Applying Ecological Principles to Land Management

Springer
New York
Berlin
Heidelberg
Barcelona
Hong Kong
London
Milan
Paris
Singapore
Tokyo

Virginia H. Dale Richard A. Haeuber
Editors

Applying Ecological Principles to Land Management

Foreword by Richard T.T. Forman

With 56 Illustrations, 5 in Full Color

Virginia H. Dale
Environmental Sciences Division
Oak Ridge National Laboratory
Oak Ridge, TN 37831-6036, USA
dalevh@ornl.gov

Richard A. Haeuber
1712 Johnson Avenue, NW
Washington, DC 20009, USA
haeuber.richard@epamail.epa.gov

Cover illustration: An overlook in the Great Smoky Mountains National Park with Mount Leconte in the distance illustrates the five ecological principles for land use and management discussed in this book: time, place, disturbance, species, and landscape. The distinct horizontal line between the canopies of the tulip poplar and the mixed hardwood forest marks the edge of areas that were once farms (now succeeding to a tulip-poplar-dominated forest) and illustrates the importance of *time* since human activity. The gradient in vegetation from the lowlands where the mixed hardwood forest dominates, to the higher elevation, where the spruce-fir forest occurs, demonstrates the *place* principle. The *disturbance* principle is depicted by the light gray patches within the high-elevation spruce-fir forest on Mount Leconte where the introduced balsam woolly adelgid has killed the native Fraser fir trees. The *species* principle is illustrated by the light patch of rhododendron monoculture or "slick" on the lower left side of Mount Leconte. The *landscape* principle is conveyed by the mix of vegetation types and histories that occur across this perspective.

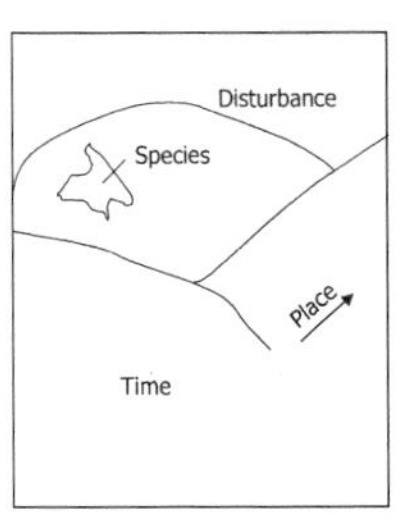

Library of Congress Cataloging-in-Publication Data
Applying ecological principles to land management/edited by Virginia H. Dale, Richard A. Haeuber.
p. cm.
Includes bibliographical references (p.).
ISBN 0-387-95099-0 (alk. paper)—ISBN 0-387-95100-8 (softcover: alk. paper)
1. Land use—Environmental aspects. 2. Land use—Planning. 3. Applied ecology. I. Dale, Virginia H. II. Haeuber, Richard A.
HD108.3 .A66 2001
333.73′16—dc21 00-061266

Printed on acid-free paper.

Production coordinated by Chernow Editorial Services, Inc., and managed by Francine McNeill; manufacturing supervised by Jacqui Ashri.
Typeset by Scientific Publishing Services (P) Ltd., Madras, India.
Printed and bound by Maple-Vail Book Manufacturing Group, York, PA.
Printed in the United States of America.

9 8 7 6 5 4 3 2 1

ISBN 0-387-95099-0 SPIN 10773621 (hardcover)
ISBN 0-387-95100-8 SPIN 10773809 (softcover)

Springer-Verlag New York Berlin Heidelberg
A member of BertelsmannSpringer Science+Business Media GmbH

Foreword

How much of the world should be planned? Or designed? Or managed? These tough questions produce few answers.

For several years I have assigned an article on a planned area in Texas (USA) for my students to read. "The Woodlands" began developing after the 1960s environmental consciousness raising, yet before the modern perspective of landscape ecology. Highlights of the planned area include prevention of flooding, nature's fingers around every home, distinctive neighborhoods, bikeways everywhere, shopping–cultural centers, and impressive aesthetics. One year my class included a resident of The Woodlands. He hated the place. People have to leave a thicket of scrubby oaks and briars in their front- and backyards; only plants on an approved list can be planted; the tidy shopping–cultural centers are designed for adults who frown on kids' activities; backyards are spaces deadened by the sounds of adjacent golfers; and much more. Everything is regular, predictable, controlled, manicured, boring. The planner's hand reaches everywhere, a permanent pervasive force. No place for a Texan. At the end of my course, which also studies natural areas and unplanned human areas, the student revealed a fuller picture of his planned landscape with an equal balance of pros and cons. The whole class listened carefully, and the memory is vivid.

People often say that land needs more planning and better management. I agree. Yet, is today's planning and management what it should be? Civilizations overwhelmingly brand the land with straight lines and regular geometry. Irregularity, randomness, curviness, aggregations, variance, and fine texture tend to be squelched. Think how this linearizing transformation alters ecological flows or natural processes across the land. Nature is truncated.

So, yes, plan all the land, but at a broad or coarse scale. Highlight key areas for ground-water protection, concentrations of interior species, wildlife movement, transportation, neighborhoods, cultural–shopping areas, remote roadless tracts, and the like. Design bits of the land for artistic expression, inspiration, model examples, dense population living, and so forth. And yes, manage all the land in an ongoing way, but with adaptive and multiscale perspectives. Adapt management actions by regularly incorporating new knowledge. Integrate top-down, middle-level, and bottom-up approaches to effectively incorporate both local and regional perspectives. The operational scale of planning and management today grows inexorably toward the landscape. A forested, agricultural, dry, or suburban landscape, or a major portion thereof, has similar characteristics throughout its area and is large enough to normally persist over generations. At the same time, the landscape is small enough that individual residents can care about it as a whole, as well as see the products of their efforts.

But what is the overall objective of this land planning and management? I would describe it as creating a mesh of nature and people where both thrive over the long term. Nature, of course, can be expressed in different ways or subdivided into categories, such as clean water, rich habitat and species diversity, wildlife movement, and natural soils. Similarly, the human dimension can be subdivided into components, including social patterns, health, transportation, economics, government, and aesthetics. However, focusing broadly on the spatial arrangement of nature and people provides insight into how the land is degrading in landscape after landscape around the globe. Fortunately, though, the landscape perspective also provides ready handles for planners and managers to make a difference and offers better communication with policy makers and the public.

Ecological principles emerge as a key to the nature-and-people solution. This book starts by highlighting major ecological principles. These unambiguous statements articulated by a group of leading ecologists and researchers from related disciplines represent a distillation of a century of ecological research and thought. Read them carefully. They should be readily repeatable by any good ecologist, planner, or manager.

Even more useful are the guidelines presented for using the principles. Planners and managers sometimes yearn to think long term, yet usually are forced to take the short view of one to several years. These succinctly stated guidelines take the long view, a boon for sustainability thinking.

Without clear articulation of ecological principles, it is easy for society to ignore or pay lip service to ecology. This book provides a foundation not to be ignored. Several chapters indicate that more principles should follow. Those that have emerged from landscape ecology, which I estimate underlies about three-quarters of land planning and management, are especially needed.

Principles should be phrased for the policy maker and the public. Who would be stimulated by words like patches, edaphic, biodiversity, bioreserves, ecosystem management, and sustainability? Instead, for these audiences, why not try ribbons of life, emerald necklace, steppingstones, primeval spot, and big blobs of luxuriant nature?

The book then uncovers an array of eye-popping case studies, which basically occurred before this articulation of principles. Authors of the case studies use the principles to effectively evaluate planning and management actions. They also pinpoint strategies and tools for achieving land-use goals. The successes and failures that emerge promise fascinating reading and wonderful insight.

A motif appears. Ecological principles alone cannot effectively plan or manage the world. Indeed, they cannot even sustain nature. Cultural principles must be included. I would add a link, that people without nature is also unsustainable. This overall feedback is not surprising, because the objective is to sustainably mesh nature and people so they both thrive.

Few persons, even among social scientists, geographers, and landscape ecologists, seem equipped to accomplish this challenge. I have a friend who wants to manage the world, but history does not offer good precedents. Maybe we will have to plan and manage the world by committee. Or maybe the world should remain unplanned and unmanaged. None of these options are appealing. The current trajectory of land degradation suggests that there must be a better way.

Paraphrasing Eugene P. Odum's father, "Poverty is no friend of the environment, but neither is unbridled growth." Many suggest that meshing ecology and economics is the key. But economics focuses on consumption, considers land as an investment commodity to buy and sell, ignores many natural resources, and often changes overnight. I would rather mesh ecology and culture, in its core sense of aesthetics, language, traditions, morals, and learning passed from generation to generation. Culture and ecology of a landscape or region tend to change gradually over decades or generations. Culture cares, plans, and manages. Nature survives, even thrives. They are tightly linked.

Providing for nature and culture avoids the single-issue view. Many people focus overwhelmingly on water quality, health, soil conservation, aesthetics, biological diversity, transportation, wildlife populations, or a sense of community. Individually these are laudable societal goals, but none alone is a basis for sustainably meshing nature and culture. Who will solve this challenge overhanging us all?

This book offers specific, clearly stated ecological principles. It also offers a dozen major case studies relating land planning and management to the principles. Such a combination is dynamite; it should not be lost. Read this and discover. Wisdom lies in these pages, a scarce and powerful attribute.

Richard T.T. Forman
Harvard University
May 2000

Preface

The many ways that people use and manage land are a primary cause of land-cover change around the world. Thus, land use and land management increasingly represent a fundamental source of change in the global environment. Despite their global importance, however, many decisions about the management and use of land are made with scant attention to ecological impacts. Through their knowledge of the functioning of the Earth's ecosystems, ecologists can make an important contribution to decisions on land use and management.

In response to this need, the Governing Board of the Ecological Society of America (ESA) established a committee to examine land-use decisions and the ways in which ecologists might help inform those decisions. The white paper developed by the ESA Land Use Committee focuses on five ecological principles for land use and management. The ideas in that paper generated a series of symposia focused on applying the five principles to various experiences with land use, land management, and conservation. These symposia were held at the August 1998 annual meeting of ESA, the January 1999 meeting of the American Association for the Advancement of Science, the March 1999 meeting of the American Association of Geographers, and the May 1999 meeting of the National Center for Environmental Decision Making. Along with several invited contributions, the papers presented at those symposia developed into this book.

This book marks a growing commitment among the ecological science and other research communities to marshal the best possible knowledge and understanding to inform policy and management issues. As reflected in the diverse background of the chapter authors, this commitment encompasses and integrates a broad base of research approaches, perspectives, and disciplines. Thus, this volume exemplifies a growing trend within the natural science community to join with researchers from many disciplines to interpret and communicate research findings to resource managers and decision makers. A critical challenge in this effort involves developing the science and tools appropriate for addressing specific land-management and land-use questions and to communicate the answers in a productive and useful manner. We hope that our effort will stimulate others to think about how ecological principles can be effectively employed in resource management. Such endeavors may further promote the professional training and interactions necessary to make interdisciplinary collaboration, as well as communication among researchers and managers, more accepted in the future.

We could not have completed this book without much support and assistance. As chair of the ESA Governing Board in 1997, when the white paper was first requested, Gordon Orians both encouraged this activity and gave it free reign. Under the leadership of Mary Barber, ESA's Sustainable Biosphere Initiative supported many parts of the effort. The work of ESA's Land Use Committee and travel to the symposia were supported by the Environmental Protection Agency and the National Aeronautics and Space Administration. Chapter authors assisted this effort greatly by participating in the peer review process. In addition, chapters were reviewed by Warren Webb, Robert O'Neill, and Dale Huff. Virginia appreciates the support from the Environmental Sciences Division at Oak Ridge National Laboratory and especially from her family who never complained about the early morning hours she spent at the keyboard. Rick gratefully acknowledges the support of the staff, management, and many members of ESA. Without their help over many years, he would have had neither the knowledge nor the opportunity to contribute to this project. More important, he thanks Amy, who cheerfully indulged the many weekend hours spent in bringing this project to fruition.

Virginia H. Dale and Richard A. Haeuber
July 2000

Contents

Contributors

Cynthia A. Botteron
SWARM-AAAS, Fort Collins, CO 80524, USA

Patrick S. Bourgeron
Institute of Arctic and Alpine Research, University of Colorado, Boulder, CO 80309-0450, USA

Bennett A. Brown
Animas Foundation, Animas, NM 88020, USA

Sandra Brown
Winrock International, Corvallis, OR 97330, USA

Mark Clark
Department of Animal Ecology, Iowa State University, Ames, IA 50011, USA

Richard M. Cruse
Department of Agronomy, Iowa State University, Ames, IA 50011, USA

Virginia H. Dale
Environmental Sciences Division, Oak Ridge National Laboratory, Oak Ridge, TN 37831-6036, USA

Brent Danielson
Department of Animal Ecology, Iowa State University, Ames, IA 50011, USA

Joseph Eilers
E & S Environmental Chemistry, Inc., Corvallis, OR 97330, USA

Kathryn Freemark
Environmental Canada, Canadian Wildlife Service, Hull, Quebec K1A 0H3, Canada

Susan Galatowitsch
Department of Horticultural Science, University of Minnesota, St. Paul, MN 55108, USA

Stanley V. Gregory
Department of Fisheries and Wildlife, Oregon State University, Corvallis, OR 97331, USA

Richard A. Haeuber
1712 Johnson Avenue, NW, Washington, DC 20009, USA

Andrew J. Hansen
Department of Biology, Montana State University, Bozeman, MT 59717-0346, USA

Jonathan B. Haufler
Boise Cascade Corporation, Boise, ID 83728, USA

N. Thompson Hobbs
Colorado Division of Wildlife and Natural Resource Ecology Laboratory, Colorado State University, Fort Collins, CO 80523, USA

David W. Hulse
Department of Landscape Architecture, Institute for a Sustainable Environment, University of Oregon, Eugene, OR 97403, USA

Hope C. Humphries
Institute of Arctic and Alpine Research, University of Colorado, Boulder, CO 80309-0450, USA

Nancy J. Huntly
Department of Biological Sciences, Idaho State University, Pocatello, ID 83209-8007, USA

Mark E. Jensen
Ecological Applications Service Team, USDA Forest Service Northern Region, Missoula, MT 59807, USA

Brian J. Kernohan
Boise Cascade Corporation, Boise, ID 83728, USA

Susan McDowell
U.S. Environmental Protection Agency, Philadelphia, PA 19103-2029, USA

Robert J. Naiman
College of Ocean and Fishery Sciences, University of Washington, Seattle, WA 98195-2100, USA

Joan Nassauer
Department of Landscape Architecture, School of Natural Resources and Environment, University of Michigan, Ann Arbor, MI 48109-1115, USA

Stephen Polasky
Department of Applied Economics, 1994 Buford Avenue, University of Minnesota, St. Paul, MN 55108-6040, USA

William E. Riebsame
Department of Geography, University of Colorado, Boulder, CO 80309, USA

Jay J. Rotella
Department of Biology, Montana State University, Bozeman, MT 59717, USA

Emily W.B. Russell
Department of Geological Sciences, Rutgers University, Newark, NJ 07102, USA

Mary Santelmann
Department of Geosciences, Oregon State University, Corvallis, OR 97331-8564, USA

Carl Steinitz
Department of Landscape Architecture, Harvard Graduate School of Design, Cambridge, MA 02138, USA

David M. Theobald
Natural Resource Ecology Laboratory, Colorado State University, Fort Collins, CO 80523, USA

Monica G. Turner
Department of Zoology, University of Wisconsin, Madison, WI 53706, USA

Kellie Vache
Department of Bioresource Engineering, Oregon State University, Corvallis, OR 97331, USA

Tom J. Valone
Department of Biology, St. Louis University, St. Louis, MO 63103, USA

Steven D. Weiss
Department of Literature and Philosophy, Georgia Southern University, Statesboro, GA 30460-8023, USA

Denis White
U.S. Environmental Protection Agency, Corvallis, OR 97333, USA

Junjie Wu
Department of Agricultural and Resource Economics, Oregon State University, Corvallis, OR 97331, USA

Part I

Introduction

1
Ecological Guidelines for Land Use and Management

Virginia H. Dale, Sandra Brown, Richard A. Haeuber, N. Thompson Hobbs, Nancy J. Huntly, Robert J. Naiman, William E. Riebsame, Monica G. Turner, and Tom J. Valone

1.1 Introduction

During the past few millennia, humans have emerged as the major force of change around the globe. The large environmental changes wrought by our actions include modification of the global climate system, reduction in stratospheric ozone, alteration of the earth's biogeochemical cycles, changes in the distribution and abundance of biological resources, and decreasing water quantity and quality (Meyer and Turner 1994; IPCC 1996; Mahlman 1997; Vitousek et al. 1997). One of the most pervasive aspects of human-induced change involves the widespread transformation of land through efforts to provide food, shelter, and products for our use. Land transformation is perhaps the most profound result of human actions because it affects so many of the planet's physical and biological systems (Kates et al. 1990). In fact, land-use changes directly impact the ability of the earth to continue providing the goods and services upon which humans depend.

Unfortunately, potential ecological consequences are not always considered in making decisions regarding land use. Moreover, the unique perspective and body of knowledge offered by ecological science rarely are brought to bear in decision-making processes on private lands. In response to this need, the Ecological Society of America established a committee to examine the ways that land-use decisions are made and the ways that ecologists could help inform those decisions. This chapter reports the principles and guidelines developed by that committee. The full report is in Dale et al. (2000). This chapter presents ecological principles that are critical to sustaining the structure and function of ecosystems in the face of rapid land-use change and discusses their implications. It also offers guidelines for using these principles in making decisions regarding land-use change. Throughout, the chapter offers specific examples to illustrate decision-making processes, relevant ecological principles, and guidelines for making choices about land use at spatial scales ranging from the individual site to the landscape. The rest of the chapters in this volume further flesh out these principles and guidelines through specific case studies.

1.1.1 Trends and Patterns of Land-Use Change

Changes in the cover, use, and management of land have occurred throughout history in most parts of the world as population has changed and human civilizations have risen and fallen (e.g., Perlin 1989; Turner et al. 1990). Over the centuries, two important trends are evident: the total land area dedicated to human uses (e.g., settlement, agriculture, forestry, and mining) has grown dramatically, and increasing production of goods and services has intensified both use and control of the land (Richards 1990). At the end of the twentieth century, almost all of the earth's habitable surface is dedicated to human use, mostly for production of food and fiber. Some is used for conservation, but even that area is mapped, zoned, and controlled.

The major anthropogenic causes of change in land cover and land use include population and associated infrastructure; economic factors, such as prices and input costs; technological capacity; political systems, institutions, and policies; and sociocultural factors, such as attitudes, preferences, and values (Kates et al. 1990; Liu et al. 1993; Turner et al. 1993; Riebsame et al. 1994; Diamond and Noonan 1996). Human population growth can be considered an ultimate cause for many land-use changes (Fig. 1.1). However, population expansion is affected by many factors, such as political dynamics and policy decisions that influence local and regional trends in urbanization and colonization. Moreover, local demography and variability

FIGURE 1.1. Human settlement in the Brazilian Amazon replaces the native tropical forest.

in per capita resource consumption can modify the effects of population. In Brazil, for example, one of the highest rates of deforestation currently occurs in the state of Rondônia, where a high rate of land-cover change results from road establishment and paving and government policies that have allowed colonists to immigrate and clear forests so farms can be established. The rate of natural resource exploitation also depends on technological advances in resource extraction and enhancement, such as logging, mining, hydroelectric power, fertilizers, pesticides, and irrigation. The relative importance of these factors varies with the situation and the spatial scale of analysis.

1.1.2 Challenges of Ecologically Sustainable Land Use

A critical challenge for land use and management involves reconciling conflicting goals and uses of the land. The diverse goals for use of the land include resource-extractive activities, such as forestry, agriculture, grazing, and mining; infrastructure for human settlement, including housing, transportation, and industrial centers; recreational activities; services provided by ecological systems, such as flood control and water supply and filtration; support of aesthetic, cultural, and religious values; and sustaining the compositional and structural complexity of ecological systems. These goals often conflict with one another, and difficult land-use decisions may develop as stakeholders pursue different land-use goals. For example, conflicts often arise between those who want to extract timber and those who are interested in the scenic values of forests. Local versus broad-scale perspectives on the benefits and costs of land management also provide different views of the implications of land actions. Understanding how land-use decisions affect the achievement of these goals can help achieve balance among the different goals. The focus of this book is on the last goal, sustaining ecological systems, for land-use decisions and practices rarely are undertaken with ecological sustainability in mind. Sustaining ecological systems also indirectly supports other values, including ecosystem services, cultural and aesthetic values, recreation, and sustainable extractive uses of the land.

To meet the challenge of sustaining ecological systems, an ecological perspective must be incorporated into land-use and land-management decisions. Specifying ecological principles and understanding their implications for land-use and land-management decisions are essential steps on the path toward ecologically-based land use, for the resulting guidelines translate theory into practical steps for land managers. Ecological principles and guidelines for land use and management elucidate the consequences of land uses for ecological systems. Thus, a major intent of this chapter is to set forth ecological principles relevant to land use and management and to develop them into guidelines for use of the land.

Trends in land-use change have increased the interdependence of ecological and human systems because there are few ecological systems that are not directly or indirectly impacted by human actions. However, the consequences of land-use decisions for both humans and ecological systems often are not felt immediately. Planning is needed to avert long-term or broad-scale harmful ecological effects resulting from unwise land-use choices. Therefore, planning should be based on a sound ecological basis.

1.2 Ecological Principles for Land Use and Management

The major lessons of ecological science for land management can be summarized in numerous ways. This chapter organizes ecological information into five principles that have implications for land management. The principles deal with *time, place, species, disturbance,* and the *landscape*. The principles are presented as separate entities, although they interact in many ways. They are translated into specific guidelines in a later section.

1.2.1 Time Principle

Ecological processes function at many time scales, some long, some short; and ecosystems change through time. Metabolic processes occur on the scale of seconds to minutes, decomposition occurs over hours to decades, and soil formation occurs at the scale of decades to centuries. Additionally, ecosystems characteristically change from season to season and year to year in response to variations in weather as well as showing long-term successional changes (Odum 1969). Early successional communities often are dominated by a few short-lived and relatively small species that grow rapidly and decompose readily after death. In contrast, later successional communities tend to be dominated by a mixture of longer-living species and contain higher standing crops of vegetation that both grow and decompose more slowly. Human activities that alter community composition or biogeochemical cycles can change the pace or direction of ecosystem succession and thus have effects lasting decades to centuries.

The time principle has several important implications for land use. First, the current composition, structure, and function of an ecological system are, in part, a consequence of historical events or conditions that occurred decades to centuries before. An ecosystem may have species or soil characteristics that reflect legacies from past land use (Foster 1992; Motzkin et al. 1996). Therefore, historical information may be needed to understand the nature of the ecosystem, including its responses to changes in use or other perturbations, and current land uses may limit those choices that are available in the future.

Second, the full ecological effects of human activities often are not seen for many years because of the time it takes for a given action to propagate through components of the system. For example, changes in nutrient inputs may alter plant growth rates and species composition (Inouye and Tilman

1995), but these changes in a plant community also affect higher trophic levels, nutrient pools, and soil organic matter, and these latter effects can develop much more slowly.

Third, the imprint of land use may persist on the landscape for a long time, constraining future land use for decades or centuries. For example, the pattern imposed on a forested landscape by extensive clear-cutting may persist for many decades after all harvesting stops (Wallin et al. 1994). Establishing roads and controlling fire or flood regimes have similarly long-lasting effects. Rapid return to previous ecological conditions often does not occur and should not be expected.

Finally, both the variation and the change that characterize ecosystem structure and process mean that the long-term effects of land use or management may be difficult to predict. This problem is exacerbated by the tendency to overlook low-frequency ecological disturbances, such as 100-year flooding or storm events (Dale et al. 1998) or processes that operate over periods longer than human life spans (e.g., forest succession).

1.2.2 Species Principle

Particular species and networks of interacting species have key, broad-scale ecosystem-level effects. These focal species affect ecological systems in diverse ways. Indicator species are important because their condition is indicative of the status of a larger functional group of species, reflective of the status of key habitats, or symptomatic of the action of a stressor. Keystone species have greater effects on ecological processes than would be predicted from their abundance or biomass alone (Power et al. 1996). Ecological engineers (e.g., the gopher tortoise [*Gopherus polyphemus*] or beaver [*Castor canadensis*]) alter the habitat and, in doing so, modify the fates and opportunities of other species (Jones et al. 1994; Naiman and Rogers 1997). Umbrella species either have large area requirements or use multiple habitats and thus overlap the habitat requirements of many other species. Link species exert critical roles in the transfer of matter and energy across trophic levels or provide critical links for energy transfer within complex food webs. Trophic cascades occur when changes in the abundance of a focal species or guild of organisms at one trophic level propagate across other trophic levels, resulting in dramatic changes in biological diversity, community composition, or total productivity. Such cascades often affect many species, including those with which the guild does not interact directly (Power 1992; Polis and Winemiller 1996). For instance, changes in the abundance of top predatory fishes may change phytoplankton composition and alter the productivity of a lake (Carpenter and Kitchell 1988; Carpenter 1992). In addition, some species (such as threatened and endangered species, game species, sensitive species, and those that are vulnerable to society because of their rarity) also require attention because of public interest in them.

The impacts of changes in the abundance and distribution of focal species are diverse. For example, keystone species affect ecosystems through such processes as competition, mutualism, dispersal, pollination, and disease and by modifying habitats and abiotic factors. Because effects of keystones are diverse and involve multiple steps, they are often unexpected despite their fundamental importance to biological diversity and ecosystem dynamics (Paine 1969, 1995; Power et al. 1996). The removal of a keystone species can radically change the diversity and trophic dynamics of a system. Changes in land use that affect keystone species may spread well beyond the boundaries of a land-use unit, extending land covers that are inhospitable to some species and favorable to others, adding barriers to movement or dispersal, introducing new predators or competitors, or changing the existing trophic or competitive dynamics. While keystone species have been found in all ecosystems, they are difficult to identify, and their effects are difficult to predict prior to a change in their abundance (Power et al. 1996).

A nonnative species can assume a focal-species role when introduced into an ecosystem and produce numerous effects on the ecosystems. Nonnative species have altered community composition and ecosystem processes via their roles as predators, competitors, pathogens, or vectors of disease and through effects on water balance, productivity, and habitat structure (Drake et al. 1989). However, determining whether a particular nonnative species will become a focal species when introduced to a new ecosystem is very difficult (Drake et al. 1989). Changes in land use often affect the establishment of nonnative species. For example, both agriculture and grazing typically introduce and spread nonnative species. In these situations, the nonnative species often are used at very high densities and can significantly alter environmental conditions, thereby reducing the abundance of native species. Furthermore, changes in the *pattern* of land cover can promote the establishment of nonnative species, for example, by creating corridors of disturbed habitat that alter movement patterns (Getz et al. 1978; DeFerrari and Naiman 1994).

Changes in species composition and diversity can result from land use through alterations to such ecosystem properties as stream turbidity (which often occurs with increased soil erosion), nutrient cycling, or productivity (e.g., Rosenzweig and Abramsky 1993). Additions of water or fertilizer also typically alter and often reduce biodiversity (Harms et al. 1987; Warner 1994; Naiman et al. 1995). The effects of land use on species composition have implications for the future productivity of ecological systems. Low-diversity systems are likely to experience large variations in realized productivity through time, as species differ in their productive potential under different conditions of weather or resource supply (Mitchell 1984; McNaughton 1993; Rosenzweig and Abramsky 1993; Tilman 1996). Moreover, low species diversity can cause resources to be used less fully than they might be in a more ecologically diverse community of primary producers (Ewel et al. 1991). Herbivores can affect the productivity of the plant communities they graze (McNaughton 1979, 1993), and top predators can initiate trophic cascades that influence productivity by changing the primary-producer community (Carpenter 1992). In fact, the introduction of predators has been used as a remediation tool to control agricultural pests and undesired algae in lakes.

1.2.3 Place Principle

Local climatic, hydrologic, edaphic, and geomorphologic factors as well as biotic interactions strongly affect ecological processes and the abundance and distribution of species at any one place. Local environmental conditions reflect location along gradients of elevation, longitude, and latitude and the multitude of microscale physical, chemical, and edaphic factors that vary within these gradients. These factors constrain the locations of agriculture, forestry, and other land uses, as well as provide the ecosystem with a particular "look." A Great Basin site looks different and has a different landscape structure from that of the Sonoran Desert, an arid montane grassland, an eastern deciduous forest, or the Great Plains. Moreover, local environmental conditions constrain the patterns of land use and the styles of architecture and development that work most efficiently and that are aesthetically pleasing.

Alternatively, the constraints of place provide opportunities to use ecological patterns and processes as models for efficient and sustainable land use. Rates of key ecosystem processes, such as primary production and decomposition, are limited by soil nutrients, temperature, water availability, and the temporal pattern of availability of these factors as mediated by climate and weather (Chabot and Mooney 1985; Givnish 1986; Frank and Inouye 1994). Thus, only certain ranges of ecological-process rates can persist in a locale without continued management inputs (e.g., irrigation of crops growing in a desert). Chronic human intervention may broaden these ranges but cannot entirely evade the constraints of place. For instance, enhanced productivity on desert uplands can be supported over the short term by additions of water; however, higher productivity generally cannot be sustained in arid land soils over the long run because of the degrading effects of high evapotranspiration rates and resulting salinization.

Agricultural production requires favorable conditions of temperature, soil, nutrients, and water, key limiting factors for plant growth and productivity. The temporal pattern of these factors is a consequence of climate and weather, restricting the location of agriculture and the suitability of particular crops. Using plants appropriate for a particular place and situating agricultural and natural patches of vegetation in an appropriate landscape context can extend sustainable agricultural land use, reduce the impacts of agriculture on adjacent areas, and more efficiently use resources. Many uses of land have failed because species composition and ecosystem processes have not been appropriately matched with the local physical, chemical, and climatic conditions. For example, the Dust Bowl in the central United States resulted from unsustainable use of Great Plains arid grassland for dryland row-crop agriculture (Sears 1980; Glantz 1994; Diamond 1997).

Naturally occurring patterns of ecosystem structure and function provide models for sustainable and ecologically sound agriculture (Carroll et al. 1990; Soulé and Piper 1992). Only those species whose adaptations suit the environmental constraints particular to an area will thrive there. For instance, arid regions cannot support plants that are unable to survive high heat and low water availability. Precipitation constrains choice of species for landscape plantings as well as for managed agricultural, forestry, or grazing systems. It also makes some places more important than others for conservation of species and ecosystems. Species lost as a consequence of land-cover changes or with increases in land-use intensity may not be easily restored or replaced. Agricultural plants may be selected to mimic the structure, physiology, growth, and flowering/fruiting phenology of local communities as a mechanism to match productive potential with local patterns of conditions suitable for production. Similarly, the use of multiple crops, typical of traditional low-technology agriculture, has attracted recent attention as a way of increasing the efficiency of resource use, stabilizing production across years of varying weather conditions, and reducing the impact of herbivores.

Land uses that cannot be maintained within the constraints of place will be costly when viewed from long-term and broad-scale perspectives. Only certain patterns of land use, settlement and development, building construction, or landscape design are compatible with local and regional hydrologic, geomorphic, and biogeochemical cycles. In terrestrial systems, land-use and land-management practices that lead to soil loss reduce the long-term potential productivity of a site. Additions of water and nutrients may exceed levels that can be used directly by primary producers, given the natural limitations of species and climate. The excess water and nutrients from enriched systems may move into adjacent areas and influence ecosystems by such processes as runoff. Similarly, sustainable settlement is limited to suitable places on the landscape. For instance, houses or communities built on transient lake-shore dunes, major flood plains, eroding seashores, or sites prone to fires are highly vulnerable to loss over the long term. Ideally, the land will be used for the purpose to which it is best suited. It does not make sense to put cities on prime farmland, requiring that more moderately productive farmland be used to provide the same quantity of food production. However, socioeconomic and political pressures have strong influences on land-use decisions.

1.2.4 Disturbance Principle

The type, intensity, and duration of disturbance shape the characteristics of populations, communities, and ecosystems. Disturbances are events that disrupt ecological systems. Disturbances may occur naturally (e.g., wildfires

[Romme and Despain 1989], storms [Boose et al. 1994; Lugo and Scatena 1996], or floods [Poff et al. 1997]) or be induced by human actions, such as clearing for agriculture, clear-cutting in forests, building roads, or altering stream channels. The effects of disturbances are controlled in large part by their intensity, duration, frequency, timing, and spatial impacts (the size and shape of the area affected) (Sousa 1984; Pickett and White 1985; Pickett et al. 1987*a*; Reice 1994; Turner et al. 1997*a*). Disturbance has been shown to have many important effects on communities and ecosystems, including enhancing or limiting biological diversity (Hastings 1980; Sousa 1984); initiating succession (Cowles 1911; Watt 1947; Pickett et al. 1987*a,b*; Glenn-Lewin and van der Maarel 1992); causing inputs or losses of dead organic matter and nutrients that affect productivity and habitat structure (Peet 1992; Scatena et al. 1996); and creating landscape patterns that influence many ecological factors, from movements and densities of organisms to functional attributes of ecosystems (Turner 1987; Turner et al. 1994, 1997*b*; Forman 1995) (Fig. 1.2). Disturbance and succession impose both spatial and temporal heterogeneity on ecological systems. Additionally, disturbances can have secondary effects, such as fragmentation caused by road development, plowing, or clear-cutting (Franklin and Forman 1987; Roland 1993).

Land-use changes that alter natural disturbance regimes or initiate new disturbances are likely to cause changes in species' abundance and distribution, community composition, and ecosystem function (Yarie et al. 1998). In addition, the susceptibility of an ecosystem to other disturbances may be altered. For example, forest fragmentation may enhance the susceptibility of the remaining forest to a variety of other disturbances, including windthrow, pest epidemics, invasion by nonnative species, and nest parasitism (Franklin and Forman 1987).

Land managers and planners should be aware of the ubiquity of disturbance in nature. Disturbances that are both intense and infrequent, such as hurricanes or 100-year floods, will continue to produce "surprises" (Turner et al. 1997*a*; Turner and Dale 1998). As discussed above for the time principle, communities and ecosystems change, with or without disturbance; thus, attempts to maintain landscape conditions in a particular state will be futile over the long term. Further, attempts to control disturbances are generally ineffectual (Dale et al. 1998). In fact, suppression of a natural disturbance may have the opposite effect of that intended. For example, suppression of fire in fire-adapted systems results in the buildup of fuels and increases the likelihood of severe, uncontrollable fires. In the aftermath of the fires at Yellowstone National Park, it was recognized that the large scale of those fires was, in part, the result of previous fire-control actions that created connected patches of fire-prone forests (Schullery 1989). Similarly, flood-control efforts have facilitated development in areas that are still subject to infrequent large events (e.g., the 1993 floods in the upper Midwest), resulting in tremendous economic and ecological loss (Haeuber

FIGURE 1.2. Long-leaf pine forests depend on low-intensity fires to reduce hard wood ingrowth and eliminate competing tree species.

and Michener 1998; Sparks 1996; Sparks et al. 1998). Land-use policy that is based on the understanding that ecosystems are dynamic in both time and space can often deal with changes induced by disturbances (Turner et al. 1995; Dale et al. 1998; Haeuber and Michener 1998).

Natural disturbances can provide a model upon which to base land-use activities, but the differences between natural and human activities must be recognized. For example, timber harvest has sometimes been considered a surrogate for natural fire, but some of the ecological attributes and effects of these two disturbances are markedly different. Natural forest fire typically causes little soil disturbance, and often fine fuels are consumed, while large, coarse wood remains to decompose after the fire. Timber harvest often results in considerable soil disturbance, and fine branches may remain while large wood is removed. This removal of wood also impacts forest streams, which are strongly influenced by the physical structure and nutrient subsidies provided by dead woody debris (Harmon et al. 1986; Bilby and Bisson 1998).

Continued expansion of human settlement into disturbance-prone landscapes is likely to result in increased conflicts between human values and the maintenance of natural-disturbance regimes necessary to sustain such landscapes. For example, building homes in conifer forests that have recurrent wildfires results in conflicts that endanger human life as well as entail financial risks.

1.2.5 Landscape Principle

The size, shape, and spatial relationships of land-cover types influence the dynamics of populations, communities, and ecosystems. The spatial array of habitats or ecosystems comprises the landscape, and all ecological processes respond, at least in part, to this landscape template (Urban et al. 1987; Turner 1989; Forman 1995; Pickett and Cadenasso 1995). The kinds of organisms that can exist (including their movement patterns, interactions, and influence over such ecosystem processes as decomposition and nutrient fluxes) are constrained by the sizes, shapes, and patterns of interspersion of habitat across a landscape. Landscape fragmentation is not necessarily destructive of ecological function or of diverse biological communities, because a patchwork of habitat types often maintains more types of organisms and more diversity of ecosystem processes than does a large area of homogeneous habitat (e.g., Wilson et al. 1997). However, large decreases in the size of habitat patches or increases in the distance between habitat patches of the same type can greatly reduce or eliminate populations of organisms (Lovejoy et al. 1986; Saunders et al. 1991; Noss and Csuti 1994; Hansson et al. 1995; Fahrig 1997; Schwartz 1997) as well as alter ecosystem processes. Making a naturally patchy landscape less patchy (more uniform) may also have adverse affects.

Human-settlement patterns and individual land-use decisions often fragment the landscape or otherwise alter land-cover patterns. Effects of habitat fragmentation on species are numerous (e.g., Saunders et al. 1991; Noss and Csuti 1994; Andren 1997), and the richness of native species is almost always reduced.

Larger patches of habitat generally contain more species (and often a greater number of individuals) than smaller patches of the same habitat (Wiens 1996). Larger patches also frequently contain more local environmental variability, such as differences in microclimate, more structural variation in plants, and greater diversity of topographic positions. This variability provides more opportunities for organisms with different requirements and tolerances to find suitable sites within the patch. In addition, the edges and interiors of patches may have quite different conditions, favoring some species over others, and the abundance of edge and interior habitat varies with patch size (Temple 1986). Large patches are likely to contain both edge and interior species, whereas small patches will contain only edge species.

Habitat connectivity can constrain the spatial distribution of species by making some areas accessible and others inaccessible (Burgess and Sharpe 1981; Pulliam et al. 1992). Connectivity is a threshold dynamic, meaning that gradual reduction of habitat may have gradual effects on the presence or abundance of a species, but the effects tend to be dramatic after the threshold is passed (Andren 1997). Land-cover changes are most likely to have substantial effects when habitat is low to intermediate in abundance (Pearson et al. 1996). Under these conditions, small changes in habitat abundance may cause the connectivity threshold to be passed. The threshold of connectivity varies among species and depends on two factors: (1) the abundance and spatial arrangement of the habitat and (2) the movement or dispersal capabilities of the organism (Gardner et al. 1989; Pearson et al. 1996).

Local ecological dynamics (e.g., the abundance of organisms at a place) may be explained by attributes of the surrounding landscape as well as by characteristics of the immediate locale (e.g., Franklin 1993; Pearson 1993). Valone and Brown (1995) found that rates of immigration by, and extinction of, small rodents in habitat patches was affected by competition from other species and by habitat structure. Therefore, understanding the implications of local land-use decisions requires interpreting them within the context of the surrounding landscape.

The ecological import of a habitat patch may be much greater than is suggested by its spatial extent. Some habitats, such as bodies of water or riparian corridors, are small and discontinuous, but nevertheless have ecological impacts that greatly exceed their spatial extent (Naiman and Décamps 1997). For example, wetlands and bodies of water in general are low in spatial extent but high in their contributions to the compositional and structural complexity of an ecoregion. In addition, the presence of riparian vegetation, which may occur as relatively narrow bands along a stream or as small patches of wetland, generally reduces the amount of nutrients being transported to the stream (e.g., Peterjohn and Correll 1984; Charbonneau and Kondolf 1993; Detenbeck et al. 1993; Soranno et al. 1996; Weller et al. 1998). This filtering by the vegetation is an ecologically important function because excess nutrients that unintentionally end up in lakes, streams, and coastal waters are a major cause of water quality problems, such as acidification and eutrophication. Thus, the presence and location of particular vegetation types can strongly affect the movement of materials across the landscape and can contribute to the maintenance of desirable water quality.

1.3 Guidelines for Land Use

Ecologically based guidelines are proposed here as a way to facilitate land managers considering the ecological ramifications of land-use decisions. These guidelines are meant to be flexible and to apply to diverse land-use situations. The guidelines recognize that the same parcel of land can be used to accomplish multiple goals and require that decisions be made within an appropriate spatial and temporal context. For example, the ecological implications of a decision may last for decades or even centuries, long outliving the political effects and impacts. Furthermore, all aspects of a decision need to be considered in setting the time frame and spatial scale for impact analysis. In specific cases, the relevant guidelines can be developed into prescriptions for action. One could think of these guidelines as a checklist of factors to be considered in making a land-use decision:

- ☐ Examine the impacts of local decisions in a regional context.
- ☐ Plan for long-term change and unexpected events.
- ☐ Preserve rare landscape elements, critical habitats, and associated species.
- ☐ Avoid land uses that deplete natural resources over a broad area.
- ☐ Retain large contiguous or connected areas that contain critical habitats.
- ☐ Minimize the introduction and spread of nonnative species.
- ☐ Avoid or compensate for effects of development on ecological processes.
- ☐ Implement land-use and land-management practices that are compatible with the natural potential of the area.

Checking the applicability of each guideline to specific land-use decisions provides a means to translate the ecological principles described in the previous section into practice.

1.3.1 Examine Impacts of Local Decisions in a Regional Context

As embodied in the landscape principle, the spatial array of habitats and ecosystems shapes local conditions and responses (e.g., Risser 1985; Patterson 1987) and, by the same logic, local changes can have broad-scale impacts over the landscape. Therefore, it is critical to examine both the constraints placed on a location by the regional conditions and the implications of decisions for the larger area. This guideline dictates two considerations for planning land use: identifying the surrounding region that is likely to affect and be affected by the local project and examining how adjoining jurisdictions are using and managing their lands. Once the regional context is identified, regional data should be examined. Items to include in a regional data inventory include land-cover classes, soils, patterns of water movement, historical disturbance regimes, and habitats of focal species and other species of special concern (see Diaz and Apostol [1992] and Sessions et al. [1997] for a thorough discussion). The focal species typically represent a diversity of functional roles that are possible within a place and reflect the environmental fluctuations that provide opportunities and constraints for species. In some cases, an attribute (such as soils) can be used as a surrogate for other information that is dependent on that feature (such as vegetation). Recent technological advances (such as the development of geographic information systems (GIS) and the general availability of databases for soils, roads, and land cover on the Internet) make regional analysis a possibility even for small projects (e.g., Mann et al., 1999).

Where one has the luxury of planning land use and management in a pristine site, both local and broad-scale decisions can be considered simultaneously. Forman (1995) suggests that land-use planning begin with determining nature's arrangement of landscape elements and land cover and then considering models of optimal spatial arrangements and existing human uses. Following this initial step, he suggests that the desired landscape mosaic be planned first for water and biodiversity; then for cultivation, grazing, and wood products; then for sewage and other wastes; and finally for homes and industry. Planning under pristine conditions is typically not possible. Rather, the extant state of development of the region generally constrains opportunities for land management.

This guideline implies a hierarchy of flexibility in land uses, and it implicitly recognizes ecological constraints as the primary determinants in this hierarchy. A viable housing site is much more flexible in placement than an agricultural area or a wetland dedicated to improving water quality and sustaining wildlife. Optimizing concurrently for several objectives requires planners to recognize that some land uses have lower site flexibility than other uses. However, given that most situations involve existing land uses and built structures, this guideline calls for examining local decisions within the regional context of ecological concerns as well as in relation to the social, economic, and political perspectives that are typically considered.

Ideally, land-use models that incorporate both ecological and other concerns about impacts of land activities could be used to design and explore implications of land-use decisions in a regional context. Land-use models that truly integrate the social, economic, and ecological considerations are in their infancy, and no consensus has yet been reached about what approaches are best for this task. Therefore, many diverse approaches have been advanced (Wilkie and Finn 1988; Southworth et al. 1991; Baker 1992; Lee et al. 1992; Dale et al. 1993, 1994*a,b*; Riebsame et al. 1994; Gilruth et al. 1995; Turner et al. 1996; Wear et al. 1996). Development and use of these models have improved understanding of the relationship between the many factors that affect land-use decisions and their impacts (including human perceptions, economic systems, market and resource demands, foreign relations [e.g., trade agreements], fluctuations in interest rates, and pressure for environmental conservation and maintenance of ecosystem goods and services). Understanding this interface between causality and effect of land-use decisions is a key challenge facing the scientific community and planners in the coming decades.

1.3.2 Plan for Long-Term Change and Unexpected Events

The time principle indicates that impacts of land-use decisions can, and often do, vary over time. Long-term changes that occur as a response to land-use decisions can be classified into two categories: delayed and cumulative. Delayed impacts may not be observed for years or decades. An example is the composition of forest communities in New England; today, those forests differ substantially among areas that were previously woodlots, pasture, or croplands (Foster 1992). Cumulative effects are illustrated by events that together determine a unique trajectory of effects that could not be predicted from any one event (Paine et al. 1998). For example, at Walker Branch Watershed in East Tennessee, patterns of calcium cycling are determined not only by past land uses (timber harvest versus agriculture) but also by the history of insect outbreaks in the recovering forest (Dale et al. 1990).

Future options for land use are constrained by the decisions made today as well as by those made in the past. These constraints are conspicuous in forested systems, where options for areas to harvest may be limited by the pattern of available timber left from past cuts (Turner et al. 1996). In addition, areas that are urbanized are unlikely to be available for any other land uses because urbanization locks in a pattern on the landscape that is hard to reverse. This difficulty of reversal also holds for suburban sprawl and the development of vacation or retirement homes.

The concept of externalities must be considered in the context of this guideline. Land actions should be implemented with some consideration as to the physical, biological, aesthetic, or economic constraints that are placed on future uses of the land. External effects can extend beyond the boundaries of individual ownership and thus have the potential to affect surrounding owners.

Planning for the long term requires consideration of the potential for unexpected events, such as variations in temperature or precipitation patterns or disturbances. Although disturbances shape the characteristics of ecosystems, estimating the occurrence and implications of these unanticipated events is difficult. Nevertheless, land-use plans must include them. For example, the western coast of the United States has a high potential for volcanic eruption, which would have severe effects. Yet, predicting exact impacts is not possible. Climate change is occurring, but global-climate-projection models cannot determine the temperature and precipitation changes that will happen in any one place. Thus, potential impacts of land-use changes on future dynamics should be recognized but cannot be precisely specified, as yet. Similarly, land-use changes that affect natural water drainages can cause catastrophic flooding during extreme rain events (Sparks 1996). Although it will not be possible to foresee all extreme events or the effects of a land-use decision on natural variations, it is important to estimate likely changes.

Long-term planning must also recognize that one cannot simply extrapolate historical land-use impacts forward to predict future consequences of land use. The transitions of land from one use or cover type to another often are not stable from one period to another (Turner et al. 1996; Wear et al. 1996) because of changes in demographics, public policy, market economies, and technological and ecological factors. Thus, models produce projections of potential scenarios rather than predictions of future events. It is difficult to model (or even understand) the full complex of interactions among the factors that determine land-use patterns, yet models offer a useful tool to consider potential long-term and broad-scale implications of land-use decisions.

1.3.3 Preserve Rare Landscape Elements and Associated Species

Rare landscape elements provide critical habitats or ecological processes. For example, in the Southern Appalachian Mountains, 84% of the federally listed terrestrial plant and animal species occur in rare communities (Southern Appalachian Assessment 1996). While these communities occupy a small area of land, they contain features important for the region's biological diversity. Therefore, rare landscape elements need to be identified, usually via an inventory and analysis of vegetation types, hydrology, soils, and physical features that identifies the presence and location of rare landscape elements and, when possible, associated species (e.g., see Mann et al. 1999). Once the inventory is complete, effects of alternative land-use decisions on these landscape elements and species can be estimated. These effects can then be considered in view of the overall goal for the project, the distribution of elements and species across the landscape, and their susceptibility, given likely future land changes in the vicinity and region. Strategies to avoid or mitigate serious impacts can then be developed and implemented. This guideline to preserve rare landscape elements and associated species derives from both the species and place principles.

1.3.4 Avoid Land Uses That Deplete Natural Resources Over a Broad Area

Depletion of natural resources disrupts natural processes in ways that often are irreversible over long periods of time. The loss of soil via erosion that occurs during agriculture and the loss of wetlands and their associated ecological processes and species are two examples. This guideline entails prevention of the rapid or gradual diminishment of resources, such as water or soil. This task first requires the determination of resources at risk. For example, in the southwestern United States, water might be the most important resource; but elsewhere, water might not be a limiting factor, yet it may not be readily replaced. The evaluation of resources at risk is thus an ongoing process as the abundance and distribution of resources change. This guideline also calls for the deliberation of ways to avoid actions that would jeopardize natural resources. Some land actions are inappropriate in a particular setting or time, and they should be avoided. Examples of inappropriate actions are farming on steep slopes, which might produce soil loss; logging on stream sides, which may jeopardize the habitat for aquatic organisms; and planting hydrophilic plants in areas that require substantial watering (e.g., lawns grown in arid areas).

1.3.5 Retain Large Contiguous or Connected Areas That Contain Critical Habitats

Large areas are often important to maintaining key organisms and or supporting ecosystem processes (e.g., Brown 1978; Newmark 1995). Habitats are places on the landscape that contain the unique set of physical and biological conditions necessary to support a species or guild. Thus, the features of a habitat must be interpreted in the context of the species or guild that defines them. Habitat becomes critical to the survival of a species or population when it is rare or disconnected. Thus, this guideline derives from both the place and landscape principles. Size and connectivity of patches provide ecological benefits. The presence of animals in an area can be predicted by the size of their home range and their ability to cross gaps of inhospitable habitat (Dale et al. 1994*b*; Mladenoff et al. 1995). However, habitat connectivity is not always a positive attribute for species and ecosystems. Land uses that serve as barriers to species' movement can have long-term negative effects on populations (e.g., Merriam et al. 1989); but, at the same time, corridors can facilitate the spread of nonnative species or diseases (see the next guideline). Additionally, habitats do not need to be in natural areas to provide benefits for wildlife. For example, golf courses in the southeastern United States often contain enough long-leaf pine (*Pinus palustris*) to provide habitat for the endangered red cockaded woodpecker (*Pocoides borealis*).

Again, the importance of spatial connections depends on the priorities and elements of a situation. A first step in implementing the guideline is to examine the spatial connectivity of key habitats in an area, determining which patches are connected and whether the connectivity has a temporal component. Second, opportunities for connectivity must be promoted. Sometimes, those opportunities complement other planning needs. For instance, corridors along streams must be protected during timber extraction to provide benefits for aquatic species (Naiman and Décamps 1997).

The term "connected" also should be defined in a manner specific to the situation. In some cases, two areas that are divided by a land-cover type may be artificially connected. For example, the habitat of panther (*Felis concolor coryi*) that is bisected by roads in Florida is now connected by tunnels under the highway (Foster and Humphrey 1995). For other species, such as meadow voles (*Microtus pennsylvanicus*), roads themselves serve as corridors (see Getz et al. 1978). In other cases, areas of similar habitat need not be directly adjacent but need only to be within the dispersal distance of the species of concern (e.g., migratory birds returning to nesting grounds [Robinson et al. 1995]). The connections provided by linear land-cover features, such as roads, may have both positive and negative effects (Forman and Alexander 1998), and thus the broad-scale impacts of these features require careful consideration.

1.3.6 Minimize the Introduction and Spread of Nonnative Species

The species principle indicates that nonnative organisms often have negative effects on such ecological processes as reproduction, growth, mortality, competition, predation, and herbivory (Fig. 1.3). Thus, land-use decisions must consider the potential for the introduction and spread of nonnative species. Land planning should consider vehicle movement along transportation routes, the planting of native species, and control of pets. For example, transportation routes have been very important in the spread of the spores of the pathogen *Phytophthora lateralis,* which kills Port Orford cedar (*Chamaecyparis lawsoniana*), an important timber species of southwestern Oregon (Harvey et al. 1985; Zobel et al. 1985). The USDA Forest Service has found that cleaning trucks or minimizing traffic during wet periods can dramatically reduce the transport of this pathogen between forests. Similarly, the spread of gypsy moth (*Lymantria dispar*) is correlated with overseas transportation of the eggs, larvae, and adults in the cargo holds of ships (Hofacker et al. 1993) or along roads at low elevations, where egg sacs are attached to vehicles (Sharov et al. 1997) or outdoor furniture. The great potential for vehicular transport of nonnative species was demonstrated by a case in which material was collected from the exterior surface of an automobile following a drive through central Europe; the collected matter represented 124 plant species and exhibited a high proportion of foreign propagules (Schmidt 1989). The introduction of aquatic organisms transported incidentally with shipping traffic is a comparable example for aquatic ecosystems. Many of these introductions have had devastating effects. Waterways for shipping have impacts on the movement of introduced species not unlike those of roadways.

FIGURE 1.3. The introduced invasive species kudzu (*Pueraria thunbergiana*) grows over trees and eventually kills them.

Often, growing native species reduces the need for planting nonnative species, particularly in urban, suburban, or other developed areas. The planted native species can then provide propagules that may disperse and establish. As an added benefit, the native species are adapted to the local conditions and frequently become established more readily and require less maintenance than nonnatives. Native species are also adapted to long-term variations in climate or disturbance regimes to which nonnative species often succumb. Terrestrial environmental conditions associated with native vegetation may also deter the spread of nonnatives. For example, in small forest islands interspersed among alien-dominated agroecosystems in Indiana, even the smallest forest remnants retained interior habitat conditions sufficient to resist invasion by the available nonnative plant species (Brothers and Spingharn 1992). Introduced agricultural crops often result in less sustainable farming practices than does the use of native crops, as has been observed in the Brazilian Amazon (Soulé and Piper 1992).

The control of pets is an essential aspect of reducing introductions. As suburbanization expands, one of the major effects on native fauna is the introduction of exotic pets. The mosquitofish (*Gambusia affinis*), swordtail (*Xiphophorus helleri*), and other species used as pets and then released into the wild have had a dramatic impact on the native fauna (Gamradt and Kats 1996). In addition, cats (*Felis cattus* or *Felis domestica*) kill birds and small mammals (Dunn and Tessaglia 1994). In Australia, conservationists have worked with developers and the public to ban dogs from suburban development projects that contain koala habitat because dogs strongly contribute to koala mortality in developed areas.

1.3.7 Avoid or Compensate for Effects of Development on Ecological Processes

Negative impacts of development might be avoided or mitigated by some forethought. To do so, potential impacts need to be examined at the appropriate scale. At a fine scale, the design of a structure may interrupt ecoregional processes. For example, dispersal patterns may be altered by a road, migrating birds may strike the reflective surfaces of a building, or fish may be entrained in a hydroelectric generator. At a broad scale, patterns of watershed processes may be altered, for example, by changing drainage patterns as part of the development.

Therefore, how proposed actions might affect other systems (or lands) should be examined. For example, landslides are generally site-specific so that development of places with a high potential for landslides should be avoided. Also, human uses of the land should avoid structures and uses that might have a negative impact on other systems; at the very least, ways to compensate for those anticipated effects should be determined. It is useful to look for opportunities to design land use to benefit or enhance the ecological attributes of a region (Fig. 1.4). For example, golf courses can be designed to serve as wildlife habitat (Terman 1997), or traffic in rural areas can be concentrated on fewer and more strategically placed roads, resulting in decreased traffic volumes and flows within the region as a whole and less impact on wildlife (Jaarsma 1997).

1.3.8 Implement Land-Use and Land-Management Practices That Are Compatible with the Natural Potential of the Area

The place principle implies that local physical and biotic conditions affect ecological processes. Therefore, the natural potential for productivity and for nutrient and water cycling partially determine the appropriate land-use and land-management practices for a site. Land-use practices that fall within these limits are usually cost-effective in terms of human resources and future costs caused by unwarranted changes on the land. Nevertheless, supplementing the natural resources of an area by adding nutrients through fertilization or water via irrigation is common. Even with such supplements, however, the natural limitations of the site must be recognized for cost-effective management.

Implementing land-use and land-management practices that are compatible with the natural potential of the area requires that land managers have

FIGURE 1.4. Areas adjacent to air strips can provide habitat for rare plants and birds, such as this site at Arnold Air Force Base in Tennessee.

an understanding of the site potential. Traditional users of the land (e.g., native farmers) typically have a close relationship with the land. As farming and other resource extraction activities become larger and more intensive, the previous close association that managers had with the land is typically lost. Still, land-management practices like no-till farming can be employed to reduce soil erosion or other resource losses. However, many land uses simply ignore site limitations or externalize site potential. For example, building shopping malls on prime agricultural land does not make the best use of the site potential. Also, establishing farms where irrigation is required or lawns where watering is necessary assumes that site constraints will be surmounted. Ultimately, land use and the products that can be derived from the land will be limited by the natural potential of the site.

1.4 Conclusions

The next step is to examine how these principles apply to particular land-use and land-management situations. The remainder of this volume tackles that task. The chapters are organized into four categories. Part II presents a diversity of examples of applying the ecological principles and guidelines. The case studies include both private and public lands, management issues from around the world, and various land-management goals (e.g., conservation of species or habitat, protection of historically important sites, and sustainable use of natural resources with limited harvesting). Some of the chapters explicitly consider how human activities influence the ability to apply the ecological principles to land management. Part III of the book contains four chapters that use different techniques to consider future conditions of the land. These alterative futures reflect the ecological guidelines to varying degrees. The explicit presentation of future conditions allow the land owner to understand how their decisions may affect future conditions. Part IV, Making Decisions About the Land, explores ways that people and groups make decisions and how the affects of these decisions are assessed. The book ends with a chapter that explores the future of ecological perspectives in decision making. It considers opportunities and constraints to including ecological guidelines in making decisions about the land. Together, the ideas, examples, and tools in the following chapters provide illustrations of how ecology can be integrated into land management.

1.5 Summary

Throughout history, the many ways that people use and manage land have emerged as a primary cause of land-cover change around the world. Thus, land use and land management increasingly represent a fundamental source of change in the global environment. Despite their global importance, however, many decisions about the management and use of land are made with scant attention to ecological impacts. Thus, ecologists' knowledge of the functioning of the earth's ecosystems is needed to broaden the scientific basis of decisions on land use and management. In response to this need, the Ecological Society of America established a committee to examine the ways that land-use decisions are made and the ways that ecologists could help inform those decisions. This chapter reports the principle and guidelines developed by that committee. The full report is in Dale et al. (2000).

Five principles of ecological science have particular implications for land use and can assure that fundamental processes of the earth's ecosystems are sustained. These ecological principles deal with time, species, place, disturbance, and the landscape. The recognition that ecological processes occur within a *temporal* setting and change over time is fundamental to analyzing the effects of land use. In addition, individual *species* and networks of interacting species have strong and far-reaching effects on ecological processes. Furthermore, each *site* or region has a unique set of organisms and abiotic conditions influencing and constraining ecological processes. *Disturbances* are important and ubiquitous ecological events whose effects may strongly influence population, community, and ecosystem dynamics. Finally, the size, shape, and spatial relationships of habitat patches on the *landscape* affect the structure and function of ecosystems. The responses of the land to changes in use and management by people depend on expressions of these fundamental principles in nature.

These principles dictate several guidelines for land use. The guidelines give practical rules of thumb for incorporating ecological principles into land-use decision making. These guidelines suggest that land managers should

- ☐ Examine impacts of local decisions in a regional context.
- ☐ Plan for long-term change and unexpected events.
- ☐ Preserve rare landscape elements and associated species.
- ☐ Avoid land uses that deplete natural resources.
- ☐ Retain large contiguous or connected areas that contain critical habitats.
- ☐ Minimize the introduction and spread of nonnative species.
- ☐ Avoid or compensate for the effects of development on ecological processes.
- ☐ Implement land-use and land-management practices that are compatible with the natural potential of the area.

Decision makers and citizens are encouraged to consider these guidelines and to include ecological perspectives in choices on how land is used and managed. The guidelines also suggest the applied and basic research required to develop the science needed by landowners and managers if ecologically based land-use approaches are to be implemented successfully.

Acknowledgments. Much of this paper is based on the report of the Land Use Committee of the Ecological Society of America published as Dale et al. (2000). Concepts in that paper are included here with the permission of the Ecological Society of America, Publisher of Ecological Applications. The submitted manuscript has been written by a contractor of the U.S. Government under contract No. DE-AC05-96OR22464. Accordingly, the U.S. Government retains a nonexclusive, royalty-free license to publish or reproduce the published form of this contribution, or to allow others to do so, for U.S. Government purposes.

References

Andren, H. 1997. Population response to landscape changes depends on specialization to different landscape elements. Oikos **80**:193–196.

Baker, W.L. 1992. Effects of settlement and fire suppression on landscape structure. Ecology **73**:1879–1887.

Bilby, R.E., and P.A. Bisson. 1998. Function and distribution of large woody debris. Pages 324–346 *in* R.J. Naiman and R.E. Bilby, editors. River ecology and management. Springer-Verlag, New York, New York, USA.

Boose, E.R., D.R. Foster, and M. Fluet. 1994. Hurricane impacts to tropical and temperate forest landscapes. Ecological Monographs **64**:369–400.

Brothers, T.S., and A. Spingharn. 1992. Forest fragmentation and alien plant invasion of central Indiana old-growth forests. Conservation Biology **6**:91–100.

Brown, J.H. 1978. The theory of insular biogeography and the distribution of boreal birds and mammals. Great Basin Naturalist Memoirs **2**:209–227.

Burgess, R.L., and D.M. Sharpe, editors. 1981. Forest island dynamics in man-dominated landscapes. Springer-Verlag, New York, New York, USA.

Carpenter, S.R. 1992. Destabilization of planktonic ecosystems and blooms of blue-green algae. Pages 461–481 *in* J. F. Kitchell, editor. Food web management. Springer-Verlag, New York, New York, USA.

Carpenter, S.R., and J.F. Kitchell. 1988. Consumer control of lake productivity. Bioscience **38**:764–769.

Carroll, C.R., J.H. Vanderneer, and P.M. Rosset, editors. 1990. Agroecology. McGraw-Hill, New York, New York, USA.

Chabot, B.F., and H.A. Mooney, editors. 1985. Physiological ecology of North American plant communities. Chapman & Hall, New York, New York, USA.

Charbonneau, R., and G.M. Kondolf. 1993. Land use change in California, USA: nonpoint source water quality impacts. Environmental Management **17**:453–460.

Cowles, H.C. 1911. The causes of vegetation cycles. Botanical Gazette **51**:161–183.

Dale, V.H., L.K. Mann, R.J. Olson, D.W. Johnson, and K.C. Dearstone. 1990. The long-term influence of past land use on the Walker Branch forest. Landscape Ecology **4**:211–224.

Dale, V.H., R.V. O'Neill, M. Pedlowski, and F. Southworth. 1993. Causes and effects of land-use change in central Rondônia, Brazil. Photogrammetric Engineering and Remote Sensing **56**:997–1005.

Dale, V.H., R.V. O'Neill, F. Southworth, and M. Pedlowski. 1994*a*. Modeling effects of land management in the Brazilian amazonian settlement of Rondônia. Conservation Biology **8**:196–206.

Dale, V.H., S.M. Pearson, H.L. Offerman, and R.V. O'Neill. 1994*b*. Relating patterns of land-use change to faunal biodiversity in the central Amazon. Conservation Biology **8**:1027–1036.

Dale, V.H., A. Lugo, J. MacMahon, and S. Pickett. 1998. Ecosystem management in the context of large, infrequent disturbances. Ecosystems **1**:546–557.

Dale, V.H., S. Brown, R.A. Haeuber, N.T. Hobbs, N. Huntly, R.J. Naiman, W.E. Riebsame, M.G. Turner, and T.J. Valone. 2000. Ecological principles and guidelines for managing the use of land. Ecological Applications **10**:639–670.

DeFerrari, C., and R.J. Naiman. 1994. A multiscale assessment of exotic plants on the Olympic Peninsula, Washington. Journal of Vegetation Science **5**:247–258.

Detenbeck, N.E., C.A. Johnston, and G.J. Niemi. 1993. Wetland effects on lake water quality in the Minneapolis/St. Paul metropolitan area. Landscape Ecology **8**:39–61.

Diamond, H.L., and P.F. Noonan. 1996. Land use in America: report of the sustainable use of land project. Island Press, Covelo, California, USA.

Diamond, J. 1997. Guns, germs, and steel: the fates of human societies. Norton, New York, New York, USA.

Diaz, N., and D. Apostol. 1992. Forest landscape analysis and design. Eco-TP-043-92. USDA Forest Service, Pacific Northwest Region, Portland, Oregon, USA.

Drake, J.A., H.A. Mooney, F. di Castri, R.H. Groves, F.J. Kruger, M. Rejmanek, and M. Williamson, editors. 1989. Biological invasions: a global perspective. John Wiley and Sons, Chichester, England.

Dunn, E.H., and D.L. Tessaglia. 1994. Predation of birds at feeders in winter. Journal of Field Ornithology **65**:8–16.

Ewel, J.J., M.J. Mazzarino, and C.W. Berish. 1991. Tropical soil fertility changes under monocultures and successional communities of different structure. Ecological Applications **1**:289–302.

Fahrig, L. 1997. Relative effects of habitat loss and fragmentation on population extinction. Journal of Wildlife Management **61**:603–610.

Forman, R.T.T. 1995. Land mosaics: the ecology of landscapes and regions. Cambridge University Press, Cambridge, England.

Forman, R.T.T., and L.E. Alexander. 1998. Roads and their major ecological effects. Annual Review of Ecology and Systematics **29**:207–231.

Foster, D.R. 1992. Land-use history (1730–1990) and vegetation dynamics in central New England, USA. Journal of Ecology **80**:753–772.

Foster, M.L., and S.R. Humphrey. 1995. Use of highway underpasses by Florida panthers and other wildlife. Wildlife Society Bulletin **23**:95–100.

Frank, D.A., and R.S. Inouye. 1994. Temporal variation in actual evapotranspiration of terrestrial ecosystems: patterns and ecological implications. Journal of Biogeography **21**:401–411.

Franklin, J.F. 1993. Preserving biodiversity: species, ecosystems, or landscapes? Ecological Applications **3**:202–205.

Franklin, J.F., and R.T.T. Forman. 1987. Creating landscape patterns by forest cutting: ecological consequences and principles. Landscape Ecology **1**:5–18.

Gamradt, S.C., and L.B. Kats. 1996. Effect of introduced crayfish and mosquitofish on California newts. Conservation Biology **10**:1155–1162.

Gardner, R.H., R.V. O'Neill, M.G. Turner, and V.H. Dale. 1989. Quantifying scale-dependent effects of animal movement with simple percolation models. Landscape Ecology **3**:217–228.

Getz, L.L., F.R. Cole, and D.L. Gates. 1978. Interstate roadsides as dispersal routes for *Microtus pennsylvanicus*. Journal of Mammalogy **59**:208–212.

Gilruth, P.T., S.E. Marsh, and R. Itami. 1995. A dynamic spatial model of shifting cultivation in the highlands of Guinea, West Africa. Ecological Modeling **79**:179–197.

Givnish, T.J., editor. 1986. On the economy of plant form and function. Cambridge University Press, New York, New York, USA.

Glantz, M.H., editor. 1994. Drought follows the plow. Cambridge University Press, New York, New York, USA.

Glenn-Lewin, D.C., and E. van der Maarel. 1992. Patterns and processes of vegetation dynamics. Pages 11–59 *in* D.C. Glenn-Lewin, R.K. Peet, and T.T. Veblen, editors. Plant succession. Chapman & Hall, New York, New York, USA.

Haeuber, R.A., and W.K. Michener. 1998. Policy implications of recent natural and managed floods. BioScience **48**:765–772.

Hansson, L., L. Fahrig, and G. Merriam, editors. 1995. Mosaic landscapes and ecological processes. Chapman & Hall, New York, New York, USA.

Harmon, M.E., J.F. Franklin, F.J. Swanson, P. Sollins, S.V. Gregory, J.D. Lattin, N.H. Anderson, S.P. Cline, N.G. Aumen, J.R. Sedell, G.W. Lienkaemper, K. Cromack, Jr., and K.W. Cummins. 1986. Ecology of coarse woody debris in temperate ecosystems. Advances in Ecological Research **15**:133–302.

Harms, W.B., A.H.F. Stortelder, and W. Vos. 1987. Effects of intensification of agriculture on nature and landscape in the Netherlands. Pages 357–379 *in* M.G.

Wolman and F.G. Fourier, editors. Land transformation in agriculture. John Wiley and Sons, New York, New York, USA.

Harvey, R.D., J.S. Hadfield, and H. Greenup. 1985. Port-Orford-cedar root rot on the Siskiyou National Forest in Oregon. U.S. Department of Agriculture, Forest Service, Forest Insect and Disease Management, Pacific Northwest Region, Portland, Oregon, USA.

Hastings, A. 1980. Disturbance, coexistence, history, and competition for space. Theoretical Population Biology **18**:363–373.

Hofacker, T.H., M.D. South, and M.E. Mielke. 1993. Asian gypsy moths enter North Carolina by way of Europe: a trip report. Gypsy Moth News **33**: 13–15.

Inouye, R.S., and D. Tilman. 1995. Convergence and divergence of old-field vegetation after 11 years of nitrogen addition. Ecology **76**:1872–1887.

IPCC (Intergovernmental Panel on Climate Change). 1996. Climate change 1995. Impacts, adaptations, and mitigation of climate change: scientific-technical analyses. Cambridge University Press, Cambridge, England.

Jaarsma, C.F. 1997. Approaches for the planning of rural road networks according to sustainable land-use planning. Landscape and Urban Planning **39**:47–54.

Jones, C.G., J.H. Lawton, and M. Shachak. 1994. Organisms as ecosystem engineers. Oikos **69**:373–386.

Kates, R.W., W.C. Clark, V. Norberg-Bohm, and B.L. Turner II. 1990. Human sources of global change: a report on priority research initiatives for 1990–1995. Discussion Paper G-90-08. Global Environmental Policy Project, John F. Kennedy School of Government, Harvard University, Cambridge, Massachusetts, USA.

Lee, R.G., R.O. Flamm, M.G. Turner, C. Bledsoe, P. Changler, C. DeFerrari, R. Gottfried, R.J. Naiman, N. Schumaker, and D. Wear. 1992. Integrating sustainable development and environmental vitality. Pages 499–521 *in* R.J. Naiman, editor. New perspectives in watershed management. Springer-Verlag, New York, New York, USA.

Liu, D.S., L.R. Iverson, and S. Brown. 1993. Rates and patterns of deforestation in the Philippines: application of geographic information system analysis. Forest Ecology and Management **57**:1–16.

Lovejoy, T.E., R.O. Bierregard, A.B. Rylands, J.R. Malcolm, C.E. Quintela, L.H. Harper, K.S. Brown, Jr., A.H. Powell, A.V.H. Powell, H.O.R. Schubert, and M.B. Hays. 1986. Edge and other effects of isolation on Amazonian forest fragments. Pages 257–285 *in* M.E. Soulé, editor. Conservation biology: the science of scarcity and diversity. Sinauer Associates, Sunderland, Massachusetts, USA.

Lugo, A.E., and F.N. Scatena. 1996. Background and catastrophic tree mortality in tropical moist, wet, and rain forests. Biotropica **28**:585–599.

Mahlman, J.D. 1997. Uncertainties in projections of human-caused climate warming. Science **278**:1416–1417.

Mann, L.K., A.W. King, V.H. Dale, W.W. Hargrove, R. Washington-Allen, L. Pounds, and T.A. Ashwood. 1999. The role of soil classification in geographic information system modeling of habitat pattern: threatened calcareous ecosystems. Ecosystems **2**:524–538.

McNaughton, S.J. 1979. Grassland–herbivore dynamics. Pages 46–81 *in* A.R.E. Sinclair and M. Norton-Griffiths, editors. Serengeti: dynamics of an ecosystem. University of Chicago Press, Chicago, Illinois, USA.

McNaughton, S.J. 1993. Biodiversity and function of grazing ecosystems. Pages 361–383 *in* E.D. Schulze and H.A. Mooney, editors. Biodiversity and ecosystem function. Springer-Verlag, New York, New York, USA.

Merriam, G., M. Kozakiewiez, E. Tsuchiya, and K. Hawley. 1989. Barriers as boundaries for metapopulations and demes of *Peromyscus leucopus* in farm landscapes. Landscape Ecology **2**:227–235.

Meyer, W.B., and B.L. Turner II, editors. 1994. Changes in land use and land cover: a global perspective. Cambridge University Press, Cambridge, England.

Mitchell, R. 1984. The ecological basis for comparative primary productivity. Pages 13–53 *in* R. Lowrance, B.R. Stinner, and G.J. House, editors. Agricultural ecosystems. Wiley Interscience, New York, New York, USA.

Mladenoff, D.J., T.A. Sickley, R.G. Haight, and A.P. Wydeven. 1995. A regional landscape analysis of favorable gray wolf habitat in the northern Great Lakes region. Conservation Biology **9**:279–294.

Motzkin, G., D. Foster, A. Allen, J. Harrod, and R. Boone. 1996. Controlling site to evaluate history: vegetation patterns of a New England sand plain. Ecological Monographs **66**:345–365.

Naiman, R.J., and H. Décamps. 1997. The ecology of interfaces: riparian zones. Annual Review of Ecology and Systematics **28**:621–658.

Naiman, R.J., and K.H. Rogers. 1997. Large animals and system-level characteristics in river corridors. BioScience **47**:521–529.

Naiman, R.J., J.J. Magnuson, D.M. McKnight, and J.A. Stanford, editors. 1995. The freshwater imperative: a research agenda. Island Press, Washington, D.C., USA.

Newmark, W.D. 1995. Extinction of mammal populations in western North American National Parks. Conservation Biology **9**:512–525.

Noss, R.F., and B. Csuti. 1994. Habitat fragmentation. Pages 237–264 *in* G.K. Meffe and C.R. Carroll, editors. Principles of conservation biology. Sinauer Associates, Sunderland, Massachusetts, USA.

Odum, E.P. 1969. The strategy of ecosystem development. Science **164**:262–270.

Paine, R.T. 1969. A note on trophic complexity and community stability. American Naturalist **103**:91–93.

Paine, R.T. 1995. A conversation on refining the concept of keystone species. Conservation Biology **9**:962–964.

Paine, R.T., M.J. Tegner, and A.E. Johnson. 1998. Compounded perturbations yield ecological surprises: everything else is business as usual. Ecosystems **1**:535–545.

Patterson, B.D. 1987. The principle of nested subsets and its implications for biological conservation. Conservation Biology **1**:323–334.

Pearson, S.M. 1993. The spatial extent and relative influence of landscape-level factors on wintering bird populations. Landscape Ecology **8**:3–18.

Pearson, S.M., M.G. Turner, R.H. Gardner, and R.V. O'Neill. 1996. An organism-based perspective of habitat fragmentation. Pages 77–95 *in* R.C. Szaro, editor. Biodiversity in managed landscapes: theory and practice. Oxford University Press, New York, New York, USA.

Peet, R.K. 1992. Community structure and ecosystem function. Pages 103–151 *in* D.C. Glenn-Lewin, R.K. Peet, and T.T. Veblen, editors. Plant succession-theory and prediction. Chapman & Hall, New York, New York, USA.

Perlin, J. 1989. A forest journey: the role of wood in the development of civilization. Harvard University Press, Cambridge, Massachusetts, USA.

Peterjohn, W.T., and D.L. Correll. 1984. Nutrient dynamics in an agricultural watershed: observations on the role of a riparian forest. Ecology **65**:1466–1475.

Pickett, S.T.A., and P.S. White, editors. 1985. The ecology of natural disturbance and patch dynamics. Academic Press, New York, New York, USA.

Pickett, S.T.A., S.C. Collins, and J.J. Armesto. 1987*a*. Models, mechanisms, and pathways of succession. Botanical Review **53**:335–371.

Pickett, S.T.A., S.C. Collins, and J.J. Armesto. 1987*b*. A hierarchical consideration of causes and mechanisms of succession. Vegetation **69**:109–114.

Pickett, S.T.A., and M.L. Cadenasso. 1995. Landscape ecology: spatial heterogeneity in ecological systems. Science **269**:331–334.

Poff, N.L., J.D. Allan, M.B. Bain, J.R. Karr, K.L. Prestegaard, B.D. Richter, R.E. Sparks, and J.C. Stromberg. 1997. The natural flow regime. BioScience **47**:769–784.

Polis, G.A., and K.O. Winemiller. 1996. Food webs: integration of patterns and dynamics. Chapman & Hall, New York, New York, USA.

Power, M.E. 1992. Top-down and bottom-up forces in food webs: do plants have primacy? Ecology **73**:733–746.

Power, M.E., D. Tilman, J.A. Estes, B.A. Menge, W.J. Bond, L.S. Mills, G. Daily, J.C. Castilla, J. Lubchenco, and R.T. Paine. 1996. Challenges in the quest for keystones. Bioscience **46**:609–620.

Pulliam, H.R., J.B. Dunning, and J. Liu. 1992. Population dynamics in complex landscapes: a case study. Ecological Applications **2**:165–177.

Reice, S.R. 1994. Nonequilibrium determinants of biological community structure. American Scientist **82**:424–435.

Richards, J.F. 1990. Land transformations. Pages 163–178 *in* B.L. Turner II, W.C. Clark, R.W. Kates, J.F. Richards, J.T. Matthews, and W.B. Meyer, editors. The Earth as transformed by human action: global and regional changes in the biosphere over the past 300 years. Cambridge University Press, New York, New York, USA.

Riebsame, W.E., W.J. Parton, K.A. Galvin, I.C. Burke, L. Bohren, R. Young, and E. Knop. 1994. Integrated modeling of land use and cover change. Bioscience **44**:350–356.

Risser, P.G. 1985. Toward a holistic management perspective. BioScience **35**:414–418.

Robinson, S.K., F.R. Thompson III, T.M. Donovan, D.R. Whitehead, and J. Faaborg. 1995. Regional forest fragmentation and the nesting success of migratory birds. Science **267**:1987–1990.

Roland, J. 1993. Large-scale forest fragmentation increases the duration of tent caterpillar outbreak. Oecologia **93**:25–30.

Romme, W.H., and D.G. Despain. 1989. Historical perspective on the Yellowstone fires of 1988. BioScience **39**:695–699.

Rosenzweig, M.L., and Z. Abramsky. 1993. How are diversity and productivity related? Pages 52–65 *in* R. Ricklefs and D. Schluter, editors. Species diversity in ecological communities: historical and geographical perspectives. Chicago University Press, Chicago, Illinois, USA.

Saunders, D.A., R.J. Hobbs, and C.R. Margules. 1991. Biological consequences of ecosystem fragmentation: a review. Conservation Biology **5**:18–32.

Scatena, F.N., S. Moya, C. Estrada, and J.D. Chinea. 1996. The first five years in the reorganization of aboveground biomass and nutrient use following Hurricane Hugo in the Bisley Experimental Watersheds, Luquillo Experimental Forest, Puerto Rico. Biotropica **28**:424–440.

Schmidt, W. 1989. Plant dispersal by motor cars. Vegetation **80**:147–152.

Schullery, P. 1989. The fires and fire policy. BioScience **39**:686–694.

Schwartz, M.W., editor. 1997. Conservation in highly fragmented landscapes. Chapman & Hall, New York, New York, USA.

Sears, P.B. 1980. Deserts on the march. University of Oklahoma Press, Norman, Oklahoma, USA.

Sessions, J., G. Reeves, K.N. Johnson, and K. Burnett. 1997. Implementing spatial planning in watersheds. Pages 271–283 *in* K.A. Kohm and J.F. Franklin, editors. Creating a forestry of the 21st century. Island Press, Washington, D.C., USA.

Sharov, A.A., A.M. Liebhold, and E.A. Roberts. 1997. Correlation of counts of gypsy moths (*Lepidoptera, Lymantriidae*) in pheromone traps with landscape characteristics. Forest Science **43**:483–490.

Soranno, P.A., S.L. Hubler, S.R. Carpenter, and R.C. Lathrop. 1996. Phosphorus loads to surface waters: a simple model to account for spatial pattern of land use. Ecological Applications **6**:865–878.

Soulé, J., and J. Piper. 1992. Farming in nature's image. Island Press, Washington, D.C., USA.

Sousa, W.P. 1984. The role of disturbance in natural communities. Annual Review of Ecology and Systematics **15**:353–391.

Southern Appalachian Assessment. 1996. The Southern Appalachian assessment: summary report. USDA Forest Service. Washington, D.C., USA.

Southworth, F., V.H. Dale, and R.V. O'Neill. 1991. Contrasting patterns of land use in Rondônia, Brazil: simulating the effects on carbon release. International Social Sciences Journal **130**:681–698.

Sparks, R.E. 1996. Ecosystem effects: positive and negative outcomes. Pages 132–162 *in* S.A. Changnon, editor. The great flood of 1993: causes, impacts, and responses. Westview Press, Boulder, Colorado, USA.

Sparks, R.E., J.C. Nelson, and Y. Yin. 1998. Naturalization of the flood regime in regulated rivers. BioScience **48**:706–720.

Temple, S.A. 1986. Predicting impacts of habitat fragmentation on forest birds: a comparison of two models. Pages 301–304 *in* J. Verner, M.L. Morrison, and C.J. Ralph, editors. Wildlife 2000: modeling habitat relationships of terrestrial vertebrates. University of Wisconsin Press, Madison, Wisconsin, USA.

Terman, M.R. 1997. Natural links: naturalistic golf courses as wildlife habitat. Landscape and Urban Planning **38**:183–197.

Tilman, D. 1996. Biodiversity: population versus ecosystem stability. Ecology **77**:350–363.

Turner, B.L., II, W.C. Clark, R.W. Kates, J.F. Richards, J.T. Matthews, and W.B. Meyer, editors. 1990. The Earth as transformed by human action: global and regional changes in the biosphere over the past 300 years. Cambridge University Press, New York, New York, USA.

Turner, M.G., editor. 1987. Landscape heterogeneity and disturbance. Springer-Verlag, New York, New York, USA.

Turner, M.G. 1989. Landscape ecology: the effect of pattern on process. Annual Review of Ecology and Systematics **20**:171–197.

Turner, B.L., II, R.H. Moss, and D.L. Skole. 1993. Relating land use and global land-cover change: a proposal for an IGBP-HDP core project. HDP Report No. 5, International Geosphere-Biosphere Programme, Stockholm, Sweden.

Turner, M.G., and V.H. Dale. 1998. What have we learned from large, infrequent disturbances? Ecosystems **1**:493-496.

Turner, M.G., W.H. Hargrove, R.H. Gardner, and W.H. Romme. 1994. Effects of fire on landscape heterogeneity in Yellowstone National Park, Wyoming. Journal of Vegetation Science **5**:731–742.

Turner, M.G., R.H. Gardner, and R.V. O'Neill. 1995. Ecological dynamics at broad scales. BioScience **45**(Supplement):S29–S35.

Turner, M.G., D.N. Wear, and R.O. Flamm. 1996. Land ownership and land-cover change in the Southern Appalachian Highlands and the Olympic Peninsula. Ecological Applications **6**:1150–1172.

Turner, M.G., V.H. Dale, and E.E. Everham III. 1997*a*. Fires, hurricanes, and volcanoes: comparing large-scale disturbances. BioScience **47**:758–768.

Turner, M.G., W.H. Romme, R.H. Gardner, and W.W. Hargrove. 1997*b*. Effects of patch size and fire pattern on early post-fire succession on the Yellowstone Plateau. Ecological Monographs **67**:411–433.

Urban, D.L., R.V. O'Neill, and H.H. Shugart. 1987. Landscape ecology. BioScience **37**:119–127.

Valone, T.J., and J.H. Brown. 1995. Effects of competition, colonization, and extinction on rodent species diversity. Science **267**:880–883.

Vitousek, P.M., H.A. Mooney, J. Lubchenco, and J.M. Melillo. 1997. Human domination of Earth's ecosystems. Science **277**:494–504.

Wallin, D.O., F.J. Swanson, and B. Marks. 1994. Landscape pattern response to changes in pattern generation rules: land-use legacies in forestry. Ecological Applications **4**:569–580.

Warner, R.E. 1994. Agricultural land use and grassland habitat in Illinois: future shock for Midwestern birds? Conservation Biology **8**:147–156.

Watt, A.S. 1947. Pattern and process in the plant community. Journal of Ecology **35**:1–22.

Wear, D.N., M.G. Turner, and R.O. Flamm. 1996. Ecosystem management with multiple owners: landscape dynamics in a Southern Appalachian watershed. Ecological Applications **6**:1173–1188.

Weller, D.E., T.E. Jordan, and D.L. Correll. 1998. Heuristic models for material discharge from landscapes with riparian buffers. Ecological Applications **8**:1156–1169.

Wiens, J.A. 1996. Wildlife in patchy environments: metapopulations, mosaics, and management. Pages 53–84 *in* D.R. McCullough, editor. Metapopulations and wildlife conservation. Island Press, Washington, D.C., USA.

Wilkie, D.S., and J.T. Finn. 1988. A spatial model of land use and forest regeneration in the Ituri Forest of northeastern Zaire. Ecological Modeling **41**:307–323.

Wilson, C.J., R.S. Reid, N.L. Stanton, and B.D. Perry. 1997. Effects of land-use and tsetse fly control on bird species richness in southwestern Ethiopia. Conservation Biology **11**:435–447.

Yarie, J.K., L. Viereck, K. Van Cleve, and P. Adams. 1998. Flooding and ecosystem dynamics along the Tanana River. BioScience **48**:690–695.

Zobel, D.B., L.F. Roth, and G.H. Hawk. 1985. Ecology, pathology, and management of Port-Orford-cedar (*Chamaecyparis lawsoniana*). General Technical Report PNW-184. U.S. Department of Agriculture, Forest Service, Portland, Oregon, USA.

Part II

Applying the Principles and Guidelines

2
Effects of Land-Use Change on Wildlife Habitat: Applying Ecological Principles and Guidelines in the Western United States

N. THOMPSON HOBBS and DAVID M. THEOBALD

We discuss ecological principles relevant to impacts of land-use change on wildlife in the Rocky Mountain West. This discussion proceeds in three parts. We first review the history of declines in abundance and distribution of black-tailed prairie dog (*Cynomys ludovicianus*) in the Rocky Mountains and Great Plains to illustrate how failure to apply ecological principles to land use can do lasting harm to natural and human systems. The long-term decline of prairie-dog populations has resulted in the loss of an important agent of natural disturbance, which led to shifts in plant community structure and composition over large areas, and has created effects that reverberate through food webs. We discuss how anthropogenic sources of changes in land cover have fragmented the naturally heterogeneous pattern of distribution of prairie dog populations. We next focus on current stressors in the region associated with growth of the human population, making the case that the region is likely to continue the rapid pace of change seen during the last decade. Finally, we discuss guidelines developed by Dale et al. (Chapter 1) in relation to impacts of development on wildlife in the Rocky Mountain West. We emphasize the need to inform local decisions with the best available ecological data and cite examples from our work in Colorado to illustrate efforts to incorporate guidelines in land-use planning by counties and municipalities.

The landscapes of the western United States support abundant and diverse wildlife. Wildlife populations contribute in important ways to the economy of the region, and they meaningfully enhance the quality of life of its citizens (Decker and Swanson 1973; Kaush 1996; Bullock et al. 1998). The human population of the region has grown rapidly during the last decade, and this growth is projected to continue well into the next millennium, largely as a consequence of immigration (Reibsame et al. 1996). The rapid expansion of the human population threatens habitat for many species of wildlife and this threat creates a fundamental paradox. Wildlife and wildlands are features of the region that attract immigrants and, paradoxically, are resources that are most likely to be impacted by land-use change needed to accommodate growth (Reibsame et al. 1996). Thus, wildlife and wildlife habitat have a reciprocal relationship with land-use change: They are a driver of shifts in land use and they are a resource sensitive to those changes. It follows that interactions between wildlife and people offers a useful way to examine ecological principles relevant to land-use change. Moreover, the widely expressed desire to preserve and protect wildlife motivates the application of ecological guidelines to inform human decisions on the use of land.

This chapter has three main sections. In the first, we offer a specific case history to illustrate the urgent need to apply ecological principles to land-use change. This history traces the deterministic decline in the abundance and distribution of a keystone species, the black-tailed prairie dog (*Cynomys ludovicianus*). We touch on each of the principles described in Chapter 1 and show how failure to attend to that principle led to lasting harm for natural and human systems that depend on the ecological services provided by prairie dogs. In the second section, we describe current patterns of land-use change in the West and discuss the implications of this change for conservation of wildlife. This section documents the continuing pace of land-use change and amplifies the need for applying principles outlined in the first section. In the final section, we look to the future by discussing the application of ecological guidelines on land use to enhance conservation of wildlife in the West. In all three sections we offer concrete examples of the general concepts described in Chapter 1.

2.1 Ecological Principles of Wildlife, and Land Use: A Case History

2.1.1 The Place Principle: Local Climatic, Hydrologic, Edaphic, and Geomorphologic Factors as Well as Biotic Interactions Strongly Affect Ecological Process and the Abundance and Distribution of Species at Any One Place

The eastern front of the Rocky Mountains is composed of a diverse set of habitats supporting more than 400 species of vertebrates. Specific conditions in the biotic and abiotic environment determine areas of the landscape that are suitable for any given species. Prairie dogs establish colonies in areas of grassland with soil conditions suitable for burrowing (Fitzgerald et al. 1994). As described in the subsequent sections, the presence of prairie dogs on the landscape is fundamentally important to sustaining food webs, maintaining disturbance regimes, and conserving the biotic diversity of the region (Sharps and Uresk 1990; Miller et al. 1994).

2.1.2 The Time Principle: Ecological Processes Function at Many Time Scales—Some Long, Some Short—and Ecosystems Change Through Time

Analyzing changes in ecological states and processes over time is fundamental to understanding anthropogenic effects of changes in land use. To that end, describing changes in the distribution and abundance of organisms occurring in parallel with shifts in land use can reveal important impacts of human decisions on the operation of natural systems. During the last century, the abundance of black-tailed prairie dogs has declined throughout their native range, and these reductions can be attributed in large measure to human choices on land use and management. Prairie dog populations were historically distributed throughout grasslands of western North America and numbered to as many as 5 billion individuals in the early 1900s (Long 1998). Current populations are believed to be less than 3% of historic levels (Miller et al. 1994; Wuerthner 1996). Hundreds of thousands of square kilometers that once supported abundant networks of "dog-towns" now have only small remnants of prairie dog activity (Sharps and Uresk 1990). As a result, prairie dogs are now listed in Category 2 under the Endangered Species Act (Miller et al. 1994; Wuerthner 1996). Category 2 species are believed to be in serious jeopardy of extinction.

The danger of extinction can be traced to direct effects of human actions on vital rates of populations and to indirect effects on availability of suitable habitat. Because prairie dogs are believed to compete with domestic livestock for herbaceous forage, and their burrows are viewed as physical hazards to cattle and horses, ranchers throughout the region sought to "manage" prairie dogs as "noxious pests" (Dold 1998). Poisoning and shooting eradicated many populations (Sharps and Uresk 1990; Roemer and Forest 1996). Populations were also decimated by introduction of an exotic pathogen from Europe, bubonic plague (Barnes 1993; Fitzgerald et al. 1994). The greatest cumulative impact, however, resulted from the conversion of native grasslands to other uses–grassland to crops under the plow earlier in the century and grassland to residential development under the bulldozer during the last 20 years (Dold 1998). Effects of land management and shifts in land use have accumulated slowly over time; largely as a result of many decisions made singly by individual landowners.

Deterministic declines in populations, like those seen in prairie dogs, result from a fundamental mismatch in time scales. The scale of human decisions, such as the choices to eradicate species and convert land, extends well beyond the scale of life histories of organisms. That is, the processes of births and deaths of individuals occur within processes operating over longer time spans—the conversion of prairie to agriculture and development. These multiscale processes interact with the time scale of human perception. Steady declines in the abundance of an organism may occur too slowly to be perceived by people. Differences in scales of processes and perception can cause fundamental harm to important components of ecosystems, such as prairie dogs (Lee 1993).

2.1.3 The Species Principle: Particular Species and Networks of Interacting Species Have Key, Broad-Scale Ecosystem–Land Effects

Changes in the abundance and distribution of individual species can exert effects that reverberate in unexpected ways throughout ecosystems. It has been estimated that prairie dogs are in some way connected to the livelihoods of some 170 other species of vertebrates, including a wide variety of predators and herbivores (Sharps and Uresk 1990; Lee 1993). For example, prairie dogs are the primary item of prey for black-footed ferrets as well as several species of raptors. The loss of prairie dog populations ripples through food webs, and these effects are believed to be directly responsible for the near extinction of black-footed ferrets and have exacerbated effects of other anthropogenic stressors on raptor populations (Long 1998).

2.1.4 The Disturbance Principle: The Type, Intensity, and Duration of Disturbance Shape the Characteristics of Populations, Communities, and Ecosystems

Disturbance is a ubiquitous and potent influence on the dynamics of natural systems. Mammals can be important agents of disturbance, and their effects on the physical environment can cause profound changes in the structure and function of ecosystems (see reviews in Jones et al. 1994; Brown 1995; Hobbs 1996). Prairie dogs modify the physical environment by creating networks of burrows. They modify the biotic environment by intense grazing in the area within and around those networks. These modifications cause effects that extend across ecosystems at multiple scales (Sharps and Uresk 1990; Miller et al. 1994).

At the broadest scale, loss of habitat disturbance by prairie dogs is believed to be responsible for the extensive invasion of woody shrubs into grasslands during the later part of the 1900s (Weltzin et al. 1997). Widespread eradication of prairie dogs eliminated a significant constraint to woody plant establishment, and the loss of this control has accelerated the transition of grassland to shrublands and woodlands. It follows that land management designed to remove an ostensible competitor with livestock appears to have "contributed significantly to development of another management problem that is now a major detriment to sustainable livestock production" (Weltzin et al. 1997).

At intermediate scales, prairie dog towns create spatial heterogeneity in plant communities by altering plant phenology and growth form (Detling and Painter 1983; Cid et al. 1991). Grazing by prairie dogs maintains vegetation in an immature state, which results in enhanced forage quality. This enhancement provides important nutritional alternatives for large mobile herbivores like bison and cattle (Coppock et al. 1980, 1983*a,b*; Whicker and Detling 1988; Sharps and Uresk 1990). At fine scales, physical alteration of the environment creates nesting habitat for other species like burrowing owls (Plumpton and Lutz 1993; Desmond et al. 1995; Desmond and Savidge 1996).

These multiscale effects offer evidence that prairie dogs function as keystones in the habitats of the West. It follows that land-use change that causes direct harm to keystone species like the prairie dogs can indirectly harm many other species that depend on the ecological services they provide.

2.1.5 The Landscape Principle: The Size, Shape, and Spatial Relationships of Land-Cover Types Influence the Dynamics of Populations, Communities, and Ecosystems

Historically, colonies of prairie dogs were distributed extensively across the grasslands of the Great Plains and the eastern front of the Rocky Mountains (Miller et al. 1994). Because soil conditions favorable for burrowing determine the suitability of habitat for prairie dogs, heterogeneity in soil conditions created a naturally "fragmented" distribution of colonies. This natural fragmentation has been intensified by land-use change (Daley 1992). Agricultural and residential development have compressed the extent of colonies (i.e., reduced patch size) and have expanded the area of unsuitable habitat separating them (Daley 1992). Fragmentation (natural and anthropogenic) exerts opposing effects on the viability of prairie dog populations. Viability is harmed by reductions in population size and by loss of genetic diversity resulting from reduced rates of gene flow among spatially isolated populations (Daley 1992). In contrast, spatial separation tends to isolate epidemics of plague that otherwise might affect much larger numbers of animals in spatially contiguous colonies.

2.1.6 Landscape Principles and the Intelligent Design of Land Use

Deliberate human enterprises require design. Designing a change in land use usually emphasizes human needs and efficiencies–the layout of an agricultural field or a housing development is most likely based on economic imperatives and constraints. The prairie dog example illustrates that ecological imperatives and constraints exist alongside economic ones. Principles of economy can contribute to the intelligent design of economic systems. However, intelligent design of the use of land must embrace principles of ecology as well as economy. Failure to do so can result in unexpected, sometimes unpleasant, surprises for people and natural systems (e.g., Weltzin et al. 1997).

2.2 Patterns of Landscape Change in the Rocky Mountain West

In the previous section, we offered a history of impacts of land-use change on a keystone species to develop the idea that failing to apply ecological principles to human decisions can produce far-reaching effects. In this section, we focus on the present to show that the rate of land change in the region is accelerating, amplifying the need for choices that consider impacts on natural systems.

The Rocky Mountain states (Arizona, Colorado, Idaho, Montana, Nevada, New Mexico, Utah, and Wyoming) are experiencing rapid landscape transformation, primarily as a result of conversion from agricultural to residential land use (Reibsame et al. 1996). This widespread shift in land-use change is driven by rates of population growth double or even triple those seen in the other regions of the United States during 1990 to 1996. Population growth in the Rocky Mountains averaged 2.1% growth annually compared to 0.9% for the remainder of the United States (Census Bureau 1997). Nineteen of the 281 Rocky Mountain counties exceeded 6% annual growth rate. The Rocky Mountain states are projected to grow by nearly 6.5 million additional people by the year 2025 (Census Bureau 1997), so that in the span of one generation, this region faces the addition of the combined 1996 populations of Colorado, Montana, and New Mexico. Although periods within the past 150 years have had equal or greater rates of growth, the magnitude and extent of the landscape change caused by this population growth are unprecedented. One clear indication of this unprecedented change is urban sprawl, where the density of a city decreases while the overall population increases (Diamond and Noonan 1996).

Though much of the recent growth is focused on large urban areas, rural areas have also exhibited remarkable growth. Resource-based economies operating in landscapes relatively free of development continue to be defining characteristics of this region (Riebsame et al. 1997). The amount of land in agricultural use has increased steadily since the late 1800s, reaching a peak in the 1960s. However, since 1978, the Rocky Mountain states have lost over 3400 km^2 of agricultural land annually (Census of Agriculture 1993). Much of this loss is the result of conversion of forest, rangeland, and pasture to development, particularly development occurring at relatively

TABLE 2.1. Total acreage and trends of development by category in Colorado during 1960–2020.

Development category	Housing density (acres/unit)	Area (sq mi) 1960	1990	2020	Annual growth rate (%) 1960–1990
Urban	<2	166	519	681	7.1
Suburban	2–10	164	483	702	6.4
Exurban	10–40	1,080	3,935	6,491	8.8
Ranch	40–80	1,084	2,903	3,869	5.5
Rural	>80	53,730	49,549	46,034	−0.2

Source: U.S. Bureau of Census Block-group statistics.

low density. For example, in Colorado almost 4000 square miles of landscape was developed at exurban densities (1 house per 10–40 acres) during 1960–1990, while less than 450 square miles was developed at urban densities (1 house per <2 acres) (Table 2.1).

Although much of the region is publicly owned, development nonetheless exerts a strong, fragmenting influence on the landscape. Private land dissects the public land ownership pattern, follows the dendritic valley and stream-bottom patterns, and extends upslope to the forest fringe. As a result of this pattern, 80% of the forested land in the Rocky Mountain region is within 3 km of private land (Theobald 1999). Moreover, private land includes many critically important resources for wildlife, including riparian zones, nesting habitat, wetlands, winter ranges, and migration corridors. Changes in the use of private land adjacent to forested areas influence the ecological functioning and impact the management of ecological processes (such as forest fire suppression) (Theobald 1999). Consequently, sustaining the region's natural character and values depends in important ways on considering the effects of development on wildlife habitat and biotic communities in land-use decisions.

2.3 Applying Ecological Guidelines to Land-Use Decisions in the West

In the absence of efforts to conserve wildlife populations and habitats, the rapid changes in land use outlined above will almost certainly create problems such as those illustrated in the prairie dog case history. Alternatively, land-use change shaped by guidelines incorporating the essential features of those principles can meet the needs of people for economic vitality, while, at the same time, meeting the needs of wildlife for functioning ecosystems. Here we discuss guidelines developed by Dale et al. (Chapter 1) in relation to impacts of development on wildlife in the Rocky Mountain West.

2.3.1 Examine Impacts of Local Decision in a Regional Context

John Wesly Powell, a geographer who explored and mapped much of the West, recognized the importance of aligning political boundaries within ecological units like watersheds. Despite this recognition, the West today is divided politically in ways that fail to consider the structure and function of natural landscapes. Thus, although rapid land-use change is occurring at regional scales in the Rocky Mountain West, the decisions that affect the temporal trajectory and spatial pattern of change are most often made locally (Hauber and Hobbs, Chapter 12). This disparity between the scale of change and the scale of decisions influencing change creates problems in conserving wildlife and wildlife habitat because the area influenced by these decisions fails to match the areas where conservation must occur.

This problem can be remedied, at least in part, by considering regional needs and objectives in developing local conservation plans. It follows that knowledge of the status of species beyond political boundaries is a prerequisite for informed decisions at the local level. Without looking beyond the boundaries of counties or municipalities, it is likely that conservation efforts will be wasted in protecting species locally that are regionally abundant or that opportunities will be forgone to conserve species locally that are regionally rare. For example, there are many counties along the Front Range in Colorado where prairie dogs are sufficiently abundant to be considered a pest that stands in the way of development. This local view needs to be shaped by the regional declines in their abundance and distribution to recognize the need for local conservation.

2.3.2 Plan for Long-Term Change and Unexpected Events

The Time Principle, described above, reveals that the temporal scale of human perception includes a much narrower extent than the temporal scale of many important ecological processes. A consequence of this mismatch is that human decisions often fail to consider effects that may require many years to express themselves. A corollary idea is that these decisions often neglect unusual events, those that are not likely to be experienced during the lives of decision makers.

Planners consider 100-year flood regimes in mapping areas of the landscape that are too hazardous to allow development. In a similar way, conservation plans must consider the full range of natural variability experienced by species and natural communities. They must consider the possibility of change that operates at rates slow enough to be difficult to detect over the scale of perception of decision makers. There are a couple of good examples. Species that are not currently rare and are not disappearing rapidly, but experience steady, deterministic declines in abundance and distribution may be overlooked in shot-term planning. Prairie dogs are an excellent example. Decline in their abundance has been slow but relentless. An additional example of slow but unrelenting change can be seen in the depletion of groundwater suppliers in many areas of the West. Human demands on aquifers for agriculture exceed the ability of groundwater to recharge them. As a result, these water sources are likely to be depleted within the next 50–100 years, which will eliminate wetland habitats critical to the persistence of many wildlife species. This time span makes the problem seem remote, but its solution can be achieved only by action now.

To that end, we developed maps of "habitat at risk" to help local governments identify needs for wildlife habitat protection within their jurisdiction based on long-term projections of development pressure. We identified areas with undeveloped native vegetation that are in the path of development (Fig. 2.1; see color insert). Areas with urban/built-up or cropland cover were assumed to have already undergone substantial land-cover conversion and were excluded from the analysis. Risk level corresponded to the magnitude of change between 1960 and projected (2020) levels of development.

2.3.3 Preserve Rare Landscape Elements and Associated Species

The Place Principle is responsible for the long-standing observation of ecologists that the diversity of wildlife species found on a landscape depends in a fundamental way on the diversity of habitats available to them. Habitat diversity is sensitive to the presence or absences of rare components of landscapes. Elimination of these rare components can reduce habitat and species diversity in ways far disproportionate to their area. It follows that conversion of uncommon landscape elements to human uses can cause precipitous declines in biotic diversity. A more optimistic view is that land-use change can proceed with relatively minor impacts on diversity if that change occurs in the most common habitat types.

This view can form the basis for land-use plans that conserve the full range of wildlife species found in a landscape. Plans should encourage development in common habitats and discourage development in uncommon ones. The needs of people can be accommodated alongside the needs of wildlife if people are willing to adjust their distribution to avoid habitats that are in short supply on the landscape.

We developed maps of rare landscape elements as part of a master land-use plan for Larimer County Colorado (Fig. 2.2; see color insert). Two types of maps focusing on rare elements were used in this planning process. The first displayed conservation sites mapped by the Colorado Natural Heritage Program (Fig. 2.2). Conservation sites are areas of the landscape containing rare and sensitive plants and animals. The boundaries of the sites are delineated to include the ecological processes needed to sustain the rare elements within them. In addition, we used land-cover maps derived from satellite imagery to delineate vegetation types contributing less than 3% of the area of the county. These maps (along with others) were used to decide which areas of the county would require conservation plans prior to approval of development proposals.

2.3.4 Avoid Land Uses That Deplete Natural Resources

Sustaining human and natural economies depends on the preservation of capital supporting production and renewal. In economic systems, this capital includes physical plant, human resources, and finance. In ecological systems, analogous capital is composed of natural resources. Land uses that consume natural resources at a rate exceeding their rate of renewal will deplete those resources and, in the absence of replacement from external sources, will lead to degradation of the natural system. Examples of such depletion include erosion of soil resulting from inappropriate tillage, and loss of wetlands resulting from misplaced development or excessive irrigation.

2.3.5 Retain Large Contiguous Areas That Contain Critical Habitats and Maintain Connections Among These Areas

Long-term trends in the abundance and distribution of wildlife species are strongly influenced by the area of habitat available to them. This simple truth is responsible for the widespread observation that wildlife abundance declines when development perforates and dissects an otherwise intact natural landscape. It also explains, at least in part, why large predators are particularly sensitive to the impacts of development. Predators require a much greater area of habitat per individual than herbivores do, and large predators require greater area than small ones. Because large intact patches of habitat tend to support more individuals in a population, those populations are more likely to persist than populations in small, fragmented habits. The relationship among persistence, abundance, and habitat area is believed to underlie the widespread observation that the diversity of species tends to increase in a nonlinear way with the area of habitat (Brown 1971; Gilpin and Diamond 1980; Harris 1985). It follows from this relationship that sustaining the abundance and diversity of wildlife in the West will depend on conserving large, intact patches of habitat wherever possible.

We developed a statewide map of "patch quality" in Colorado for use in land-use decisions by local governments (Fig. 2.3). The patch quality map identifies large intact patches of vegetation and weights them by relative rarity of species' habitat. The patch quality map was created by counting the number of species that have potentially suitable habitat within each patch on a vegetation map. We then weighted each patch by the ratio of the patch

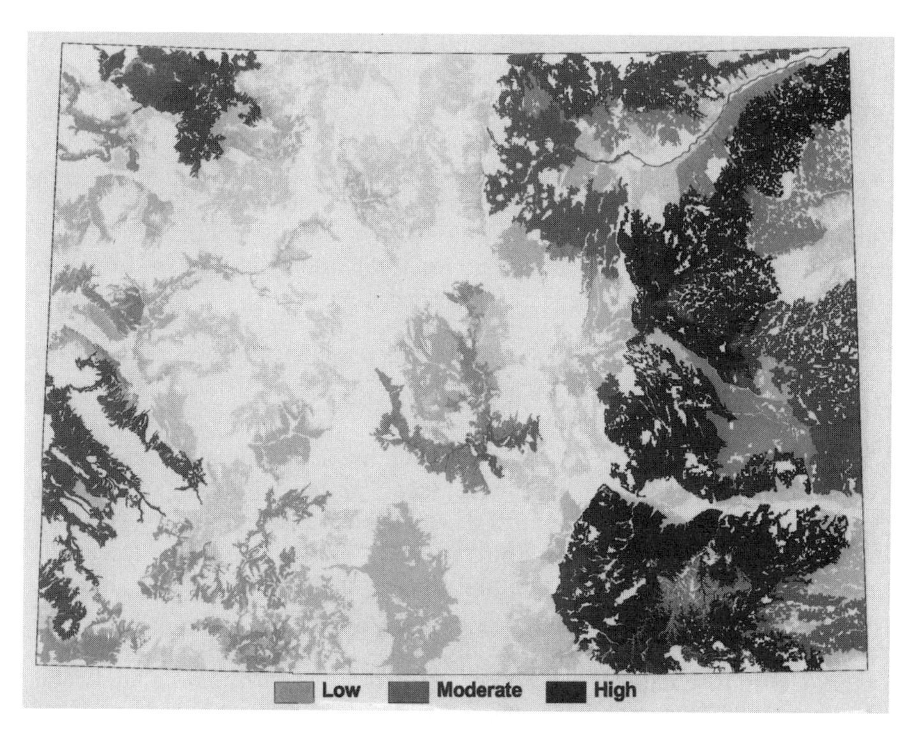

FIGURE 2.3. We developed maps of patch quality to help local governments and land trusts identify priorities for habitat protection. The patch-quality map was created by counting the number of species that have potentially suitable habitat within each patch on a vegetation map. We then weighted each patch by the ratio of the patch area to the total area of potential suitable habitat for each species.

area to the total area of potential suitable habitat for each species. Small patches are given greater weight if they comprise a large proportion of a species' habitat. Larger patches have higher values relative to their size.

2.3.6 Minimize Introduction and Spread of Nonnative Species

Introduction of exotic species can cause extinction of native species of wildlife by altering relationships among competitors and between predators and prey (Short and Smith 1994; Kirch 1996; Marschall and Crowder 1996; May and Norton 1996; Turner 1996; Calver et al. 1998; Sinclair et al. 1998). Reid and Miller (1989) estimated that introductions of nonnative species were implicated in more extinctions of terrestrial and aquatic vertebrates than any other single cause. Moreover, exotic invasion can accelerate habitat loss caused by other anthropogenic stresses. A particularly useful example of this acceleration is found in riparian zones of the Western United States. Damming rivers to impound water for agriculture and residential use has reduced riparian habitat throughout the Rocky Mountains. Simultaneously, invasion of exotic trees and shrubs in remaining riparian areas has further reduced suitability of habitat for many species (DeBolt and McCune 1995; Tabacchi et al. 1996; Ellis et al. 1997; Germaine et al. 1998; Lynn et al. 1998). It follows that wildlife conservation plans should encourage use of native species for landscaping and restoration. Exotic species should be actively controlled.

2.3.7 Compensate for the Effects of Development on Ecological Processes

Historically, conserving wildlife populations and habitats focused on proscribing uses of land that were incompatible with the needs of a particular species or community. Thus, habitat protection was virtually synonymous with "leaving nature alone." Traditional management of national parks, wildlife refugees, and other conservation areas emphasized this non-interventionist approach. In contrast, a contemporary view recognizes that conservation areas are imbedded in context that often differs markedly in its composition and behavior from the surrounding landscape (Botkin 1990; Pickett et al. 1992). Many of the internal dynamics of conservation areas, areas ostensibly "protected," respond in a sensitive way to these external conditions (Pickett et al. 1992). It follows that a central challenge in efforts to conserve wildlife populations and habitats depends on compensating for effects of development surrounding conservation areas (Halvorson and Davis 1996).

An excellent example of the need to compensate for effects of a changing context resulting from development is the loss of natural disturbance that occurs when conservation areas are embedded in a developed landscape. We studied effects of a changing large-scale context on fine-scale landscape structure of the Rawhide Wildlife Management Area in southern Wyoming (Miller et al. 1995). The North Platte River bisects the area, supporting riparian zones with an overstory of plains cottonwood (*Populus* spp.). Midgrass prairie predominates the adjacent uplands. Although the area has been "protected" from development for more than 30 years, it has nonetheless changed dramatically as result of extensive impoundment of water along sections of the Platte upstream from the area. As a consequence of construction of upstream dams, the river within the Rawhide area declined in wetted area by 75% between 1937 and 1990. More importantly, the area of the landscape proportion occupied by cottonwood (*Populus* spp.) stands with less than 30% canopy closure increased, while stands with greater than 70% canopy closure decreased during this period, indicating a shift from young, dense stands to older, more open stands. The absence of periodic flooding from a free-flowing river prevented recruitment of young cottonwood. These changes illustrate that despite regulating land use at the site to preserve riparian habitat for wildlife, the effects of changes in the landscape surrounding the site caused marked change in those habitats.

2.3.8 Implement Land-Use and Management Practices Compatible with the Natural Potential of the Area

One of the defining characteristics of the West is an abundance of public land, land that provides a variety of environmental values, not the least of which is habitat for diverse species of wildlife (Wolfe et al. 1996). However, many wildlife species in the West depend on habitat that is privately owned. For example, migratory ungulates spend the summer at high elevation on public land, but require lower elevation range in mountain valleys during the winter. For the most part, these winter ranges are found on private land (Baumeister 1996). Historically, the primary land use in these valleys was ranching. A ranching-based economy supports human populations at relatively low density, and well-managed grazing is compatible with the habitat needs of many species of wildlife (Wolfe et al. 1996). However, ranches are being subdivided for residential development throughout the region (Reibsame et al. 1996) and these subdivisions degrade the quality and amount of habitat available for wildlife (Henderson and O'Herren 1992; Theobald et al. 1996). This trend has fostered political alliances between environmentalists who value wildlife habitat and open space, and land-owners who value ranching as a way of life. These alliances are working with local governments to develop land-use plans compatible with the natural potentials of the area to support economies based on livestock grazing and wildlife-oriented recreation.

2.4 Conclusions: Next Steps for Implementing Guidelines

We use the case history of the decline of prairie dogs in the West to illustrate far-reaching impacts that can occur as a result of failure to consider ecological principles in decisions on land use. These impacts include loss of biological diversity and a variety of ecosystem services. Avoiding these impacts will require application of ecological guidelines to shape land-use decisions.

Our experience in applying ecological research to support decisions on land-use planning in the West (Dale et al. 2000) suggests that several challenges must be overcome to assure that the guidelines described above are meaningfully applied in established decision-making processes. First, guidelines must be effectively translated into specific prescriptions for action, usually at the local level. This translation requires two things: ecological expertise and ready access to local data on wildlife and natural communities. This is difficult because the diffuse nature of land-use planning

(Hauber and Hobbs, Chapter 12) means that expertise and data will be spread thin across many different decision-making processes. A potential remedy to this difficulty is offered by web-based information systems that allow local citizens and governments to gain rapid access to ecological data and analysis. We have developed such a system for Colorado (see *http://ndis.nrel.colostate.edu*).

Second, process research and modeling must overcome the traditional reluctance of ecologists to work on human-dominated systems. A recent literature search revealed that fewer than 15% of articles published in ecological journals during the last decade dealt with effects of residential development on natural systems (Hobbs, unpublished data). Given that development of land for housing is one of the major agents of landscape change occurring in the United States (Dale et al. 2000), ecologists shift research to understand the effects of this change.

Finally, implementation of ecological guidelines on land use depends on applying the proper levers of policy and management. Linkages between land-use guidelines and policy alternatives are not well developed (see Duerksen et al. 1997). Nonetheless, the fundamental problem of implementation reduces to a fairly simple truth. The utility of guidelines will depend in a fundamental way on convincing private landowners that implementation of land-use guidelines is in their best interest. Until that occurs, the best we can hope for is a minor tidying of an expanding infrastructure.

References

Barnes, A.M. 1993. A review of plague and its relevance to prairie dog populations and the black-footed ferret. U.S. Fish & Wildlife Service Biological Report **13**:28–37.

Baumeister, T. 1996. Sustaining elk on the Rocky Mountain Front: the importance of private lands. Intermountain Journal of Science **2**:40–41.

Botkin, R.D. 1990. Discordant Harmonies. Oxford University Press, Oxford, England.

Brown, J.H. 1971. Mammals on mountaintops: nonequilibrium insular biogeography. The American Naturalist **105**:467–478.

Brown, J.H. 1995. Organisms as engineers: a useful framework for studying effects on ecosystems? Trends in Ecology and Evolution **10**:51–52.

Bullock, C.H., D.A. Elston, and N.A. Chalmers. 1998. An application of economic choice experiments to a traditional land use: deer hunting and landscape change in the Scottish Highlands. Journal of Environmental Management **52**:335–351.

Calver, M.C., D.R. King, D.A. Risbey, J. Short, and L.E. Twigg. 1998. Ecological blunders and conservation: the impact of introduced foxes and cats on Australian native fauna. Journal of Biological Education **32**:67–72.

Census of Agriculture, U.S. 1993. Final county file. The Bureau of Census, Washington, D.C., USA.

Census Bureau, U.S. 1997. County population estimates.

Cid, M.S., J.K. Detling, A.D. Whicker, and M.A. Brizuela. 1991. Vegetational responses of a mixed-grass prairie site following exclusion of prairie dogs and bison. Journal of Range Management **44**.

Coppock, D.L., J.K. Detling, and M.I. Dyer. 1980. Interaction among bison, prairie dogs, and vegetation in Wind Cave National Park.

Coppock, D.L., J.K. Detling, J.E. Ellis, and M.I. Dyer. 1983*a*. Plant–herbivore interactions in a North American mixed-grass prairie, I: effects of black-tailed prairie dogs on intraseasonal aboveground plant biomass and nutrient dynamics and plant species diversity. Oecologia **56**:1–9.

Coppock, D.L., J.E. Ellis, J.K. Detling, and M.I. Dyer. 1983*b*. Plant–herbivore interactions in a North American mixed-grass prairie, II: responses of bison to modification of vegetation by prairie dogs. Oecologia **56**:10–15.

Dale, V.H., S. Brown, R.A. Haeuber, N.T. Hobbs, N. Huntly, R.J. Naiman, W.E. Riebsame, M.G. Turner, and T.J. Valone. In press. Ecological principles and guidelines for managing the use of land: a report from the Ecological Society of America. Ecological Applications.

Daley, J.G. 1992. Population reductions and genetic variability in black-tailed prairie dogs. Journal of Wildlife Management **56**:212–220.

DeBolt, A.M., and B. McCune. 1995. Ecology of Celtis reticulata in Idaho. Great Basin Naturalist **55**:237–248.

Decker, E., and G.A. Swanson. 1973. Wildlife and the environment: The Governor's Conference: Environmental Resources Center, Colorado State University.

Desmond, M.J., and J.A. Savidge. 1996. Factors influencing burrowing owl (*Speotyto cunicularia*) nest densities and numbers in western Nebraska. American Midland Naturalist **136**:143–148.

Desmond, M.J., J.A. Savidge, and T.F. Seibert. 1995. Spatial patterns of burrowing owl (*Speotyto cunicularia*) nests within black-tailed prairie dog (*Cynomys ludovicianus*) towns. Canadian Journal of Zoology **73**:1375–1379.

Detling, J.K., and E.L. Painter. 1983. Defoliation responses of western wheatgrass populations with diverse histories of prairie dog grazing. Oecologia **57**: 65–71.

Diamond, H.L., and P.F. Noonan. 1996. Land use in America. Island Press, Washington, D.C., USA.

Dold, C. 1998. Making room for prairie dogs. Smithsonian **28**:60–68.

Duerksen, C.J., D.L. Elliot, N.T. Hobbs, E. Johnson, and J.R. Miller. 1997. Habitat protection planning: where the wild things are. American Planning Association Planning Advisory Service Report Number 470/471.

Ellis, L.M., C.S. Crawford, and M.C. Molles, Jr. 1997. Rodent communities in native and exotic riparian vegetation in the Middle Rio Grande Valley of central New Mexico. Southwestern Naturalist **42**:13–19.

Fitzgerald, J.P., C.A. Meaney, and D.M. Armstrong. 1994. The mammals of Colorado. Denver Museum of Natural History and University Press of Colorado. Denver, Colorado, USA.

Germaine, S.S., S.S. Rosenstock, R.E. Schweinsburg, and W.S. Richardson. 1998. Relationships among breeding birds, habitat, and residential development in Greater Tucson, Arizona. Ecological Applications **8**:680–691.

Gilpin, M.E., and J.M. Diamond. 1980. Subdivision of nature reserves and the maintenance of species diversity. Nature **285**:567–568.

Halvorson, W.L., and G.E. Davis, editors. 1996. Science and ecosystem management in the national parks. University of Arizona, Tucson, Arizona, USA.

Harris, L. 1985. The fragmented forest: island biogeography and the preservation of biotic diversity.

Henderson, R.E., and A. O'Herren. 1992. Winter ranges for elk and deer: victims of uncontrolled subdivisions? Western Wildlands **18**:20–25.

Hobbs, N.T. 1996. Modification of ecosystems by ungulates. Journal of Wildlife Management **60**:695–713.

Jones, C.G., J.H. Lawton, and M. Shachak. 1994. Organisms as ecosystem engineers. Oikos **69**:373–386.

Kaush, A. 1996. Sustaining wildlife values on private lands: a survey of state programs for wildlife management on private lands in California, Colorado, Montana, New Mexico, Oregon, Utah and Washington. Transactions of the North American Wildlife and Natural Resource Conference **61**:267–273.

Kirch, P.V. 1996. Late Holocene human-induced modifications to a central Polynesian island ecosystem. Proceedings of the National Academy of Sciences of the United States of America **93**:5296–5300.

Lee, K.N. 1993. Compass and gyroscope: integrating science and politics for the environment. Island Press, Washington, D.C., USA.

Long, M.E. 1998. The vanishing prairie dog. National Geographic **279**:116–130.

Lynn, S., M.L. Morrison, A.J. Kuenzi, J.C.C. Neale, B.N. Sacks, R. Hamlin, and L.S. Hall. 1998. Bird use of riparian vegetation along the Truckee River, California and Nevada. Source Great Basin Naturalist **58**:328–343.

Marschall, E.A., and L.B. Crowder. 1996. Assessing population responses to multiple anthropogenic effects: a case study with brook trout. Ecological Applications **6**:152–167.

May, S.A., and T.W. Norton. 1996. Influence of fragmentation disturbance on the potential impact of feral predators on native fauna in Australian forest ecosystems. Wildlife Research **23**:387–400.

Miller, B., G. Ceballos, and R. Reading. 1994. The prairie dog and biotic diversity. Conservation Biology **8**:677–681.

Miller, J.R., T.T. Schulz, N.T. Hobbs, K.R. Wilson, D.L. Schrupp, and W.L. Baker. 1995. Changes in the landscape structure of a southeastern Wyoming riparian zone following shifts in stream dynamics. Biological Conservation **72**:371–379.

Pickett, S.T.A., V.T. Parker, and P.L. Fiedler. 1992. The new paradigm in ecology: implications for conservation biology above the species level. Pages 66–88 *in* P.L. Fiedler and S.K. Jain, editors. Conservation biology: the theory and practice of nature conservation, preservation, and management. Chapman & Hall, New York, New York, USA.

Plumpton, D.L., and R.S. Lutz. 1993. Nesting habitat use by burrowing owls in Colorado. Journal of Raptor Research **27**:175–179.

Reibsame, W.E., H. Gosnell, and D.M. Theobald. 1996. Atlas of the new west: portrait of a changing region. Norton, New York, New York, USA.

Reid, W.V., and K.R. Miller. 1989. Keeping options alive: the scientific basis for conserving biological diversity. World Resources Institute, Washington, D.C., USA.

Roemer, D.M., and S.C. Forest. 1996. Prairie dog poisoning in Northern Great Plains: an analysis of programs and policies. Environmental Management **20**:349–359.

Sharps, J.C., and D.W. Uresk. 1990. Ecological review of black-tailed prairie dogs and associated species in western South Dakota, USA. Great Basin Naturalist **50**:339–346.

Short, J., and A. Smith. 1994. Mammal decline and recovery in Australia. Journal of Mammalogy **75**:288–297.

Sinclair, A.R.E., R.P. Pech, C.R. Dickman, D. Hik, P. Mahon, and A.E. Newsome. 1998. Predicting effects of predation on conservation of endangered prey. Conservation Biology **12**:564–575.

Tabacchi, E., A.M. Planty Tabacchi, M.J. Salinas, and H. Decamps. 1996. Landscape structure and diversity in riparian plant communities: a longitudinal comparative study. Regulated Rivers Research & Management **12**:367–390.

Theobald, D.M. In press. Fragmentation by inholdings and exurban development. Pages 155–174 *in* Forest fragmentation in the central Rocky Mountains. University Press of Colorado, Boulder, Colorado, USA.

Theobald, D.M., H. Gosnell, and W.E. Riebsame. 1996. Land-use and landscape change in the Colorado Mountains, II: a case study of the East River Valley. Mountain Research & Development **16**:407–418.

Turner, I.M. 1996. Species loss in fragments of tropical rain forest: a review of the evidence. Journal of Applied Ecology **33**:200–209.

Weltzin, J.F., S. Archer, and R.K. Heitschmidt. 1997. Small-mammal regulation of vegetation structure in a temperate Savanna. Ecology **78**:751–763.

Whicker, A.D., and J.K. Detling. 1988. Ecological consequences of prairie dog disturbances. Bioscience **38**:778–784.

Wolfe, M.L., G.E. Simonds, R. Danvir, and W.J. Hopkin. 1996. Integrating livestock production and wildlife in a sagebrush-grass ecosystem. United States Forest Service General Technical Report **285**:73–77.

Wuerthner, G. 1996. Ghost Towns. National Parks **70**:30–35.

3
Nature Reserves and Land Use: Implications of the "Place" Principle

ANDREW J. HANSEN and JAY J. ROTELLA

Many nature reserves are undergoing human-induced change despite our best attempts to keep them natural. Some of this change is due to the fact that the boundaries of nature reserves do not include all of the "places" in the landscape that are needed for ecosystem function and that are used by native species. The ecological principle of place emphasizes that ecological processes and organisms reflect the biophysical stage on which they occur. Abiotic factors, such as topography, climate, soil, and hydrology influence rates of processes, such as ecological productivity and disturbance regimes. The population status of organisms reflects this milieu of physical and biological interactions. Nature reserves whose boundaries exclude key biophysical settings are most apt to lose native species and change from their pre-EuroAmerican settlement condition. We offer four points of view for judging whether reserves include the "right" biophysical settings—those that will allow the reserve to function well. These involve disturbance initiation and run-out zones, life-history requirements of organisms, population source and sink areas, and climate change. A case study of the Greater Yellowstone Ecosystem illustrates that management conflicts can arise when administrative boundaries conflict with ecological boundaries. The place principle offers a basis for managing nature reserves and surrounding lands to maintain adequate function. We explore guidelines for selecting new reserves and managing existing reserves. Consideration of these guidelines should help managers maintain well-functioning nature reserves in the upcoming century of global change.

National parks, wilderness areas, and other nature reserves are a cornerstone of conservation in many nations. As the term nature reserve implies, these tracts are reserved primarily for natural ecosystems and native organisms. They are often managed with relatively little human intervention to let natural factors regulate ecological processes and organisms (Boyce 1998). Beyond serving as primary habitats for species that cannot tolerate more intense human land use, nature reserves are considered as benchmarks or reference systems that can be used to better understand human impacts on areas outside of reserves (Sinclair 1998).

Despite our attempts to protect these areas from human impacts, however, many nature reserves are undergoing human-induced change (U.S. General Accounting Office 1994; Murray 1996; Landres et al. 1998*a*). Key natural disturbances are changing in intensity, size, and frequency, bringing about novel vegetation dynamics. Exotic weeds and diseases are invading nature reserves and exerting negative effects on native species. Air and water pollutants are entering reserves from exogenous sources. Some native species are suffering population reductions and extinctions in nature reserves (Newmark 1987, 1995). These changes are likely due to several factors, among which are land-use practices on surrounding ownerships (Wilcove and May 1986; Knight and Landres 1998; Woodroffe and Ginsberg 1998).

Many reserves are surrounded by gradients in land use, including public lands used for resource extraction, private lands dedicated to agriculture and rural residential development, and suburban and urban areas (Knight and Landres 1998). In recent decades, many people have relocated to the lands surrounding nature reserves (Propst et al. 1998). Consequently, land use has become more extensive (expanding into natural habitats) and more intensive (exerting greater human impact) on these surrounding lands (Hansen and Rotella 2000). What might be the connections between this development on adjacent private lands and the changes observed in nature reserves?

In this chapter, we suggest that the ecological principle of "place" (Dale et al. Chapter 1; Hansen and Rotella 1999) can help us understand some of the interactions between nature reserves and use of private lands. We first explore the implications of the place principle for understanding change in nature reserves. We then provide a case study from the Greater Yellowstone Ecosystem. Finally, implications for conservation and management are discussed.

3.1 The "Place" Principle and Nature Reserves

3.1.1 Biophysical Factors and Ecosystems

Environmental conditions such as temperature, soil nutrients, and proximity to water vary from place to place. Biological activities are promoted by some levels of these abiotic factors and constrained at other levels. Consequently, spatial variation in abiotic factors causes ecosystem processes and organisms to vary in space (Hansen and Rotella 1999; Dale et al., Chapter 1). Ecological productivity is high in some landscape settings and low in others. Viable populations of a species may be able to persist in some places but not others. Human communities may have the benefit of high levels of ecological services in particular locations but have to pay to import these services in other locations. The ecological principle of place emphasizes that the potential for natural ecosystems and human communities varies in space. Knowledge of this spatial variation can improve our ability to manage landscapes sustainably (Harris et al. 1996).

This variation in abiotic factors is often not randomly distributed. Rather, heterogeneity in abiotic factors varies predictably with topography, latitude, distance from coastlines, and other factors (Whittaker 1960). For example, temperature often decreases and precipitation increases at increasing elevations. Similarly, soils often grade from coarse texture with low water-holding capacity on ridge tops to fine textured with higher water-holding capacity in valley bottoms. These predictable patterns of abiotic factors lead to corresponding patterns in ecological processes.

This interplay of abiotic factors, ecological processes, and organisms can be described by drawing analogy to the theater (Hutchinson 1965; Harris et al. 1996). If organisms represent actors that interact to produce the drama of ecological dynamics, then the landscape represents the stage on which the drama unfolds. The distribution of biophysical conditions across the landscape influences the behavior of the organisms and the nature of ecological processes, much as the theater and stage sets influence the integrity of a play. Each theater and stage set offers particular constraints and opportunities to the performers. Similarly, the distribution of environmental conditions across landscapes shapes the behaviors of organisms and ecosystems. Much of the challenge of conservation and landscape management is to maintain, restore, or create the suite of biophysical patterns that will support the ecological dynamics that will best achieve management objectives.

3.1.2 "Right" Places for Nature Reserves

The place principle has important implications for nature reserves. Specifically, the principle implies that reserves can function properly only if they contain the right places. By "right places" we mean the locations that contain the range of abiotic conditions and habitats that are required to maintain natural ecological processes and native organisms and communities. If some of these places are omitted from inclusion in the reserve, natural disturbance regimes, ecological productivity, and species dynamics all may change to levels outside of the pre-EuroAmerican Settlement range of variation. In some cases, the boundaries of nature reserves were drawn in ways that omitted key landscape settings. This is akin to performing a play on only a portion of the stage. Just as the play would change as portions of the stage were removed, ecosystem dynamics change as portions of the landscape are made unavailable.

What are the "right places" to include in nature reserves? The simple answer is that nature reserves should include a variety of biophysical settings. The importance of particular combinations of biophysical settings needs to be judged relative to which landscape attributes exert strong influence over ecosystem function and organism performance. Here we judge the right places, or suites of biophysical factors, from four points of view: organism life histories, disturbance regimes, population source and sink dynamics, and climate change.

3.1.2.1 Life Histories of Organisms

Most species occur in particular biophysical settings. The suite of biophysical conditions within a nature reserve sets constraints on the types of species the reserve can support. If abiotic conditions are outside the tolerances of an organism, survival, growth, and reproduction may be reduced (Hansen and Rotella 1999). Abiotic factors may also influence organisms by altering rates of processes. Disturbance rates alter habitat quality for organisms, and primary productivity influences food availability at higher trophic levels.

Some species have a narrow range of requirements and occupy throughout their lifetimes a particular landscape setting with specific environmental conditions. Studies by Whittaker (1960), for example, elegantly demonstrated that tree species are found in particular locations along elevational gradients due to the effects of climate and soils. Other organisms actively move across abiotic gradients on a daily, seasonal, or life-stage basis. If resources and conditions vary in time and space, such organisms can maintain access to a particular set of conditions by moving across the varying environmental gradient. Examples are migratory birds that move seasonally to maintain access to food supplies. Other species move along abiotic gradients to gain access to two or more different habitat types (e.g., amphibians that use aquatic and terrestrial habitats in different life stages).

Nature reserves that do not include the range of habitats and conditions required by species often require more intensive management. Serengeti National Park, Tanzania, for example, contains both wet-season and dry-season habitats, allowing several species of ungulates to use natural migration routes (Sinclair 1998). Kruger Park, South Africa, in contrast, is situated only in wet-season habitats. Traditional ungulate migration routes were cut off by the park fences. Consequently, considerable human intervention has been needed to maintain these species and their habitats. Wells were drilled to provide adequate water for these herds. These animals are now largely resident near water sources and substantially impact the vegetation. Culling is now used to control herd sizes and reduce negative impacts on the vegetation.

3.1.2.2 Disturbance

Disturbances tend to be initiated in particular landscape settings and move to other locations in the landscape. Interactions between the location where disturbance gets started (initiation zones) and locations where disturbances move to (run-out zones) influence the nature of the disturbance regime in an area (Baker 1992). In southwest Montana, for example, lightning strikes occur across the landscape but more frequently ignite fires in dry valley-bottom grasslands than in moister conifer forests in the uplands (Arno and Gruell 1983). These fires then spread up slope to the conifer forests. Thus, the juxtapositioning of grasslands and conifer forests strongly influences the regional fire regime. Local disturbance regimes can best be maintained in nature reserves that include the disturbance initiation zones within their boundaries (Baker 1992). In the case of the example above, a reserve placed only in the upland conifers may suffer more or less frequent fire, depending on the management of the valley-bottom grasslands outside of the reserve.

It is also important to include disturbance run-out zones within reserve boundaries. Run-out zones may contain unique abiotic conditions and habitat patterns important to ecological processes and organisms. For example, flood severity often increases from headwaters to large flood-plains. The large scour area and bare gravel bars that form on flood plains support vegetation communities not found in other landscape settings. A reserve that does not contain this disturbance run-out zone will not include these unique riparian vegetation communities. In reserves that omit either the initiation or run-out zones, human manipulation of disturbance may be required to maintain landscape patterns and organisms (Baker 1992; Sinclair 1998).

3.1.2.3 Population Sources and Sinks

New discoveries in population dynamics suggest a less visible, but still important, way that reserve location may influence organisms. Within the range where a species is found, some locations may be population source areas and other locations population sinks (Pulliam 1988). Source areas are characterized by having birth rates that exceed death rates, while sinks are the opposite: Death rates are higher than birth rates. These differences in demographics are sometimes due to biophysical factors. Individuals may suffer less physiological stress and have higher energy availability in more equitable biophysical settings (Hansen et al. 1999). Source and sink areas may be difficult to detect because population densities can be similar in both sources and sinks (Pulliam and Danielson 1991). Surplus individuals produced in source areas may immigrate to the sink areas and maintain relatively high densities there. In this case, individuals may persist in the sink areas only if population source areas are maintained. Thus, populations in nature reserves that include only sink habitats depend on source populations in other areas for their continued existence. Such sink populations may suffer extinction if population sources outside the reserve are degraded. Sinclair (1998) suggested that most of Serengeti National Park is a sink for the lion population and that the species is maintained there because of connectivity with the Ngorongoro Conservation Area, which is a population source area for lion.

3.1.2.4 Climate Change

The location of nature reserves is also relevant to how well reserves will fair under potential future climate change. Halpin (1997) predicted that 47–77% of biosphere reserves globally will undergo a change in ecoclimatic zone under a doubling of atmospheric CO_2. The magnitude of predicted changes varies among reserves. Those at higher latitudes are more likely to change than those in equatorial locations. The ability of organisms to cope with climate change is also likely to vary with reserve. Organisms have a greater ability to relocate to suitable habitats in reserves that include a wide range of biophysical conditions. Also, reserves that are connected to other reserves by seminatural habitats, such as along mountain chains, are more likely to exchange organisms with other reserves under climate change. Local analyses are needed to determine how much change a given nature reserve is likely to experience and what management strategies might be used to cope with these changes (Halpin 1997).

3.1.3 Historic Criteria for Reserve Selection

Given the importance of reserve location for conservation, we might expect that nature reserves were carefully placed relative to biophysical gradients. This was not the case for many national parks and wilderness areas in the United States. Knowledge of the interactions between biodiversity, biophysical gradients, and natural disturbance was underdeveloped when most reserves were established (Craighead 1991). Moreover, conservation of biodiversity was often not a key criterion for reserve selection. Many national parks were selected because of their scenic grandeur, geologic or biological uniqueness, potential for tourism, and public ownership (Newmark 1987). In the case of Wilderness Areas, a key criterion was lack of human impact, as evidenced, for example, by an absence of roads.

Such selection criteria likely biased the location of nature reserves toward relatively harsh sites. Lands with high potential for agriculture and other intense human land uses were often claimed for private ownership and lands that remained public were relatively harsh in climate and were lower in soil fertility (Huston 1993). Even on public lands, resource extraction was often first concentrated on the most productive sites. Hence, the areas that remained roadless at the time of the Wilderness Act in 1964 were often high in elevation and extreme in climate.

This history of reserve selection suggests that more equitable biophysical settings are underrepresented in our nature reserves (Scott et al., in review). In mountainous areas, valley bottoms, and lowlands with warmer temperatures, longer growing seasons, more fertile soils, and high ecological productivity are often outside reserve boundaries. In arid systems, reserves often omit areas of higher precipitation and soil water-holding capacity.

3.1.4 Land Allocation and Biodiversity

What might be the consequences of nature reserves being placed in harsher landscape settings? As mentioned above, these more equitable settings are apt to represent important habitats, population source areas, and possible disturbance initiation or run-out areas. In reserves in more extreme biophysical settings, organisms that migrate seasonally across elevational or precipitation gradients are apt to cross administrative boundaries. Species that exhibit source–sink population dynamics may be dependent on source areas that are on private lands. Thus, nature reserves that omit equitable biophysical settings may experience changes in ecosystem function and biodiversity as lands outside of the reserve undergo human development.

The rate at which seminatural private lands outside of nature reserves have developed appears to be accelerating in recent years. Human population density is growing rapidly in rural landscapes in the United States. Some of this growth may be a function of expansion around the periphery of urban areas. Such is the case around Denver and Phoenix. Another factor is that people are increasingly attracted to live near nature reserves because opportunities for recreation and high-quality lifestyles abound in such locales (Johnson and Rasker 1995).

An ironic consequence of this rural residential development can be an erosion of the ecosystem qualities that originally attracted the new residents. One effect of this development is a reduction of the functional size of a reserve. The number of species an ecosystem can support is strongly related to its area (Huston 1994; Rosenzweig 1995). Population sizes of species with large home ranges such as grizzly bear (*Ursus arctos horribilis*) have been associated with the size of the wildlands they inhabit (Picton 1994; Woodroffe and Ginsberg 1998). Evidence of the relationship between habitat area and species viability comes from Newmark (1987), who found that extinction rates of mammals in National Parks in the western United States were correlated with park size. As semi-wildlands are developed around reserves, the area of habitat available to organisms is decreased and likelihood of extinction is increased. This rate of extinction is likely to be even greater than predicted based on species-area relationships if humans are developing the more equitable biophysical settings that are especially important habitats for many species.

Beyond converting natural habitats to human land uses, humans can have more subtle effects on ecosystem function and biodiversity. Human activities often favor invasive organisms and facilitate the spread of these weedy species from private lands into nature reserves (Landres et al. 1998*b*). Suppression of disturbance on private lands can alter the flow of natural disturbance regimes into nature reserves (Baker 1992). Land uses such as agriculture and livestock grazing can favor predators that then exert strong negative effects on prey species in adjacent wildlands (Terborgh 1989; Hansen et al. 1999).

A key implication of the ecological principle of place is that many nature reserves cannot function adequately as islands in a human-dominated matrix because they do not include key places for ecological processes and native species. The dynamics of ecosystems within reserve boundaries can be heavily influenced by the state of the ecosystem outside the boundaries. As these surrounding lands are increasingly developed by humans, ecosystem processes and organisms within the reserves will change. Fortunately, a variety of strategies can be used to minimize or cope with these boundary effects (see below).

3.2 Case Study: Biodiversity and Land Use in Greater Yellowstone

3.2.1 Biophysical Gradients and Land Allocation

The Greater Yellowstone Ecosystem (GYE) (Fig. 3.1; see color plate) represents an interesting example where nature reserve boundaries bisect key biophysical gradients. Yellowstone National Park was established in 1872, largely to protect unique hydrothermal and geological features and remnant wildlife populations (Mackintosh 1991). The original boundaries included a largely rectangular area and centered on the key geological and thermal features. Soon after its establishment, there were calls to expand the size of the park to better include key wildlife habitats (Craighead 1991). Eventually, a small area of ungulate winter range was added to the northern boundary of the park and national forests were designated on surrounding lands.

The concern over Yellowstone National Park's boundaries largely stemmed from its location on the Yellowstone Plateau. The plateau and the surrounding mountains are relatively high in elevation. The severe winters, short growing season, and relatively infertile volcanic soils greatly constrain primary productivity (Despain 1990). The length of growing season increases at lower elevations and soil fertility is relatively high in some midslope and valley-bottom settings in the area. Consequently, primary productivity is highest in the lowlands (Hansen et al. 2000). Natural disturbances like wildfire vary across this elevational gradient (Barrett 1994), and many wildlife species migrate seasonally between lowland and upland habitats (Frank 1998).

The boundaries of Yellowstone National Park cut across these gradients in climate, soils, primary productivity, and natural disturbances. Land allocation across the GYE was largely stratified by elevation. Yellowstone National Park and other nature reserves in the ecosystem are placed primarily at higher elevations (Fig. 3.2). Other public lands, such as the national forests, are at intermediate elevations. Private lands are mostly located at lower elevations. An important consequence of this pattern of land allocation is that the productive lowland habitats are mostly outside of

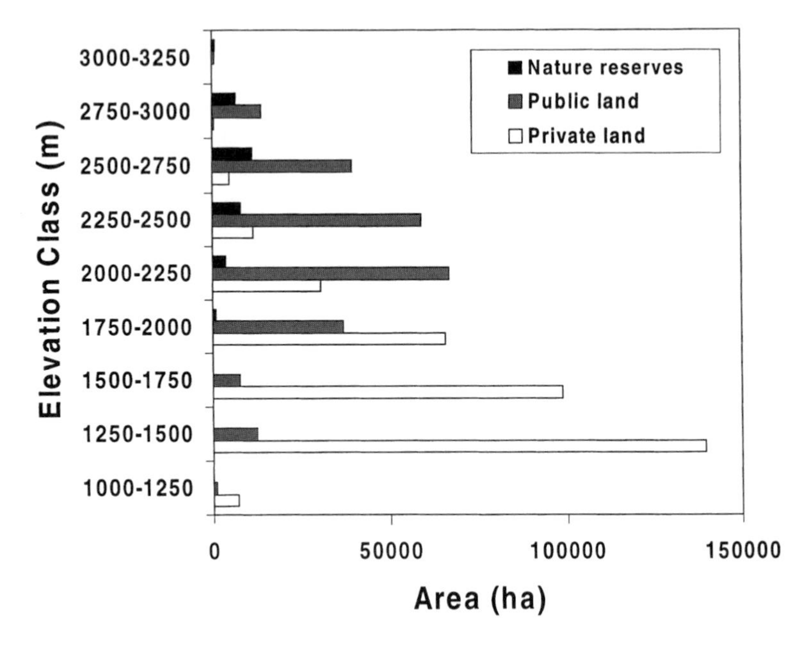

FIGURE 3.2. Distribution of land allocation across elevations for the Greater Yellowstone Ecosystem.

the nature reserves and on private lands. The term Greater Yellowstone Ecosystem was coined to emphasize that ecological processes and organisms are integrated across biophysical gradients from lowlands to the alpine zone and, consequently, coordinated management across the ownership jurisdictions is required to maintain these processes and organisms (Craighead 1991; Patten 1991).

3.2.2 Human Demography, Land Use, and Effects on Ecosystems

The northern Rocky Mountains are known for their wildness and low human-population densities. In recent decades, however, many people have moved to the private lands of the GYE. The population of the GYE has increased 55% since 1970. Of the 20 counties in the GYE, 13 were among the fastest-growing 25% of counties in the United States in the 1990s. Many of the new residents and businesses have been attracted to the GYE by outdoor recreation, scenery, and other environmental amenities (Johnson and Rasker 1995).

This population growth has had a large impact on rural private lands across the GYE. Many new residents have chosen to live outside of towns and cities. Consequently, rural residential development has been rapid in recent decades. Gallatin County, Montana, in the northwest portion of the ecosystem, is typical in these trends. The number of rural residences increased by more than fourfold between 1970 and 1995 (Hansen and Langner, in preparation).

Scientists are just beginning to examine the impact of rural residential development on biodiversity and ecosystems. Initial studies indicate substantial negative impact on native wildlife due to fragmentation of natural habitats, invasion of nonnative weeds and predators, and harassment of wildlife by pets and recreationists (Whitcomb et al. 1981; Riesbame et al. 1996; Theobald et al. 1996; Hansen et al. 1999; Hobbs and Theobald, Chapter 2).

How much has the functional size of the natural habitats of the GYE been reduced by development on private lands? Hansen and Langner (in preparation) analyzed of the rate of loss of seminatural lands in Gallatin County, Montana. They assumed that the zone of influence of a home has a radius of 2 km. The zone of influence of a home likely varies depending on the ecological process or species of interest. The radius of 2 km is conservative for some impacts; bird reproductive success here was found to be correlated with density of homes within 6 km of nests (Rotella and Hansen, unpublished data). Figure 3.3 shows the cumulative proportion of private land in

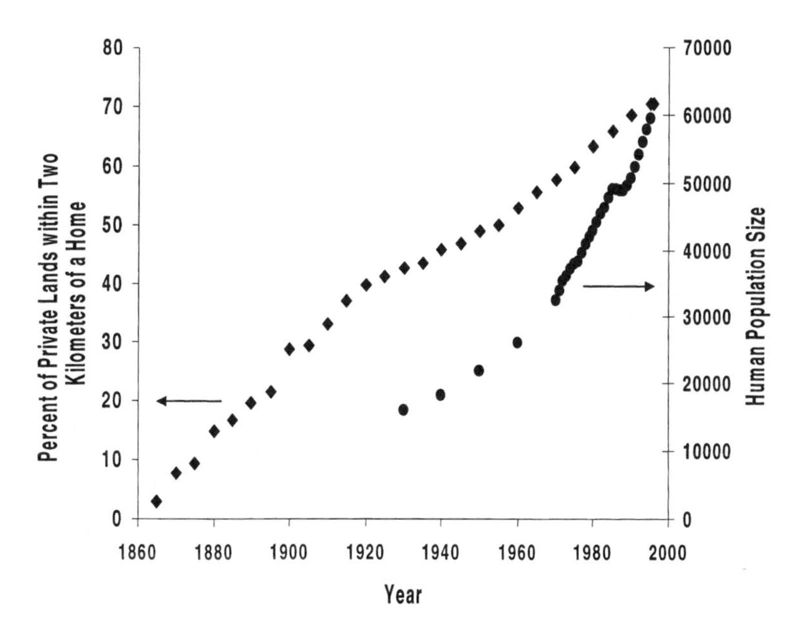

FIGURE 3.3. Land-use and human population trends in Gallatin County, MT. Cumulative proportion of private lands influenced by rural residential development from 1860 to 1995 is shown on the left axis. Human population size 1930–1995 is shown on the right axis. From Hansen and Langner (in preparation).

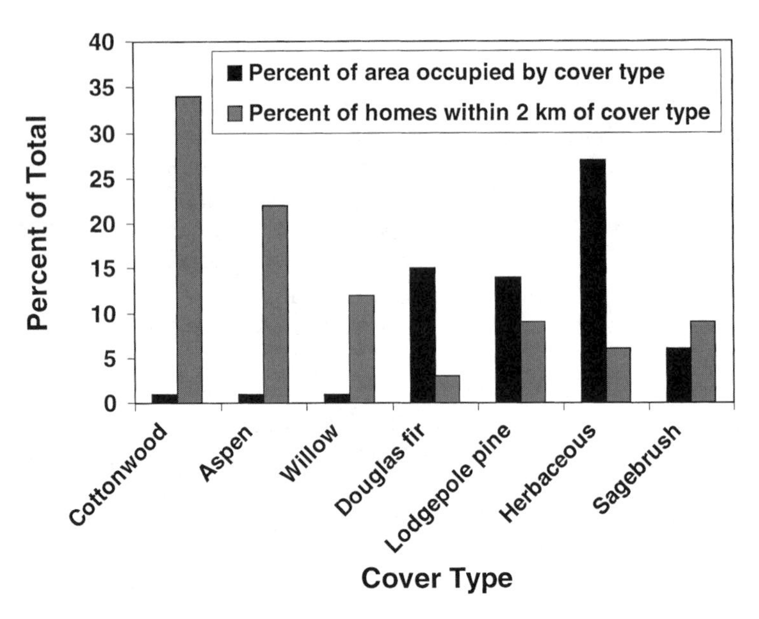

FIGURE 3.4. The distribution of rural residences relative to vegetation cover types in Gallatin County, MT. The percentage of the county occupied by each cover type is show on the left axis. The proportion of the total rural residences with 2 km of each cover type is shown on the right axis. From Hansen and Langner (in preparation).

the county that is within 2 km of one or more houses from 1860 to 1995. Notice that the increase has been almost constant over this time period at a rate of about 5% per decade. We might expect that the rate of habitat impact would decrease in modern times as new homes are placed within 2 km of existing homes. However, this rate has not decreased, largely due to the rapid increase in human population size since 1970 (Fig. 3.3). Currently, only about 30% of the private lands in the county are more than 2 km from a home. This analysis illustrates how development on private lands can erode the area of natural lands around nature reserves.

The impact of humans on many native species in Gallatin County is likely even larger than expected based on species–area relationships because the rural residential development has not been located randomly relative to biophysical factors. Rural homes in the county are disproportionately dense near deciduous habitats (cottonwood, aspen, willow) (Hansen and Langner, in preparation) (Fig. 3.4). These deciduous habitats cover only about 3% of the county and are hot spots for primary productivity and biodiversity (see below). These data suggest that rural residential development may be having a particularly strong influence on native species because it is concentrated in the equitable biophysical settings. This process is likely to continue into the future, further eroding the size and quality of habitats across the GYE.

3.2.3 Conflicts at Reserve Boundaries

Many of the conservation controversies in the GYE ultimately stem from the location of Yellowstone National Park relative to biophysical gradients. We will briefly mention examples involving natural disturbance and wildlife habitats and then provide more detail on source–sink population dynamics from our studies on birds.

Wildfire is the dominant natural disturbance in the GYE. Fire frequency and size vary with biophysical setting, ranging from relatively small, frequent fires at lower elevations to very large fires at 200- to 300-year intervals at higher elevations (Romme 1982; Barrett 1994). Neither fire-initiation nor run-out zones are entirely included in Yellowstone National Park. Fires initiated in the productive lowlands on the windward side of the park likely spread into the park in presettlement times (Hansen et al. 2000). This area is now managed as the Targhee National Forest. Extensive clear-cutting across the Targhee has dramatically reduced fuel loads and likely reduced the potential for fire to spread of into the park. Thus, the nature of the disturbance regime in the park may be altered due to land-use practices in disturbance initiation zones outside of the park. The spread of fire from the park to surrounding private lands is also a difficult management question. Current policy allows wildfires to burn in the park. However, land managers and property owners on the leeward side of the park are concerned that this policy puts their lands at increased risk of fire.

Elk, bison, and other ungulates in the GYE historically migrated between summer habitats at high elevations and winter range at lower elevations (Frank 1998). Human development on private lands in lowlands has reduced the ungulates' access to traditional winter range. This loss of winter habitat on private lands may have led to higher densities of elk on the winter range within Yellowstone National Park and exacerbated the loss of woody plants there due to herbivory. Over the last two decades the area of winter range available to the Northern Range elk herd was expanded by land acquisition and conservation easements. A greater proportion of this elk herd now winters outside of Yellowstone National Park (Lemke et al. 1998). Densities of elk on the winter range inside of the park have decreased and may allow higher growth rates of woody plants (Singer et al. 1998).

Bison illustrate most visibly, perhaps, the difficulties of managing when administrative boundaries do not coincide with ecological boundaries. Bison have increasingly migrated out of Yellowstone National Park in recent years, possibly because of increased population size and/or ease of movement along snow-cleared roads (Meagher 1989). The herd carries the pathogenic bacterium *Brucella abortus*, and livestock growers are concerned that the disease brucellosis will spread from bison to cattle. The current management policy is to slaughter diseased bison as they leave the park. This policy has been highly controversial and has led to Congressionally mandated reviews, lawsuits, and public protests (Keiter 1997).

Just as ungulates move down in elevation and leave the park, exotic species have invaded the park from the surrounding lowlands. Weedy nonnative plants are now common in the park, possibly due to dispersal from reservoirs on surrounding private lands (Kurtz 1999). These invasive plants may reduce forage quality for elk and other native herbivores and may outcompete native plant species. Similar invasions have occurred in stream systems. Human activities have favored nonnative trout in larger streams in the lowlands and these exotic fishes have displaced native cutthroat trout. The remaining cutthroat populations in headwater streams are highly fragmented and under threat of extinction (Shepard et al. 1997). Nonnative lake trout were recently discovered in Yellowstone Lake in the heart of the park and threaten the native trout population and the suite of carnivores that feed on the cutthroat. This population of cutthroat is also jeopardized by the exotic whirling disease that has been expanding in warmer waters in lowland habitats (D. Gustafson, personal communication). These introductions raise question as to whether invasions would be less likely if the lowland habitats were included in the nature reserve rather than subjected to intense land use.

Our own studies of birds in the GYE suggest that lowland habitats are especially important for the maintenance of regional populations (Hansen et al. 1999; Hansen and Rotella 1999). We sampled bird abundance and community richness across cover types and elevation zones in the northwest portion of the GYE. We found that birds were not randomly distributed among our samples. Rather, bird richness and abundance were high in sites with wet alluvial soils, equitable climate, and deciduous forest cover types—likely because these sites offer relatively high levels of food production and habitat structural complexity. We then extrapolated bird species over the study area based on these biophysical factors. Predicted bird species richness and total abundance were relatively low over most of the study area and high only in localized settings. Hot spots for bird richness were rare (2.7% of the study area), primarily at lower elevations, and the majority were on or near private lands. Only 3.0% of the total area in hot-spot habitats was in Yellowstone National Park.

We used these data as the basis for a risk assessment (Hansen et al. 1999) and found that the majority of the species most at risk of extinction were dependent on these hot-spot habitats. Reproductive success varied among the hot spots, however. Reproduction was relatively high in hot spots surrounded by seminatural lands, and simulation modeling indicated that these are population source areas. Hot-spot habitats surrounded by rural residential and agricultural land uses, however, had low reproduction and appear to be population sinks for this guild of at-risk species. These intense land uses favor higher densities of nest predators and brood parasites that enter the hot-spot habitats and reduce bird reproduction. These results suggest that the population source areas in the lowlands maintain the viability of many bird species across the region. Human development appears to have converted some of these source areas to population sinks. Further intensification of land use near hot-spot habitats could lead to extinctions of some of these species across the region, including within nature reserves.

3.2.4 Management of the GYE

Many of the conflicts described above can be resolved only through coordinated management across the many ownership jurisdictions of the GYE. While this has been a considerable challenge, there have been some successes (Glick and Clark 1998). Some important lowland habitats outside of Yellowstone National Park have been placed in a conservation status through purchase or trade of the lands or purchase of conservation easements by both governmental and private organizations. Cooperative efforts to manipulate animal populations are in place for some species. The hunting of elk that migrate outside of the park is used to manage the size of the herd to prevent an overpopulation within the park (Lemke et al. 1998). Similarly, the endangered grizzly bear is managed to minimize bear mortality on public and private lands outside the park. Reintroduction programs are being used to restore native trout populations in lowland streams. Some local governments have begun to manage rural residential development to reduce impacts on ecosystems. These initial efforts at cooperative management will have to be strengthened, however, to maintain the quality of the GYE in the face of expected future intensification of use of private lands and increases in recreation on public lands.

3.3 Guidelines for Conservation and Management

The ecological principle of place not only helps us to better understand patterns in and around nature reserves, it also provides a context for managing existing reserves and for establishing new reserves. Here, we highlight some guidelines for conservation and management of nature reserves.

The designers of new reserves have the luxury of taking advantage of the best current knowledge on ecology, socioeconomics, and other factors. Halpin (1997) summarized and evaluated considerations for reserve selection that relate to climate change. We modify the list slightly with reference to the ecological principle of place.

- *Biophysical setting*. The major conclusion of this chapter is that reserves are most likely to function well when they contain the right configuration of biophysical settings. Hence, reserve boundaries should be set with consideration of disturbance initiation and run-out zones, habitat requirements of organisms, the spatial distribution of population source and sink

areas, and gradients in biophysical factors that allow organisms to relocate under changing climate.
- *Buffer-zone flexibility.* Land allocation and management of lands surrounding the reserve should maintain options for readjusting for future change. The biosphere reserve concept offers a sociopolitical construct for achieving this flexibility.
- *Landscape connectivity.* The degree of upheaval wrought by climate change will be partially influenced by the success of organisms in dispersing to newly created suitable habitats. Maintaining connectivity among reserves will aid this dispersal. However, the effectiveness of positioning reserves along latitudinal or elevational gradients, corridors between reserves, and management of the intervening human-dominated matrix remains poorly understood (Halpin 1997).
- *Redundant reserves.* In some cases, placing two or more reserves in a particular ecosystem type may be desirable as a hedge against unforeseen change within a reserve.

Managers of existing reserves must cope with the legacy that they inherited from those that originally allocated lands in and around a reserve. The challenge is to understand current patterns of biophysical factors, organism dynamics, land use, and management approaches and to use this knowledge as a basis for achieving/maintaining the objectives of the reserve and surrounding lands. Possible steps to this end are as follows:

- *Assessment.* Quantify biophysical gradients, ecological processes, organisms, and land use to understand how well the reserve is functioning relative to its objectives. Approaches for assessment can be found in Hansen et al. (1999) and Bourgeron et al. (Chapter 13).
- *Habitat acquisition.* If key biophysical settings were not included in reserve boundaries, opportunities may exist to restore them. Means include land acquisition, conservation easements, tax incentives, education for private landowners, and local government planning and/or regulation.
- *Human intervention in reserves.* If surrounding lands necessary for reserve function cannot be conserved, human intervention within reserves may be required (Sinclair 1998). Intervention strategies may include manipulation of disturbance regimes and nutrient flows, control of overabundant native species or of nondesirable exotic species, and management of recreationists.
- *Coordinated management.* Neighboring agencies and landowners often have very different objectives and cultures. Nonetheless, coordination of management approaches across jurisdictions can greatly reduce boundary conflicts. Community conservation forums, regional management committees, or entirely new administrative structures may be needed to achieve this coordination (see also Botteron, Chapter 7, and Haufler and Kernohan, Chapter 4).

- *Monitoring*. As in any management endeavor, ongoing monitoring is needed to gauge the effectiveness of management implementations.

3.4 Conclusions

The value of nature reserves for conservation has become overwhelmingly apparent over the last century as land use has intensified on public and private lands. Many scientists and conservationists consider nature reserves to be "vignettes" of primitive times (Boyce 1998). However, we suggest that nature reserves are better seen as islands that feel the winds blowing from the sea of human-dominated landscapes around them. These winds have brought change to many nature reserves. Those changes will undoubtedly accelerate under the two vectors of global change—climate change and land-use intensification. The ecological principle of place offers a basis for better understanding interactions between reserves and their surroundings and better coping with global change. Place is the biophysical stage that determines the nature of the ecological drama in and around nature reserves. By managing the biophysical stage well, we have the best hope of maintaining well-functioning nature reserves. This management will require a new generation of approaches for cooperation among reserve managers and their neighbors.

Acknowledgments. Ute Langner designed the graphics. We also thank David Theobold, three anonymous reviewers, and editors Virginia Dale and Richard Haeuber for helpful comments on the manuscript.

References

Arno, S.F., and G.E. Gruell. 1983. Fire history at the forest-grassland ecotone in southwestern Montana. Journal of Range Management **39**:332–336.

Baker, W. 1992. The landscape ecology of large disturbances in the design and management of nature reserves. Landscape Ecology **7**:181–194.

Barrett, S. 1994. Fire regimes on andesitic mountain terrain in northeastern Yellowstone National Park, Wyoming. International Journal of Wildland Fire **4**:65–76.

Boyce, M.S. 1998. Ecological-process management and ungulates: Yellowstone's conservation paradigm. Wildlife Society Bulletin **26**:391–398.

Craighead, J.J. 1991. Yellowstone in transition. Pages 27–40 *in* R.B. Keiter and M.S. Boyce, editors. The greater Yellowstone ecosystem: redefining America's wilderness heritage. Yale University Press, New Haven, Connecticut, USA.

Despain, D. 1990. Yellowstone vegetation. Roberts Rinehart, Boulder, Colorado, USA.

Frank, D.A. 1998. Ungulate regulation of ecosystem processes in Yellowstone National Park: direct and feedback effects. Wildlife Society Bulletin **26**:410–418.

Glick, D.A., and T.W. Clark. 1998. Overcoming boundaries: the greater Yellowstone ecosystem. Pages 237–256 *in* R.L. Knight and P.B. Landres, editors, Stewardship across boundaries. Island Press, Washington, D.C., USA.

Halpin, P.N. 1997. Global climate change and natural-area protection: management responses and research directions. Ecological Applications **7**:828–843.

Hansen, A.J., and J.J. Rotella. 1999. Abiotic factors. Pages 161–909 *in* M. Hunter, editor. Managing forests for biodiversity. Cambridge University Press, London, England.

Hansen, A.J., and J.J. Rotella. 2000. Bird response to forest fragmentation. *In* R.L. Knight, F.W. Smith, S.W. Buskirk, W.H. Romme, and W.L. Baker, editors. Forest fragmentation in the Southern Rocky Mountains. Island Press, Washington, D.C., USA.

Hansen, A.J., J.J. Rotella, and M.L. Kraska. 1999. Dynamic habitat and population analysis: a filtering approach to resolve the biodiversity manager's dilemma. Ecological Applications **9**:1459–1476.

Hansen, A.J., J.J. Rotella, M.L. Kraska, and D. Brown. 2000. Spatial patterns of primary productivity in the Greater Yellowstone Ecosystem. Landscape Ecology **15**:505–522.

Harris, L.D., T.S. Hoctor, and S.E. Gergel. 1996. Landscape processes and their significance to biodiversity conservation. Pages 319–347 *in* O.E. Rhodes, R.K. Chesser, and M.H. Smith, editors. Population dynamics in ecological space and time. University of Chicago Press, Chicago, Illinois, USA.

Huston, M. 1993. Biological diversity, soils, and economics. Science **262**:1676–1680.

Huston, M.A. 1994. Biological diversity. Cambridge University Press, Cambridge, England.

Hutchinson, G.E. 1965. The ecological theater and the evolutionary play. Yale University Press, New Haven, Connecticut, USA.

Johnson, J.D., and R. Rasker. 1995. The role of economic and quality of life values in rural business location. Journal of Rural Studies **11**:405–416.

Keiter, R.B. 1997. Greater Yellowstone's bison: Unraveling of an early American wildlife conservation achievement. Journal of Wildlife Management **61**:1–11.

Knight, R.L., and P.B. Landres, editors. 1998. Stewardship across boundaries. Island Press, Washington, D.C., USA.

Kurtz, D. 1999. Geographic characterization of exotic plant species in Grand Teton National Park Thesis, Montana State University, Bozeman, Montana, USA.

Landres, P.B., R.L. Knight, S.T.A. Pickett, and M.L. Cadenasso. 1998*a*. Ecological effects of administrative boundaries. Pages 39–64 *in* R.L. Knight and P.B. Landres, editors. Stewardship across boundaries. Island Press, Washington, D.C., USA.

Landres, P.B., S. Marsh, L. Merigliano, D. Ritter, and A. Norman. 1998*b*. Boundary effects on wilderness and other natural areas. Pages 117–139 *in* R.L. Knight and P.B. Landres, eds., Stewardship across boundaries. Island Press, Washington, D.C., USA.

Lemke, T.O., J.A. Mack, and D.B. Houston. 1998. Winter range expansion by the northern Yellowstone elk herd. Intermountain Journal of Sciences **4**:1–9.

Macintosh, B. 1991. The national parks: shaping the system. Division of Publications, National Park Service, Washington, D.C., USA.

Meagher, M.M. 1989. Evaluation of boundary control for bison of Yellowstone National Park. Wildlife Society Bulletin **17**:15–19.

Murray, M.P. 1996. Natural processes: wilderness management unrealized. Natural Areas Journal **16**:55–61.

Newmark, W.D. 1987. Mammalian extinctions in western North American parks: a land-bridge island perspective. Nature **325**:430–432.

Newmark, W.D. 1995. Extinction of mammal populations in western North American national parks. Cons. Biol. **9**:512–526.

Patten, D.T. 1991. Defining the Greater Yellowstone Ecosystem. Pages 19–26 *in* R.B. Keiter and M.S. Boyce, editors. The Greater Yellowstone Ecosystem: redefining America's wilderness heritage. Yale University Press, New Haven, Connecticut, USA.

Picton, H.D. 1994. A possible link between Yellowstone and Glacier grizzly bear populations. International Conference on Bear Research and Management **6**:7–10.

Propst, L., W.F. Paleck, and L. Rosan. 1998. Partnerships across park boundaries: the Rincon Institute and Saguaro National Park. Pages 257–278 *in* R.L. Knight and P.B. Landres, editors. Stewardship across boundaries. Island Press, Washington, D.C., USA.

Pulliam, H.R. 1988. Sources, sinks, and population regulation. American Naturalist **132**:652–661.

Pulliam, H.R., and B.J. Danielson. 1991. Sources, sinks, and habitat selection: a landscape perspective on population dynamics. American Naturalist **137**:S50–S66.

Riebsame, W.E., H. Gosnell, and D. Theobald. 1996. Land use and landscape change in the Colorado Mountains, I: theory, scale and pattern. Mountain Research and Development **16**:395–405.

Romme, W.H. 1982. Fire and landscape diversity in subalpine forests of Yellowstone National Park. Ecological Monographs **52**:199–221.

Rosenzweig, M.L. 1995 Species diversity in space and time. Cambridge University Press, Cambridge, England.

Scott, J.M., F.W. Davis, G. McGhie, and C. Groves. In review. Biological reserves: do they capture the full range of America's biological diversity?

Shepard, B.B., B. Sanborn, L. Ulmer, and D.C. Lee. 1997. Status and risk of extinction for westslope cutthroat trout in the upper Missouri River Basin, Montana. N.A.J. Fish. Mgt. **17**:1158–1172.

Sinclair, A.R.E. 1998. Natural regulation of ecosystems in protected areas as ecological baselines. Wildlife Society Bulletin **26**:399–409.

Singer, F.J., L.C. Seigenfuss, R.G. Cates, and D.T. Barnett. 1998. Elk, multiple factors, and persistence of willows in national parks. Wildlife Society Bulletin **26**:419–428.

Terborgh, J. 1989. Where have all the birds gone? Princeton University Press, Princeton, New Jersey, USA.

Theobald, D.M., H. Hosnell, and W.E. Riebsame. 1996. Land use and landscape change in the Colorado Mountains, II: a case study of the East River Valley. Mountain Research and Development. **16**:407–418.

U.S. General Accounting Office. 1994. Activities outside park borders have caused damage to resources and will likely cause more. U.S. General Accounting Office GAO/RCED-94-59, Washington, D.C., USA.

Whitcomb, R.F., C.S. Robbins, J.F. Lynch, B.L. Whitcomb, K. Klimkiewicz, and D. Bystrak. 1981. Effects of forest fragmentation on avifauna of the eastern deciduous forest. Pages 125–205 *in* R.L. Burgess and D.M. Sharpe, editors. Forest island dynamics in man-dominated landscapes. Springer-Verlag, New York, New York, USA.

Whittaker, R.H. 1960. Vegetation of the Siskiyou Mountains, Oregon and California. Ecological Monographs **30**:279–338.

Wilcove, D.S., and R.M. May. 1986. National park boundaries and ecological realities. Nature **324**:206–207.

Woodroffe, R., and J.R. Ginsberg. 1998. Edge effects and the extinction of populations inside protected areas. Science **280**:2126–2128.

4
Ecological Principles for Land Management Across Mixed Ownerships: Private Land Considerations

JONATHAN B. HAUFLER and BRIAN J. KERNOHAN

Ecological objectives relating to sustainability and land management typically focus on maintaining and enhancing biological diversity and ecosystem integrity, objectives that require planning at landscape levels. Nearly all landscapes of sufficient size to address these objectives contain a diversity of landowners, including many private lands. For conservation planning to be effective in these mixed-ownership landscapes, a number of ecological principles and management considerations should be recognized. Ecological objectives need to be addressed in an ecosystem-management context in which social and economic objectives are integrated with ecological objectives. To accomplish this integration, the most effective approach focuses on a coarse filter; an approach that strives to meet ecological objectives through the identification of an appropriate mix of ecological communities correctly configured within the landscape. One effective coarse-filter approach has identified ecological site complexity as well as an understanding of historical disturbance regimes that shaped the inherent diversity and function of the ecological communities. Monitoring effectiveness of coarse-filter planning will need to be hierarchical to address all levels of biological diversity.

Private lands are critical elements in land-management planning for conservation objectives. In the Northeast and South, approximately 90% of the land is privately owned. Even in the West where Federal lands comprise higher percentages, private lands control much of the lower elevation and terrestrial/riparian interface zones (Hansen and Rotella, Chapter 3), thus representing a majority of area for many types of ecological communities. Further, private lands encompass a high percentage of habitat for many species listed as threatened or endangered under the Endangered Species Act. The Natural Heritage Data Network (1993) estimated that more than 50% of the federally listed threatened and endangered species occur only on private lands. In addition, most riverine systems pass through private land where land-management practices can influence many downstream characteristics. For all of these reasons, incorporating private lands into conservation planning efforts is essential if effective results are to be obtained.

Ecological objectives frequently identified as management goals for conservation efforts include the maintenance and enhancement of biological diversity and ecosystem integrity (Grumbine 1994). Biological diversity is defined as the variety of life and its processes, and is usually considered to have genetic, species, community, and landscape components (Keystone Center 1991). Ecosystem integrity is related to biological diversity at the community and landscape levels, and specifically addresses the ecological processes that are essential for ecosystems to function in a defined and predictable fashion. Ecosystem integrity addresses the need for proper functioning communities to support the species and genetic levels of biological diversity.

Given the importance of private lands in supporting a full complement of ecological communities and biological diversity, their inclusion in conservation planning is paramount. Private lands are affected by the same ecological principles as public lands. However, private landowners operate under a different set of management objectives than most public land-management agencies. Objectives of private land management are as diverse as the types of landowners. However, one generalization can be made: Economic considerations are primary drivers for management decisions on the majority of private lands (Sample 1994). This means that private lands require a different approach to incorporate conservation planning than public lands, which are typically managed under legislative guidelines. Recognition of these differences and the need to accommodate legitimate economic concerns of private landowners must be an important planning component. Strategies that help engage private landowners in conservation planning are of critical importance to meeting ecological objectives for the broader landscape.

In this chapter, we describe a number of ecological principles that have great relevance to private land management. Although similar, our key principles differ from those developed for a more general application (Dale et al., Chapter 1). However, the principles described by Dale et al. of time, place, species, disturbance, and landscape are embodied in our principles. The principles presented here include social and economic considerations that can significantly influence the success of land-management activities in meeting ecological objectives. We note in our ecological principles how they relate to, or differ from, those ecological principles described by Dale et al. In addition, we discuss guidelines for conservation planning within mixed ownership landscapes and some of the challenges to effective conservation planning.

4.1 Ecological Principles Applied Within Mixed Ownership Landscapes

Mixed ownership landscapes present unique challenges to land managers in terms of addressing ecological objectives. While private landowners have some responsibilities under the Endangered Species Act and Clean Water Act, unlike many federal lands, private landowners are not generally required to address biological diversity or ecosystem integrity requirements. Given that private landowners are not legally constrained to conserve biodiversity, ecological principles that emphasize incentives for voluntary and adaptable approaches to biodiversity conservation offer higher probabilities of success.

4.1.1 The Principle of Ecosystem Management

Ecosystem management is a land-planning and land-management process that integrates specific ecological, social, and economic objectives (Kaufmann et al. 1994). The challenge of ecosystem management is to meet the ecological objectives of maintaining and enhancing biological diversity and ecosystem integrity, while also integrating these objectives with social and economic objectives.

An important distinction can be made between ecosystem approaches or ecosystem-based management, and ecosystem management. Ecosystem approaches or ecosystem-based management are terms often used synonymously with ecosystem management. However, ecosystem approaches and ecosystem-based management typically do not address all three objectives (ecological, social, and economic), nor do they usually address the full attainment of the ecological objectives. For example, the U.S. Fish and Wildlife Service (1994:4) defined ecosystem approach as "Protecting or restoring the function, structure, and species composition of an ecosystem, recognizing that all components are interrelated." Ecosystem approaches and ecosystem-based management have often been used in the context of analyzing or managing a specific problem in terms of historical disturbances, multiple species interactions, ecosystem processes, or influences of surrounding landscape features. While these efforts can be effective in addressing the specific problem, and can contribute toward the conservation of biological diversity or ecosystem integrity, they usually have not addressed the overall ecological objectives, nor do they integrate the social or economic objectives. This is not to say that an ecosystem approach or ecosystem-based management should not be used; it is a clear advancement in our understanding of land management. However, it should not be confused with the need for true ecosystem-management initiatives that tackle the difficult task of integrating all three objectives at landscape scales. Even some initiatives touted as ecosystem management, such as the Forest Ecosystem Management Team (FEMAT) (1993), defined ecosystem management as "a strategy or plan to manage ecosystems for all associated organisms, as opposed to a strategy or plan for managing individual species." This definition of ecosystem management is species focused so it fails to address the full complexity of biodiversity, and does not integrate social or economic objectives. Obviously, ecosystem management has been defined in a number of ways (Christensen et al. 1996), with only some of the definitions emphasizing the importance of integration of social, economic, and ecological objectives.

To address all three objectives of ecosystem management requires consideration of large planning landscapes, almost always involving mixed ownership. For private lands to become engaged in these management efforts, the integration of all three objectives is critical. Thus, a key principle for ecological land management of private lands is that ecosystem management, using a definition of the term that emphasized integration of objectives, provides a workable approach. Unfortunately, because of the confusion over the use of terms related to ecosystem management, many private landowners are skeptical of such an approach. Only by clearly identifying the full scope of ecosystem management will private landowners be assured that their economic and social objectives will be addressed when cooperating in conservation planning initiatives.

To meet ecological sustainability objectives, environmental performance, social performance, and economic performance must all be incorporated (Fava et al. 1999). While this concept of ecosystem management does not present an ecological principle per se, it has direct bearing on obtaining ecological objectives on private lands. The specific mix of ecological, social, and economic objectives for any given landowner may vary, but it is a consideration for each landowner individually, as well as a major consideration across the overall mixed ownership landscape. Failure to integrate the social and economic objectives will usually produce an unsustainable outcome, whether ecologically, socially, or economically. Involving the private landowner in conservation planning through the consideration of social and economic objectives is discussed later in this chapter.

4.1.2 Coarse-Filter Principle

A coarse-filter approach (The Nature Conservancy 1982; Hunter et al. 1988; Schwartz 1999) to address ecological objectives involves maintaining a variety of ecosystems or ecological communities. Coarse-filter approaches will be most effective for meeting the ecological objectives of ecosystem management, especially across large, mixed-ownership landscapes. A primary reason for the effectiveness of a coarse-filter approach is that land management cannot address the complexity of needs of all species (Schwartz 1999). The Interior Columbia Basin Ecosystem Management Project (1997) identified 31,743 terrestrial species that may occur within its analysis area. Addressing the needs of this many species is not a feasible task. Furthermore, using a fine-filter (species focused) approach instead of a coarse-filter approach places landowners in the position of addressing the needs of species of concern today, with no assurances that additional species of concern will not be continually identified in the future.

The Nature Conservancy (1982) discussed the use of a coarse filter with an associated fine-filter approach. The coarse filter is used to identify the complex of ecological communities that are needed to maintain most biological diversity across a landscape. The associated fine filter is designed to pick up those rare or specialized species that might be missed by the coarse-filter approach.

The challenge of implementing an effective coarse-filter approach across large, mixed-ownership landscapes lies in the development of appropriate and compatible classification systems that can be used to describe and plan across various landowners (Haufler 1995). Classification systems must be of sufficient detail to distinguish all ecological communities that need to be represented within the landscape, but not so detailed as to preclude management feasibility (Haufler et al. 1996, 1999). While developing an appropriate coarse filter is a significant challenge to land-management efforts aimed at conserving biological diversity, it holds more promise for achieving results than efforts that focus on a fine-filter approach. Haufler et al. (1996, 1999) described a process for applying a coarse-filter approach to landscape planning (Fig. 4.1) where a species-viability method was used as a check on adequate representation of the ecological communities of the coarse filter.

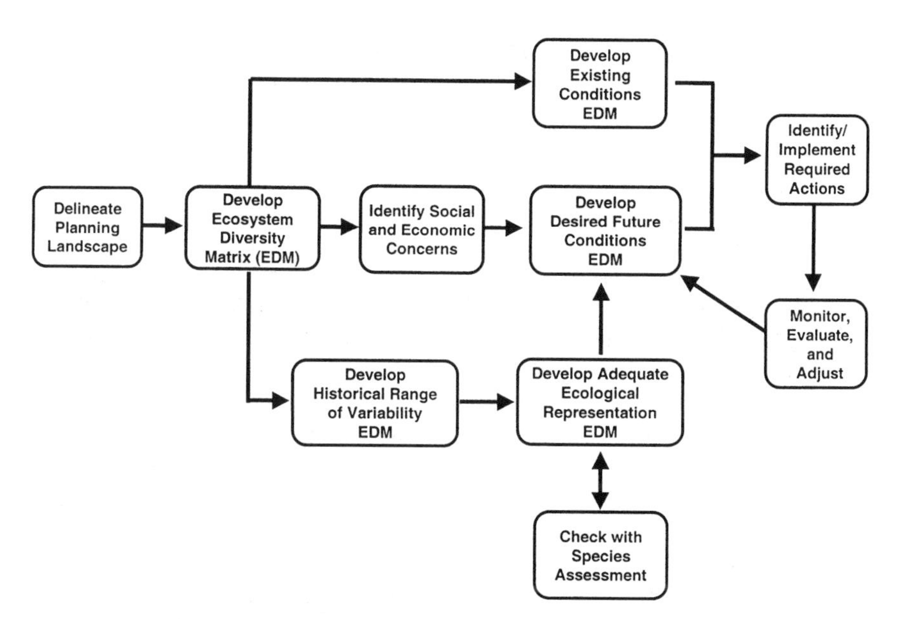

FIGURE 4.1. A diagram of a process for ecosystem management (after Haufler et al. 1999). The 10-step process is designed to address the conservation of biological diversity as well as incorporate social and economic objectives. The process uses ecosystem diversity matrices (EDMs) as a coarse filter with a species assessment as a check on its effectiveness.

These recommendations expand on the place principle described by Dale et al. (Chapter 1), where opportunities to use ecological patterns and processes as models for sustainable land use are related to community conditions. However, these recommendations differ somewhat from the species principle presented by Dale et al., who identified species influence as a critical ecological principle. While we agree that there is considerable importance in species as both a component of biological diversity and in perpetuating various ecosystem functions, we feel that effective conservation programs, especially for mixed-ownership landscapes, must be primarily based on a coarse-filter approach. In addition, a coarse-filter approach will allow for consideration of ecological integrity, as it focuses on ecosystems and processes (Schwartz 1999).

4.1.3 Ecological Site and Historical Disturbance Principle

Effective coarse filters integrate ecological site complexity, natural or historical disturbance regimes, and resulting community dynamics. To maintain or restore ecosystem integrity, managers must recognize and describe the processes and functions inherent in each ecological community. Disturbance regimes, as influenced by site characteristics such as soils, slope, aspect and elevation, historically shaped the composition, structure, and functions of ecological communities. We feel that providing for ecosystem integrity, which in turn provides the basis for meeting biological diversity objectives, will require understanding of the proper functions and processes of ecological communities under historical disturbance regimes.

Many land-classification systems in use today provide descriptions of existing conditions (such as vegetation types) (e.g., Grossman et al. 1998). These classification systems typically do not factor in underlying site complexities or disturbance regimes. Other systems have been developed that describe the complexity of ecological sites (e.g., Cleland et al. 1997; Steele et al. 1981), but do not describe existing conditions (such as successional stage) occurring on a site. All of these are important components of an effective coarse filter. Haufler (1994) and Haufler et al. (1996, 1999) described a tool termed an ecosystem diversity matrix (Fig. 4.2). An ecosystem diversity matrix, when used for terrestrial classifications, can combine both site and existing vegetation classifications. The ecosystem diversity matrix allows for an understanding and quantification of inherent ecological communities within the landscape that were produced by historical disturbance regimes. It can also be used to quantify existing conditions. Tools such as an ecosystem diversity matrix are essential for conservation planning and management across large and complex landscapes.

These recommendations are consistent with the ecological principles of Dale et al. (Chapter 1) in terms of place and disturbance. Ecological complexity of sites, as discussed above, needs to be a component of an

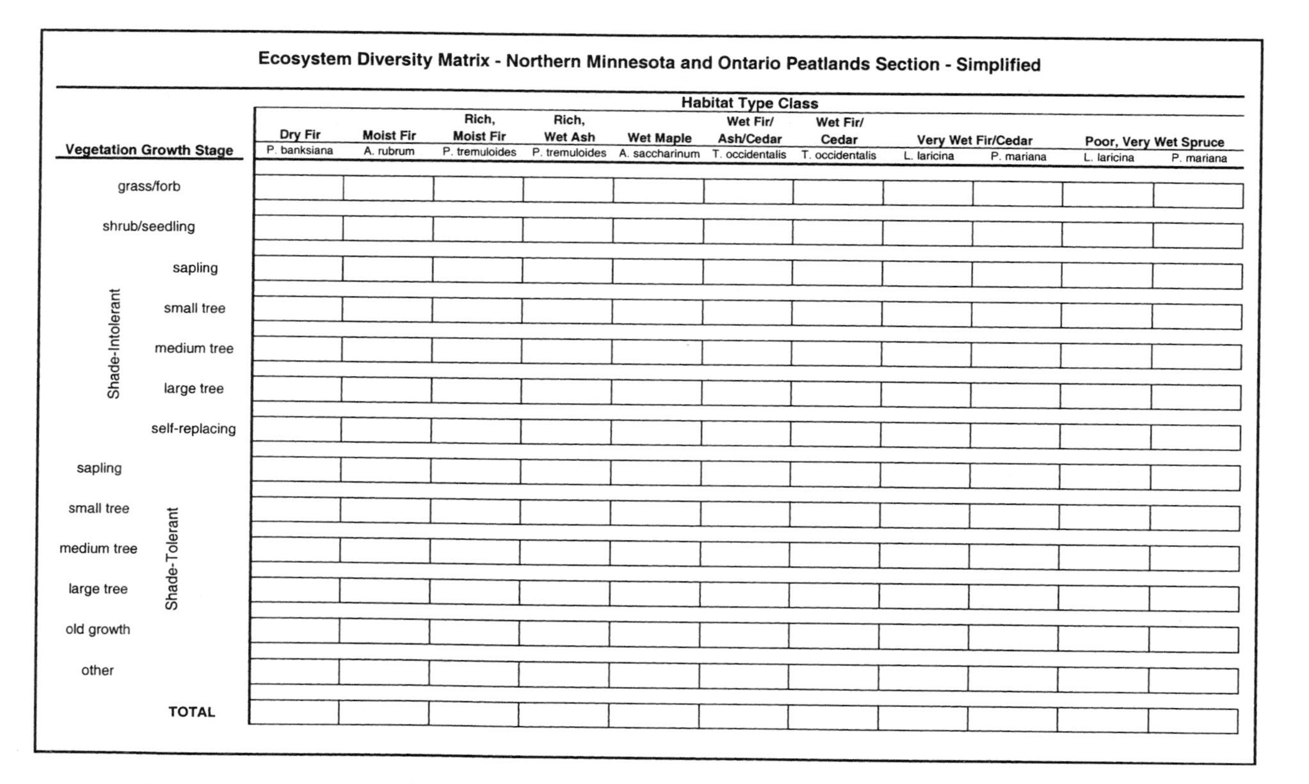

FIGURE 4.2. A simplified version of an ecosystem diversity matrix (Kernohan et al. 1999) for forested ecosystems developed for the Northern Minnesota and Ontario Peatlands Section through the Boise Cascade Corporation Ecosystem Management Demonstration Project. Primary axes of the ecosystem diversity matrix describe site potential as depicted by habitat type classes and temporal stand dynamics depicted by vegetation growth stages.

effective coarse filter. Further, the role of historical disturbance regimes in shaping ecological communities is an important component of an effective coarse filter.

The principles of site complexity and historical disturbance are not unique to private lands. These principles apply to effective coarse-filter application for any land ownership. However, because a coarse-filter approach is considered to be most effective for mixed-ownership landscapes, it does have great relevance to ecological principles for private lands alone.

4.1.4 Hierarchical Monitoring Principle

Monitoring the effectiveness of land-management planning for conservation at landscape scales will require a hierarchy of metrics at the landscape, community, species, and genetic levels operating across a variety of time scales. At the landscape level, effectiveness monitoring can be based on providing the correct or adequate mix of ecological communities within a coarse filter (Haufler et al. 1996, 1999). The communities described and quantified at the landscape level then frame the sampling units for effectiveness monitoring at the community level. Effectiveness monitoring considers whether a community (e.g., stand or stream reach) contributes to meeting the ecological objectives defined by the coarse filter. In other words, effectiveness monitoring aids in determining if a given community is functioning within acceptable bounds to meet ecosystem integrity objectives for that particular community. Measures at the community level may be evaluated relative to their historical range of variability (Morgan et al. 1994). At the species level, effectiveness monitoring evaluates individual species as a check on the coarse filter (Haufler et al. 1996, 1999). If the coarse filter is maintaining viable populations of all species inherent in the landscape, then the proper functioning of the coarse filter is confirmed. The genetic level is the finest scale in the ecological hierarchy. Genetic parameters such as the heterozygosity of a particular subpopulation can be extremely sensitive to landscape changes. Therefore, effectiveness monitoring at the genetic level provides a means of ensuring that genetic flow and patterns of diversity persist across entire landscapes.

The hierarchy of measures described above provides a way for managers and landowners to select appropriate measures that relate to their management situation while understanding how their activities may impact other levels. For example, a small private woodlot owner may be most interested in monitoring how the specific ecological community of the woodlot is contributing to ecosystem management. By having a hierarchical system of measures, this woodlot community can be evaluated based on its composition, structure, or function and related to its contribution to the overall landscape (i.e., coarse filter). Conversely, a landowner with a pair of an endangered species occurring on his or her land can monitor the success of that pair, which in turn contributes to the overall viability of that species across the landscape. Furthermore, understanding the location and productivity of all pairs of the species would allow for a viability assessment to be conducted.

Implicit in this monitoring principle is time. Dale et al. (Chapter 1) described how time can influence ecosystem change. The important aspect of monitoring is to describe ecosystem change through repeated monitoring. To effectively implement a coarse-filter strategy, or ecosystem management in general (see principles 4.1.1 and 4.1.2) monitoring the effects of change over time is critical.

Dale et al. (Chapter 1) discussed landscape influence as a key ecological principle. We agree that landscape effects are important to understand and evaluate. We advocate use of species viability assessments linked to the coarse filter as a means of understanding both the spatial and temporal distribution of suitable habitat for a given species. While connectivity may be a concern, of greater relevance is the identification of any significant barriers to movements of species that may impede either demographic or genetic support of metapopulations.

4.2 Considerations for Private Land Applications

The ecological principles discussed above relate to conservation planning and land management regardless of land ownership. Few landscapes of sufficient size to address ecosystem management objectives occur without including private lands and their diverse ownership. To engage private landowners in conservation planning and management, several considerations are important.

4.2.1 Private Property Rights Considerations

Private property rights need to be respected. Private landowners own their lands for a variety of reasons, ranging from commodity production, use as home sites, providing aesthetics or recreation, to conservation land trusts. All private landowners have made economic investments in their land, and most expect to maintain some type of economic value or return from their land either in terms of direct income or invested value. Economic investment, along with diverse views on the extent of reasonable government intervention in land-management decisions, has led to recent debates on property rights. Most private landowners wish to contribute to a healthy environment and wish to see biological diversity maintained. However, most private landowners do not believe government agencies should tell them how to manage their lands. Effective programs for mixed-ownership landscapes must recognize and respect the objectives of private landowners and develop mechanisms for engaging landowners in cooperative efforts. Top-down regulatory approaches continue to mandate specific responses from private landowners in land-management practices, but typically result in increased animosity toward regulating agencies as well as toward the environmental feature that is the focus of the regulation. In the long run, these approaches often create barriers to ecological land-management, and certainly make cooperative land-management programs more difficult to implement.

4.2.2 Voluntary, Incentive-Based Considerations

Incentive-based, voluntary programs hold the most promise for private land conservation efforts. Efforts that build on making conservation contributions from private lands attractive rather than an economic cost or regulatory restriction will gain far more support from private landowners, and in the long run yield far greater conservation benefits. Numerous examples of these incentive-based programs exist. Vickerman (1998) discussed stewardship incentives for possible application in Oregon. She identified some of the more promising incentives as stewardship councils, watershed councils, stewardship certification, tax reforms, information for conservation planning, regulatory relief, mitigation banking, and direct financial assistance. Hocker (1996) discussed the importance and use of land trusts to land conservation. Efforts by the U.S. Fish and Wildlife Service to develop both Safe Harbor Agreements and Candidate Conservation Agreements are aimed at reducing economic or management risks associated with improving existing conditions for listed species or candidate species on private lands. These are all good examples of programs that either provide actual incentives or that reduce economic disincentives associated with some stewardship activities. Additional opportunities for expansion in these types of programs exist, and represent some of the best prospects for expansion of effective conservation planning efforts to private lands.

4.2.3 Bioreserve Disincentives

A bioreserve strategy has been promoted by many (e.g., the Wildlands Project (Noss 1994)) as a necessity for conservation planning. However a bioreserve strategy can be a disincentive to private landowners by attempting to delineate specific areas across the United States as core reserves, buffer areas, and corridors. An example of a bioreserve strategy mapped proposed locations of these areas for the state of Florida (Noss 1994), and then advocated for their protection. A bioreserve strategy strives to minimize human activities within the designated core reserves and connecting corridors. This mapping effort is conducted irrespective of the ownership of the lands being designated. Landowners within these proposed areas may be willing to discuss stewardship potential of their lands, but usually have additional land-management objectives that would not be addressed by a protected status that minimizes human activities. The bioreserve strategy polarizes these potential conflicts rather than involving private landowners and working toward compatible solutions.

Bioreserves defined and used as a conservation planning tool differ from the bioreserve strategy. Defining a bioreserve as an area with a primary objective of the maintenance or enhancement of biodiversity (Haufler 1999), rather than as a system of protected core areas, offers more flexibility in what constitutes a bioreserve, and allows for greater inclusion of landowner objectives into conservation planning. When bioreserves are defined as a planning tool, landowners will likely be more willing to enter into collaborative planning efforts than under a rigid system approached from a top-down planning effort such as the Wildlands Project. Efforts at pushing top-down strategies onto private lands will fuel private property rights issues and result in increased political and legal activity rather than constructive conservation efforts.

4.2.4 Collaboration Considerations

Collaboration is an effective process for involving private landowners in land management for conservation. The Keystone National Policy Dialogue on Ecosystem Management stated that "collaboration among organizations and individuals comprising an ecosystem management initiative is usually critical to the success of the initiative" (Keystone Center 1996). Collaboration among organizations is particularly needed when such initiatives include diverse mixed-ownership landscapes. The Keystone Dialogue further stated that "by using an effective collaborative process, the inherent conflicts caused by changing legal, economic, or ecological landscapes that can derail or distort decision making can be minimized or even eliminated" (Keystone Center 1996). Inherent conflict cannot be ignored in landscape planning efforts. Often times, the reason a collaborative effort is started is to prevent the unnecessary spread of conflict beyond that which is inherent. Therefore, to accept inherent conflict and minimize expansion of conflict becomes the goal of collaborative planning for conservation. To facilitate the above agenda, participation in any collaborative process should be voluntary. At a minimum, participants representing county, state, and federal governments, industrial and nonindustrial private landowners, environmental organizations, tribal governments, elected officials, and concerned citizens should be included (Kernohan and Haufler 1999). Identification of participants in collaborative planning processes is an important decision in terms of the likelihood of success. Additional criteria for effective collaborative planning processes include early involvement of all participants, identification of mutual goals and objectives, respect for each other, decision making by consensus, and building of trust (Kernohan and Haufler 1999; Keystone Center 1996; Haufler 1995).

Participant involvement early in the planning process is critical to the success of any initiative. Early involvement provides a way for participants to take ownership in the process. Once a group has been formed, mutual goals and objectives should be set and processes established for achieving those goals and objectives (Kernohan and Haufler 1999; Breckenridge 1995). Although a partnership will generate its own goals and objectives, it must be recognized that each participant has a set of objectives that may or may not coincide with others.

Collaborative processes should employ consensus decision-making. Consensus means considering all points of view while working toward a solution. The process of developing consensus will facilitate implementation of most initiatives and encourage participants to assist resource managers in generating support for the decision. Several examples of good collaborative planning efforts exist throughout the United States. The Keystone National Policy Dialogue on Ecosystem Management participated in a variety of field trips to learn more about such efforts and summarized them in their final report (Keystone Center 1996).

4.2.5 Regulation Considerations

Regulation does have a role in management of private lands for conservation. There is a need for various types of regulation to provide for the good of society. For example, regulations that prevent a single landowner from degrading the quality of waters downstream from his or her property that would otherwise preclude others from achieving their conservation objectives or other uses are appropriate. Debates will always exist over how much the rights and needs of society should be dominant or subservient to the rights of property owners. Conservation efforts would be better served to avoid these types of debates and focus on achieving the desired solutions through collaboration and cooperation. Encouraging incentive programs, removing disincentives, and building on mutual trust and respect are likely to produce far greater conservation gains in the long term than trying to force solutions through regulation. The development of collaborative solutions may take longer to put into place than some regulatory solutions, but over time are likely to produce far greater gains to conservation.

4.3 Guidelines for Land-Management Planning for Conservation on Private Lands

4.3.1 Collaboration Efforts

Based on the issues already discussed, it should be clear that private lands are critical components of conservation planning, and that the most effective approaches to meeting ecological objectives are those that encourage and reward private land involvement. While regulation will always play a role, the less that it is needed, the greater the likely gain from private lands. Although collaboration is currently a buzzword for many agencies and initiatives, programs and assistance to enhance and enable collaborative efforts are still needed. Assistance can be accomplished through establishment of sources or mechanisms for incentives, reduction of legal or monetary barriers or disincentives, development and deployment of technical assistance programs, and recognition of and commitment to collaboration from many diverse groups.

4.3.2 Incentives

Incentive programs and landowner recognition are important components of effective conservation programs. Existing incentive programs need to be highlighted so that all available "tools" or funding are available to an initiative. Some of these incentive programs can be strengthened to more effectively meet ecological objectives of conservation planning. For example, there are a number of programs that have provided matching monies for various habitat-improvement activities, such as the Stewardship Incentive Program, sometimes at relatively high percentages of total expenditure amounts. However, these programs still require funds to be matched and some landowners may lack even the small amount of money needed to match program funds. Collaborative initiatives may need to generate the matching funds, or the incentive programs may need to be altered to allow for total funding of habitat improvement or restoration efforts that are identified as key activities in a collaborative effort. Similarly, maintaining current conditions on some lands may be an important conservation activity. If a landowner has an economic objective for a site that requires alteration of the site, failing to conduct the alteration is an economic cost to the landowner. Many programs with matching funds cannot expend those funds to compensate landowners for precluded economic return from activities, even if payment for maintaining existing conditions is an important conservation action for that site. For example, the Stewardship Incentive Program is based on cost sharing for specific planned activities, but not for payment for maintaining existing conditions. The inability to receive compensation for maintaining existing conditions is an example of a current deficiency in incentive programs and an area where programs may be improved to help support collaborative conservation planning.

4.3.3 Science-Based Assistance

Private landowners generally need science-based assistance in developing and describing a landscape through a coarse filter. As previously discussed in the ecological principles section, a coarse-filter approach to conservation planning is the most effective approach for obtaining ecological objectives of ecosystem management. However, a coarse filter is not a simple approach for lay audiences to grasp and apply, let alone to develop and utilize on their own. Therefore, collaborative efforts involving private lands will usually require professional assistance from ecosystem ecologists who can assist in the development, mapping, and description of ecological communities that comprise the coarse filter. Unfortunately, implementation of effective coarse-filter strategies is a relatively new area of ecological science, so the number of ecosystem ecologists with these skills is limited and financial support for their involvement may not be available. Finding both the professional expertise needed to provide the scientific base for a collaborative effort as well as funding to support the initiative is a major challenge. Some collaborative initiatives that have lacked one or both of these often end up with a lot of good discussions among participants, but little tangible conservation progress because of the inability to understand and quantify the landscape in meaningful conservation terms.

4.3.4 Ecological, Social, and Economic Integration

Collaborative efforts must involve ecosystem management defined as the integration of social, economic, and ecological objectives, not just a focus on ecological objectives. While the goal of most conservation initiatives is to address ecological objectives, it must be recognized that attainment of these objectives can be greatly facilitated through ecosystem management that factors in all three objectives. The integration of social, economic, and ecological objectives through ecosystem management is consistent with Fava et al.'s (1999) view of sustainability. This does not mean that ecological objectives should be compromised in this integration, but rather they should be clearly defined and understood so that they can be addressed along with social and economic objectives. Innovative solutions are needed that will allow ecological objectives to be met while also meeting social and economic objectives, especially those objectives of private landowners in the planning landscape.

4.3.5 Recognition

An effective type of incentive program includes rewarding private land-conservation efforts by publicizing good achievements of landowners. This can increase private landowner involvement. For example, a new award called the National Private Land Fish and Wildlife Stewardship Award was recently developed and sponsored by the International Association of Fish and Wildlife Agencies, Wildlife Management Institute, The Wildlife Society, American Fisheries Society, and American Farm Bureau Federation. Recipients of this or similar awards will certainly be encouraged to continue their conservation efforts, while other landowners will be encouraged to obtain similar recognition for good management practices. While recognition by national professional societies is of value, even greater rewards may be obtained by encouraging awards or other recognition by landowner organizations, particularly if these organizations can be linked to professional or conservation organizations.

4.4 Case Study

4.4.1 Boise Cascade Corporation Ecosystem Management Projects: An Example of a Private Landowner-Led Collaborative Program

Applying the principles and guidelines discussed above is not an easy task given today's political environment. Recognizing the need to compete in today's ever-changing environment, Boise Cascade Corporation initiated three projects designed to develop and promote a collaborative ecosystem management process. The three projects were geographically located in central Idaho, south-central Washington, and northern Minnesota. The project in northern Minnesota will be used to illustrate these private landowner-led collaborative programs.

The challenge for Boise Cascade Corporation was to develop a process whereby multiple landowners, government and private, could actively participate in landscape-level conservation planning designed to integrate ecological, social, and economic objectives. The project focused primarily on addressing the ecological objectives, but also integrated social and economic considerations. The project followed the steps of an ecosystem-management process (Fig. 4.1) described by Haufler et al. (1996, 1999). Determining the extent of the planning landscape was the first step in the process. The landscape was defined by ecological boundaries with minimal consideration for political jurisdictions. The planning landscape defined by Boise Cascade in northern Minnesota followed the delineation of the Northern Minnesota & Ontario Peatlands section of the National Hierarchy of Ecological Units (Cleland et al. 1997). The National Hierarchy of Ecological Units classification system is hierarchical, with the section level based on broad areas of similar geomorphic processes, topography, and regional climate (Cleland et al. 1997).

Once the planning landscape was defined, potential participants were identified. Participant involvement was designed as a two-step process, beginning with the formation of an advisory group consisting of landowners, managers, and technical experts that assisted in refining the ecosystem management process and fitting it to the landscape. A second step would involve the establishment of a broader group of "stakeholders." The group of primary landowners, managers, and technical experts that participated in Boise Cascade Corporation's collaborative project included the Minnesota Department of Natural Resources, Chippewa and Superior National

Forests, Koochiching County, The Nature Conservancy, Minnesota Forest Industries, USFS North Central Forest Experiment Station, University of Minnesota, and the Minnesota Forest Resources Council.

It was the role of this first group of landowners and managers to establish the ecological context using a coarse-filter approach. Establishing the ecological context included four steps: (1) classifying the existing landscape with regard to ecological communities, (2) determining the historical range of variability of occurrence of ecological communities for the entire landscape, (3) identifying minimum threshold levels of ecological communities to meet biological diversity and ecosystem integrity objectives, and (4) checking the coarse filter using single-species assessments (Haufler et al. 1996, 1999). The forested component of the landscape was classified using an ecosystem diversity matrix (Kernohan et al. 1999) (Fig. 4.2). Additional matrices would provide for description of riparian and wetland systems, aquatic systems, and grassland/shrub systems, if present. The ecosystem diversity matrix was used to express existing and historical conditions.

The historical ecosystem diversity matrix was used to quantify adequate ecological representation (i.e., minimum threshold levels for ecological communities), which was checked with the species viability assessment (Kernohan et al. 1999; Haufler et al. 1996, 1999). Adequate ecological representation is defined as the amount and distribution of ecological units necessary to maintain viable populations of all native species (Haufler 1994). Adequate ecological representation is calculated as a fixed percent of the maximum historical range of variability for any given ecological unit. A species-specific check of adequate ecological representation is performed using a variety of habitat potential models that assess the quality and size of home ranges and their locations necessary to maintain viable populations of selected species (Roloff and Haufler 1997).

Haufler et al. (1996, 1999) described the remainder of the ecosystem management process as the integration of social and economic conditions and generation of desired future conditions. These steps in the process require the addition of the broader group of "stakeholders" to accurately convey the social and economic issues constraining the landscape. Participants in this broader stakeholders group would likely include interested citizens, environmental organizations, local businesses, the tourism industry, and policy makers. Responsibilities of the stakeholders group would include (1) developing strategies for integrating the social and economic demands within the ecological thresholds, (2) determining desired future conditions across the landscape, and (3) monitoring the effectiveness of implemented strategies using adaptive management principles.

This two-step approach to collaboration provides for the development of critical data and scientific analysis procedures for ecological objectives prior to discussions of social and economic objectives. This strategy of separating assimilation of ecological data and general stakeholder involvement provides a process for identifying threshold levels for all ecological communities (with the exception of historically rare communities) for meeting ecological objectives first. The identification of threshold levels provides a foundation from which to evaluate tradeoffs and impacts while integrating ecological, social, and economic objectives as they relate to desired future conditions.

4.5 Challenges to Implementation

The principles, considerations, guidelines, and example described herein emphasize ecological, social, economic, and political methodologies to enhance conservation planning for private lands. While there are many exciting and promising efforts underway to incorporate private lands into effective conservation planning, there are also some significant challenges that may complicate such efforts.

4.5.1 Historical Disturbance Challenges

Historical disturbances may be incompatible with private lands. Recognizing the importance of historical disturbances to maintaining ecological integrity of ecosystems does not solve the dilemma that many of these disturbances are incompatible with other land-management objectives. For example, catastrophic fire events were a primary disturbance to many forest ecosystems. However, such events are generally unacceptable, even where feasible to implement, for most private landowners. Alternative management practices that can provide most, if not all, of the conditions for ecosystem integrity are needed. More information and research on the various alternatives are needed to evaluate the full range of their effectiveness. Even on public lands, these same questions about acceptable levels of disturbance still need to be resolved.

4.5.2 Past Management Challenges

Past management practices may preclude effective restoration of many lands. Humans have altered landscapes in North America for thousands of years. The level of alteration has greatly accelerated within the last 200 years with European settlement and technological advances. Many areas have been significantly altered, including many private lands. Some of these areas may be so altered that restoring ecosystem integrity, if desired, may not be possible in the near future. For example, if a site has been plowed, the soil horizons developed over millennia may have been changed, and the resulting functions and species compositions that can be supported on the site altered. Similarly, drainage of lands that were once wetlands may have altered regional hydrology, with ramifications for the entire watershed. The significance of past management to overall conservation planning objectives may not be known. Thus, even with good collaborative efforts and willing participants, the ecological feasibility of some restoration efforts may be less than desired.

4.5.3 Exotic Species Challenges

Exotic species across mixed-ownership landscapes may not be controllable. Invasive plants or animals are causing many changes in ecosystem composition, structure, and function (Campa and Hanaburgh 1999). Most of these exotics are strong competitors, and are difficult to control or eradicate. Control and eradication can be a significant challenge on any single ownership. On mixed-ownership landscapes, all landowners must be motivated to participate in conservation efforts or control or eradication may not be possible. Addressing the threats of exotics involves the development and application of control methods, as well as effective education programs to increase awareness of the seriousness of this threat.

4.5.4 Ownership Pattern Challenges

Patchwork of ownership may continue to cause landscape fragmentation. Under historical disturbance regimes some disturbances were very big, producing large blocks of similar conditions. In mixed-ownership landscapes, creating large blocks of similar conditions, given the different existing conditions and varying landowner objectives, may not be feasible. In addition, current societal and political pressures often restrict large areas from being managed in a similar manner. The key question to address relative to these potential constraints is how big does any given block of habitat need to be to provide adequate ecological function and integrity. In many landscapes, historical patch dynamics were quite variable (Camp et al. 1997). In others, even where large blocks of similar conditions were more consistent, biological diversity and ecosystem integrity can be provided in much smaller-sized areas, as much of the research on forest fragmentation is documenting. The challenge is to identify what block sizes are needed within the coarse filter for a planning landscape and how different landowners can work together to maintain or restore functional sizes of ecosystems.

4.5.5 Existing Regulation Challenges

Mixed-ownership landscapes face challenges to conservation planning from some existing regulations. Laws such as the Sherman Anti-Trust Act are designed to protect the public from such activities as price fixing by competing companies. Unfortunately, some interpretations of the law make collaborative efforts worrisome. For example, corporate lawyers of forest products companies are fearful of any collaborative efforts involving other companies that address future land-management plans, because these could be construed as efforts at fixing the supply of timber or fiber. While there are a number of possible ways to avoid problems with anti-trust, it is an additional impediment to collaborative efforts. Similarly, laws such as the National Forest Management Act are designed to insure that planning by the U.S. Forest Service is conducted in a regulated manner with structured opportunities for public input. While the importance of structured input can be clearly shown, it also complicates the ability of the Forest Service or other similarly legislated land-management agencies from entering into active collaboration initiatives for conservation planning (Carr et al. 1998).

4.6 Conclusions

While the primary emphasis of this chapter has been on the importance of voluntary, collaborative initiatives for effective conservation planning across mixed-ownership landscapes, several other key messages can be highlighted. Landscape-level conservation planning will be most effective if it is based on a coarse-filter approach. For implementation of a coarse-filter approach to occur through collaborative efforts, technical assistance in ecosystem ecology, remote-sensing methods, geographic information system applications, and data collection and analysis will usually be needed. Often, these will require financial support. Sources of both technical and financial assistance need to be identified and made available to collaborative conservation planning initiatives.

While many effective and imaginative incentive programs already exist for private land conservation efforts, innovative new programs based on broad ecosystem management objectives or improvements in existing programs are still needed. Such programs need to be identified, funded, legislated, or promulgated as appropriate.

Legal barriers to effective collaborative efforts need to be identified and corrected in existing legislation. Alternatively, new legislation that details exclusions to the existing legislation to accommodate collaborative conservation planning efforts could be drafted and enacted.

Education and training programs that can assist collaborative conservation planning initiatives with both ecological principles and methods as well as effective processes for collaboration need to be made available. Funding sources for support of technical assistance, education programs, or data generation and analysis are also needed for effective collaborative efforts.

Private lands hold the key to long-term, effective conservation planning efforts across most mixed-ownership landscapes. Application of ecological principles through collaborative land-management efforts that incorporate landowner objectives and concerns offers the best opportunity for maintaining biological diversity and ecosystem integrity.

References

Breckenridge, L.P. 1995. Reweaving the landscape: The institutional challenges of ecosystem management for lands in private ownership. Vermont Law Review **19**:363–422.

Camp, A., C. Oliver, P. Hessburg, and R. Everett. 1997. Predicting late successional fire refugia predating European settlement in the Wenatchee Mountains. Forest Ecology and Management **95**:63–77.

Campa, H., III, and C. Hanaburgh. 1999. A management challenge now and in the future: what to do with exotic species. Pages 203–215 *in* R.K. Baydack, H. Campa, III, and J.B. Haufler, editors. Practical approaches to the conservation of biological diversity. Island Press, Washington, D.C., USA.

Carr, D.S., S.W. Selin, and M.A. Schuett. 1998. Managing public forests: understanding the role of collaborative planning. Environmental Management **22**:767–776.

Christensen, N.L., A.M. Bartuska, J.H. Brown, S. Carpenter, D. D'Antonio, R. Francis, J.F. Franklin, J.A. MacMahon, R.F. Noss, D.J. Parsons, C.H. Peterson, M.G. Turner, and R.B. Woodmansee. 1996. The report of the Ecological Society of America committee on the scientific basis for ecosystem management. Ecological Applications **6**:665–691.

Cleland, T.D., P.E. Avers, W.H. McNab, M.E. Jensen, R.G. Bailey, T. King, and W.E. Russell. 1997. National hierarchical framework of ecological units. Pages 181–200 *in* M.S. Boyce and A. Haney, editors. Ecosystem management: applications for sustainable forest and wildlife resources. Yale University Press, London, England.

Fava, J.A., S.B. Young, and A. Veroutis. 1999. Viewpoint: making sustainability accessible: use of sustainability measures. Strategic Environmental Management **1**:175–179.

Forest Ecosystem Management Team (FEMAT). 1993. Forest ecosystem management: an ecological, economic, and social assessment. U.S. Government, Washington, D.C., USA.

Grossman, D.H., D. Faber-Langendoen, A.S. Weakley, M. Anderson, P. Bourgeron, R. Crawford, K. Goodin, S. Landaal, K. Metzler, K. Patterson, M. Pyne, M. Reid, and L. Sneddon. 1998. International classification of ecological communities; terrestrial vegetation of the United States, Volume I, The national vegetation classification system: development, status, and applications. The Nature Conservancy, Arlington, Virginia, USA.

Grumbine, R.E. 1994. What is ecosystem management? Conservation Biology **8**:27–38.

Haufler, J.B. 1994. An ecological framework for forest planning for forest health. Journal of Sustainable Forestry **2**:307–316.

Haufler, J.B. 1995. Forest industry partnerships for ecosystem management. Transactions of the North American Wildlife and Natural Resources Conference **60**:422–432.

Haufler, J.B. 1999. Strategies for conserving terrestrial biological diversity. Pages 17–30 *in* R.K. Baydack, H. Campa III, and J.B. Haufler, editors. Practical approaches to the conservation of biological diversity. Island Press, Washington, D.C., USA.

Haufler, J.B., C.A. Mehl, and G.J. Roloff. 1996. Using a coarse-filter approach with a species assessment for ecosystem management. Wildlife Society Bulletin **24**:200–208.

Haufler, J.B., C.A. Mehl, and G.J. Roloff. 1999. Conserving biological diversity using a coarse-filter approach with a species assessment. Pages 107–125 *in* R.K. Baydack, H. Campa III, and J.B. Haufler, editors. Practical approaches to the conservation of biological diversity. Island Press, Washington, D.C., USA.

Hocker, J.W. 1996. Patience, problem solving, and private initiative: local groups chart a new course for land conservation. Pages 245–259 *in* H.L. Diamond and P.F. Noonan, editors. Land use in America. Island Press, Washington, D.C., USA.

Hunter, M.L., Jr., G.L. Jacobson, Jr., and T. Webb III. 1988. Paleoecology and the coarse-filter approach to maintaining biological diversity. Conservation Biology **2**:375–385.

Interior Columbia Basin Ecosystem Management Project. 1997. Upper Columbia River Basin Draft Environmental Impact Statement. Boise, Idaho, USA.

Kaufmann, M.R., R.T. Graham, D.A. Boyce, Jr., W.H. Moir, L. Perry, R.T. Reynolds, R.L. Bassett, P. Mehlhop, C.B. Edminster, W.M. Block, and P.S. Corn. 1994. An ecological basis for ecosystem management. USDA Forest Service, General Technical Report RM-246.

Kernohan, B.J., and J.B. Haufler. 1999. Implementation of an effective process for the conservation of biological diversity. Pages 233–249 *in* R.K. Baydack, H. Campa III, and J.B. Haufler, editors. Practical approaches to the conservation of biological diversity. Island Press, Washington, D.C., USA.

Kernohan, B.J., J. Kotar, K. Dunning, and J.B. Haufler. 1999. Ecosystem diversity matrix for the Northern Minnesota and Ontario Peatlands landscape: technical manual. Unpublished Report. Boise Cascade Corporation, Boise, Idaho, USA.

Keystone Center, The. 1991. Biological diversity on Federal lands. The Keystone Center, Keystone, Colorado, USA.

Keystone Center, The. 1996. The Keystone national policy dialogue on ecosystem management: final report. The Keystone Center, Keystone, Colorado, USA.

Morgan, P., G.H. Aplet, J.B. Haufler, H.C. Humphries, M.M. Moore, and W.D. Wilson. 1994. Historical range of variability: a useful tool for evaluating ecosystem change. Journal of Sustainable Forestry **2**:87–111.

Natural Heritage Data Network. 1993. Perspectives on species imperilment. The Nature Conservancy, Washington, D.C., USA.

Nature Conservancy, The. 1982. Natural heritage program operations manual. The Nature Conservancy, Arlington, Virginia, USA.

Noss, R.F. 1994. The wildlands project: land conservation strategy. Pages 233–266 *in* R.E. Grumbine, editor. Environmental policy and biodiversity. Island Press, Washington, D.C., USA.

Roloff, G.J., and J.B. Haufler. 1997. Establishing population viability planning objectives based on habitat potentials. Wildlife Society Bulletin **25**:895–904.

Sample, V.A. 1994. Building partnerships for ecosystem management on mixed ownership landscapes. Journal of Forestry **92**(8):41–44.

Schwartz, M.W. 1999. Choosing the appropriate scale of reserves for conservation. Annual Review of Ecology and Systematics **30**:83–108.

Steele, R., R.D. Pfister, R.A. Ryker, and J.A. Kittams. 1981. Forest habitat types of central Idaho. U.S. Forest Service Intermountain Forest and Range Experiment Station General Technical Report **INT-114**.

U.S. Fish and Wildlife Service. 1994. An ecosystem approach to fish and wildlife conservation. Washington, D.C., USA.

Vickerman, S. 1998. Stewardship incentives; conservation strategies for Oregon's working landscape. Defenders of Wildlife, Washington, D.C., USA.

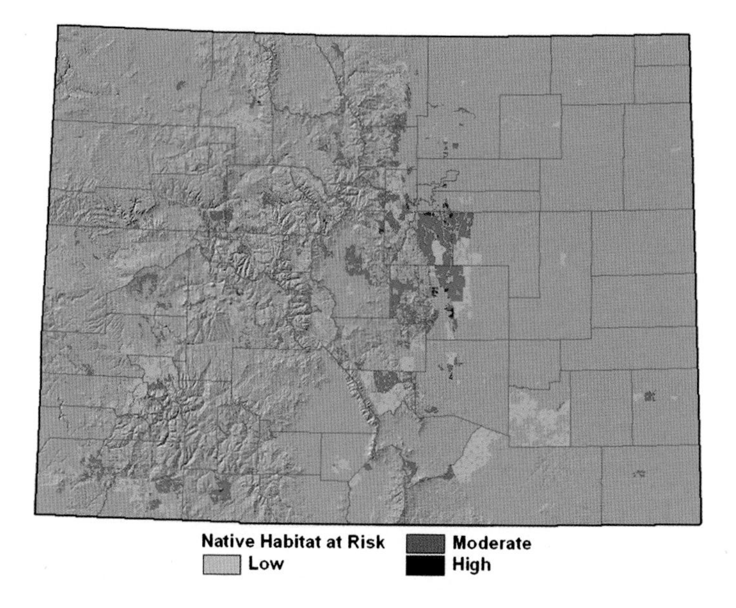

FIGURE 2.1. We mapped native habitat that was at the greatest risk of development in Colorado to help local governments develop conservation plans that included statewide needs for habitat protection. Risk level corresponded to the magnitude of change between 1960 and projected (2020) levels of development. Greater level of risk is shown by darker shades.

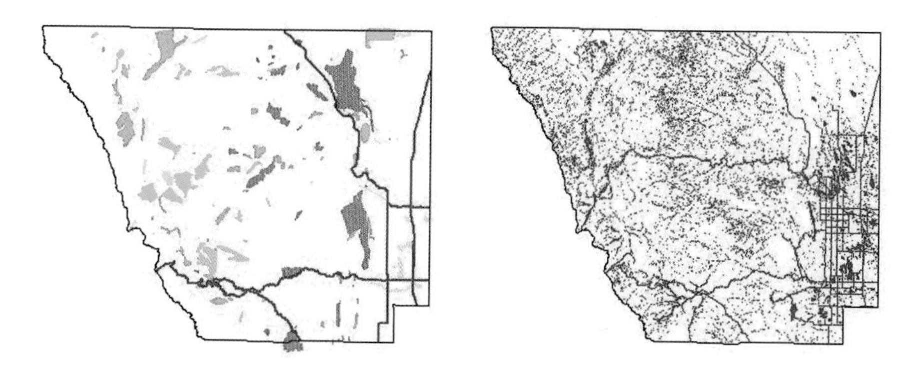

FIGURE 2.2. We developed maps of rare landscape elements as part of a master land-use plan for Larimer County, Colorado. Two types of maps were used in this planning process. The first displayed conservation sites mapped by the Colorado Natural Heritage Program (*left*). Conservation sites are areas of the landscape containing rare and sensitive plants and animals. The boundaries of the sites are delineated to include the ecological processes needed to sustain the rare elements within them. In addition, we used landcover maps derived from satellite imagery to delineate vegetation types contributing less than 3% of the area of the county (*right*). These maps (along with others) were used to decide which areas of the county would require conservation plans prior to approval of development proposals.

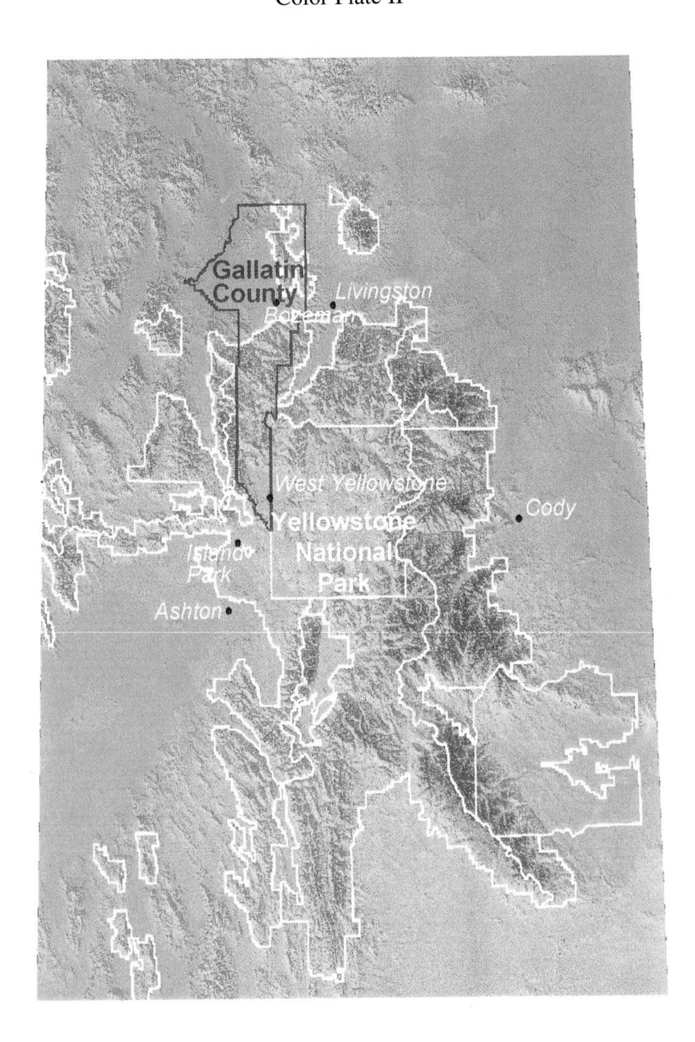

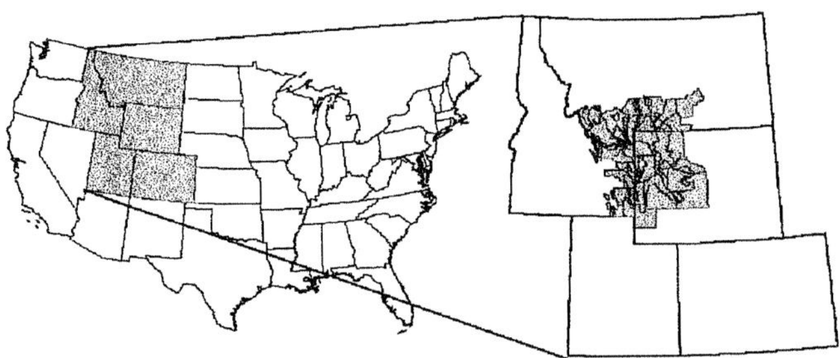

FIGURE 3.1. Shaded-relief map of the Greater Yellowstone Ecosystem. The color gradient is from brown (high elevation) to green (low elevation). White lines denote public lands. Private lands are primarily at lower elevations around the perimeter of the ecosystem. Gallatin County, MT, outlined in red, is the location of more detailed socioeconomic studies described in this chapter.

Color Plate III

FIGURE 8.4. Perspective view of (*a*) existing conditions, (*b*) Southern alternative, (*c*) Spline alternative, and (*d*) Park alternative.

FIGURE 9.2. Land-use/land-cover map of the Willamette River Basin ca. 1990 compiled from remotely sensed and ground surveys as well as census, agricultural, hydrological, and transportation data sources.

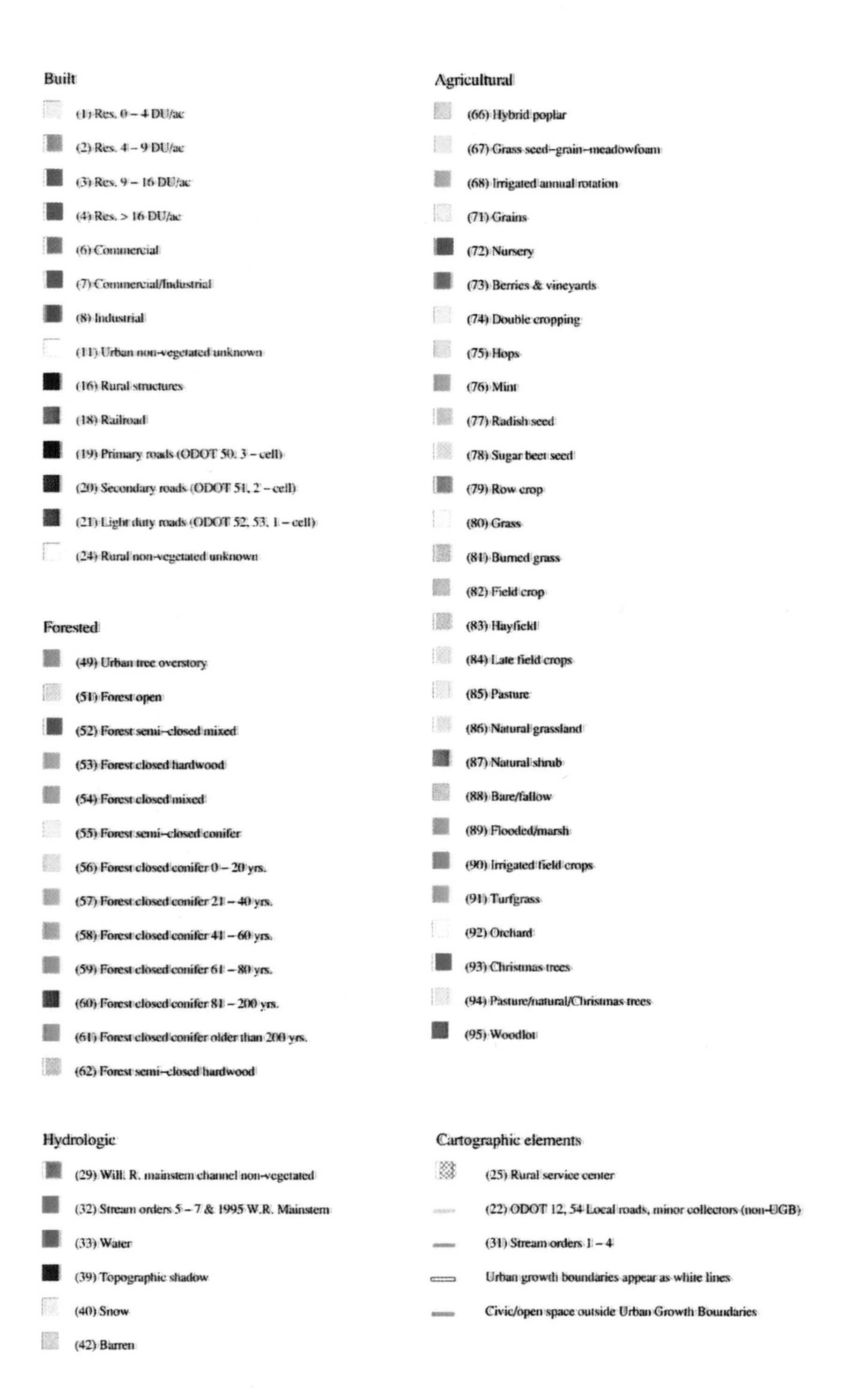

FIGURE 9.2. (*Continued*).

5
On the Road to Ecorecovery and Protection: A Case Study of Endangered Tribal Habitats and Culture in the Eastern Ghats of India

STEVEN D. WEISS

This paper provides an overview of the culture and ecology of the Eastern Ghats in southern India. Both the forests and tribal communities of this region are threatened by an increasing number of environmentally destructive commercial ventures operating in the area. A case study presents tribal habitat and communities in Andhra Pradesh that were disrupted by illegal mining operations. Linking environmental degradation and social domination, the study traces the efforts of a local nongovernment organization to empower tribal communities to protect their habitat and livelihood. Social, political, and economic issues become environmentally charged as the sustainable use and management of land depends on strengthening viable, local communities with control over their own bioresources. The study also illustrates how the promotion of ecologically sound land-use and land-management decisions in tribal communities contributes to community building and empowerment.

Perched above the bend of a mountain river in the hilly Hukkumpet section of Visakhapatnam district in Andhra Pradesh stands the small Matsyagaddam ("fish") temple. While this area has been heavily deforested, the temple is surrounded by a sacred grove of massive mango trees. A local folk tale relates how the fish temple and its companion grove came to be. According to the legend, a terrible conflict ensued upstream between the fish and snake communities that shared the river. To protect her brood, the mother fish escaped downstream and took refuge in the depths of a whirlpool at the river's bend. Not only did she look after her own, but she also rescued any hapless person who fell into the dangerous waters. One day, however, someone played a mean trick on the mother fish by throwing a bramble of thorns into the whirlpool, and she was badly injured when she tried to rescue the object. Ever since, she has refused to rescue those who fall into the whirlpool. To make amends, the local people built a temple to worship the mother fish as their goddess and surrounded the temple with a sacred grove, always to be protected. This folk tale poignantly captures the fate of the

Eastern Ghats, which is home to this temple. Extreme environmental degradation has severed the nature–human relationship in the Eastern Ghats, and if this relationship is to be restored, special efforts must be taken to conserve and protect the natural wealth of this region. This paper is about one such effort and its implications for land use and land management in India.

5.1 Linking Environmental Degradation and Social Domination

In Chapter 1, Dale et al. cite the principle human causes of changes in land cover and land use as population and its accompanying infrastructure: economic factors; technological capacity; political systems, institutions, and policies; and sociocultural factors, such as attitudes, preferences, and values. The following case study offers an analysis of these anthropogenic-generated changes in land cover and land use, which links environmental destruction and degradation to oppressive power relations and social structures. The notion that environmental and ecological destruction springs from certain patterns of social control and dominance is a central tenet of social ecology—a comprehensive environmental ethic and philosophy of nature and social existence that owes its origin to the work of one theorist, Murray Bookchin. In claiming that "the domination of nature by man [gender specific] stems from the very real domination of humans by humans," Bookchin views the human domination and degradation of nature as arising out of patterns of domination and hierarchy in society in which some human beings exercise power and control over others (1991). While it cannot be assumed that there is any kind of *necessary* relationship between the domination of humans by humans and the domination of nature (Fox 1989), social ecology offers a theoretical perspective that is particularly illuminating in understanding the underlying causes of the loss of both ecological and cultural diversity in the Indian subcontinent.

5.1.1 Hierarchies and Domination

Bookchin defines hierarchies as "cultural, traditional and psychological systems of obedience and command" (1991) that enable one group within society to promote its ends and interests at the expense of another group which it dominates and controls. Such hierarchies for Bookchin are not limited to the domination originating from economic classes or the nation-state but also include the domination "of the young by the old, of women by men, of one ethnic group by another, of 'masses' by bureaucrats who profess to speak of 'higher social interests,' of countryside by town" (1991). Social domination and hierarchies can also become internalized by the oppressed as states of consciousness that instruct them to accept their subordinate position. The challenge for Bookchin is to explain how the domination and degradation of nature "stems from" these patterns of social hierarchy and domination. Without drawing a direct causal relation between the two types of domination, Bookchin asserts that environmental destruction and degradation are more likely to occur in societies characterized by a high degree of hierarchy; social hierarchies provide the psychological motivation and the material means for dominating nature (Des Jardins 1997). The linkage between social domination and the domination and degradation of nature is particularly evident in the present case study—as it is within the Indian subcontinent at large. The domination of farmers, peasants, and rural people by urban-industrial elites, of tribals by nontribals, of lower castes by upper castes, of the poor by the economically advantaged, of the politically disenfranchised by powerful state and central government bureaucracies has allowed a privileged minority to profit from the extraction and exploitation of natural resources at the expense of a degraded environment and those who depend on it for their subsistence needs—over half of the country's total population (Gadgil and Guha 1995).

5.1.2 Sustaining Ecological and Human Systems

Focusing in Chapter 1 on the goal of sustaining ecological systems, Dale et al. call for an incorporation of an ecological perspective into land use and land management. The ecological principles and guidelines for land use and land management put forth in the chapter represent just such a perspective by delineating the consequences of land uses and practices for ecological systems. The problem posed by the current case study is how to apply these principles and guidelines to the Indian context. While it may be possible to articulate ecologically sound principles and practical guidelines for land use and management, their implementation requires that we take into account not only the pertinent environmental parameters but also the complex social, political, economic, and cultural conditions that impinge upon and determine the particular habitat in question. This approach is not at odds with the focus of Dale et al., for in articulating guideline 3 which directs land managers and policymakers to examine the impact of local decisions in a regional context, they note that, given preexisting land-use practices and realities, such an examination will also attend to the social, economic, and political perspectives pertinent to the situation. Furthermore, Dale et al. endorse the development of land-use models that integrate social, economic, and ecological considerations while broadening our understanding of the complex relationships between these factors and their impact on land-use decisions. The present case study contributes to the development of such a model by exploring the relationship between ecological and human systems within India and the ways in which they can be mutually sustained.

5.2 Ecological Decline in the Eastern Ghats

The Eastern Ghats form a long chain of broken hills that run northeast to southeast along the eastern coast of India. Approximately 800 km in length, this hill range occupies 3 districts in the state of Orissa, 13 districts in Andhra Pradesh (AP), and 7 districts in Tamil Nadu. The tallest peaks of the Eastern Ghats, which rise above 1680 m, are found in the northern section of AP that stretches into the Garjhat hills of Orissa. Groundwater is replenished in the highlands by seasonal streams flowing in surface rock pools. The Mahanadi, Krishna, Godavari, and Cauvery rivers break the chain of hills and valleys and provide drainage for the Eastern Ghats (EPTRI 1995). The many rivers and streams that flow both eastward and westward from the Eastern Ghats are crucial in maintaining the hydrologic balance of the region. The Eastern Ghats constitute the best moisture and soil traps in the southern peninsula of India: The watersheds in the valleys and plains below the hills are directly dependent on the catchments in the highlands (EPTRI 1996).

While the Eastern Ghats are a discontinuous hill range, which do not form an integrated forest habitat, they, nonetheless, represent one of India's most diverse ecological regions—and an area where industrial and commercial development has come into direct conflict with nature and tribal people. Although most physical maps of the Eastern Ghats show about 80% of the region under evergreen and deciduous forests, the situation is much different in reality. At most, 30% of the entire region can lay claim to a canopy cover of more than 40% (EPTRI 1995). According to a study conducted by the National Institute of Oceanography and the National Remote Sensing Agency, the forest cover of the Eastern Ghats is disappearing at a rate of 7% a year (Lankesh and Radhakrishna 1995). Deforestation in the Eastern Ghats has resulted in increased flooding, soil erosion, heavy siltation in dams, and adverse changes in the micro climate. All of these factors have contributed to the gradual ecological decline of the region, with deforestation and environmental degradation endangering many floral and faunal species. Either extinct or on the decline are the black buck, wild buffalo, guar, sambar, tiger, 280 species of birds (especially the great pied hornbill), golden hill gecko, Indian monitor, and python (EPTRI 1995).

According to one environmental and tribal-rights group based in Andhra Pradesh, wood-based industries within the state have contributed greatly to deforestation in the area by placing heavy demands on such forest resources as bamboo and hardwood. Most of the bamboo resources in the region have been depleted by Andhra Pradesh Paper Mills over the course of the last 30 years—a loss that has placed a serious hardship on tribals who depend on this resource for the construction of their huts. The state shares the responsibility for forest depletion, because it facilitates the purchase of wood by private industries at below market rates. In addition to overlogging, private industry further contributes to environmental degradation in the region by failing to carry out adequate afforestation programs, leaving nothing but tracts of barren land once trees have been cut. Furthermore, both forest ecology and tribal livelihoods are threatened by the state's promotion of monoculture crops on expansive teak and coffee plantations. After being expelled from government reserve forests to make way for these plantations, tribals retaliated by pulling out teak and coffee plants (D'Souza 1997). The illegal felling of trees through the collusion of government officials, contractors, and private entrepreneurs also remains a problem in the region. The loss of forest cover and forest diversity threatens both the ecological stability of the region and the livelihood of tribal communities whose daily struggle for survival becomes increasingly more difficult with the disappearance of their resource base (D'Souza 1997).

The Eastern Ghats are also under heavy threat from mining interests in the region. Extensive mining is pursued throughout Orissa, primarily in tribal areas. Approximately 600 of the 857 mines in the state are located in tribal areas. Since 6320 sq km or 85.29% of the 7388.19 sq km of wildlife sanctuaries and parks are located in tribal areas, this mining activity poses an ecological threat to the flora and fauna of the region. The intrusion of mining to tribal habitats has also resulted in the loss of tribal land and displacement of thousands of indigenous people in the Eastern Ghats (Fernandes 1996). The intensive mining of calcite, bauxite, and limestone in Orissa has even driven one tribal group, the Khonds, from their native habitat into neighboring AP, placing population pressure and ecological stress on the already degraded environment of that state.

5.2.1 Causes of Environmental Decline

The fundamental causes of environmental degradation in the Eastern Ghats and throughout India can be traced to decades, if not centuries, of unsustainable land-use and land-management practices pursued by power elites in a hierarchically ordered society. Dale et al.'s linking in Chapter 1 of the rate of natural resource exploitation to the *technological* advances in resource extraction and enhancement of logging, mining, hydroelectric power, fertilizers, pesticides, and irrigation does not fully explain the ways in which these technologies significantly contributed to environmental destruction and degradation in India (Gadgil and Guha 1995; Shiva 1989). It was their hierarchical advantage in the social, political, and economic power structure that allowed elites in both colonial and postcolonial India to disregard basic ecological principles and use these technologies to advance their own interests while passing on the costs of environmental degradation to those who directly drew their daily sustenance from the land.

Extensive deforestation began in India during the colonial period when the British placed forests under state control in the late nineteenth century and commercially exploited them as a source of revenue and timber for railway development and ship building. The British demand for teak and other timbers devastated vast stretches of forests and jungles and, in some regions, triggered cycles of flood and drought (Gadgil and Guha 1992). To better meet the demands of commercial timber and revenue, the British forest department employed two strategies. First, transportation networks were improved to make remote forests more accessible for exploitation. And, second, silvicultural techniques were introduced to transform the mixed oak–conifer forests in the Himalayas into faster-growing pure coniferous forests that were more commercially profitable. The ecological consequences of these two developments were greater deforestation and biodiversity loss. Forest ecology was further diminished as large tracts of land were cultivated for tea, coffee, and rubber plantations. The development of roads and railways to transport these commercially grown crops for export made forests all the more accessible and thus quickened their rate of exploitation (Gadgil and Guha 1992).

The colonial takeover and commodification of Indian forests meant that local communities were deprived of control over the very lands they had occupied and depended on for centuries to meet their subsistence needs. Property rights were redefined by the colonial state, and a new system of profit-driven forest management was imposed—a system that conflicted with the traditional land-use and land-management practices of indigenous peoples. Forest ecology was dramatically affected by this radical change in management systems as species which local people depended on for sustenance, fuel, fodder, and leaf manure were replaced by teak, pine, and deodar species more suitable for timber production but of little practical value to rural communities. While traditional village-based systems of community forests and pastures utilized their natural surroundings to satisfy subsistence needs, the colonial state saw the Indian countryside as a source of profit and raw materials. The loss of community control over forests also spelled the demise of indigenous systems of resource management. Over the course of centuries, a system of customs, traditions, taboos, practices, and beliefs had evolved within rural communities, which effectively functioned as an elaborate conservation scheme to manage and restrain resource use within a given locale. Local communities often placed restrictions on harvesting and hunting times; sacred ponds and groves were respected; certain species were barred altogether from hunting; certain methods of harvesting were strictly regulated or restricted, and limits were placed on yields (Shiva 1989; Fernandes 1992; Gadgil and Guha 1992; Kothari et al. 1996).

The ecological decline that accompanied the delegitimization of community control and indigenous systems of prudent resource use was accelerated in the postindependence era as the state apparatus strengthened its hold over natural resources. After 1947, a nexus of politicians, bureaucrats, and industrialists within India embarked on an ambitious development program to bootstrap a largely rural, agrarian country into a modern, industrial state. Again, it was traditional communities and their natural habitats that bore the brunt of this so-called development. Millions of peasants, farmers, and indigenous people lost their lands and livelihoods as a result of the aggressive promotion of large-scale dams and reservoirs, extensive irrigation systems, plantation forestry, monoculture cropping, and agri-business (Shiva 1989; Gadgil and Guha 1995).

5.2.2 Tribal Culture and Communities in Andhra Pradesh

With its rich mineral deposits of coal, bauxite, and limestone, the Eastern Ghats of AP are also coming under increased pressure from mining interests (Rao 1993), with environmental degradation posing a direct threat to the habitat and livelihood of the 33 tribal communities that inhabit the hills and forests of the state. The majority of these tribal groups are subsistence-economy agriculturists distinguished by a diversity of customs, socio-cultural traditions, occupations, mythologies, folk tales, and linguistic usages. Some tribal groups pursue their livelihood as food gatherers and hunters, others as cattle breeders, artisans, fisherfolk, settled cultivators, agricultural laborers, and shifting cultivators (Rao 1993). Whatever their occupational pursuits, however, the majority of these tribes depend directly on forests and natural surroundings for their very survival. In India, forests are typically tribal habitat; tribals traditionally drew 80% of their food from the forest. Tribals continue to satisfy some 50% of their food needs today from the forest, despite heavy deforestation in many regions (Fernandes 1996).

The inseparability and *interdependency* of tribals and forests is expressed in tribal culture by the extension of kinship relationships to forest flora and fauna: tribals regard the various plant and animal species of their region as their kith and kin. Totemic clans form the basis of tribal social organization, with various tribes claiming mythical affinity with special totem plants and animals. For example, tribal clans within the Visakhapatnam district of AP identify themselves with these totem objects: the snake, monkey, fish, kite, tiger, bear, and goat. Tribals developed sacred relationships with their totem objects, which they believed protected their ancestors from danger. These sacred objects are in turn protected by tribals. At least a dozen trees within the tribal areas of AP are regarded as sacred by tribals who will neither cut them nor use them for fire wood. Throughout tribal regions, groves of sacred trees can be found that are protected by tribals. Tribals also regard the very hills and forests they inhabit as sacred terrain and home to their gods and goddesses. The first fruit harvest of the season is always shared by community members after it has been ritually offered to their local gods and goddesses (Rao 1993). This rich religo-cultural tradition of totems, taboos, rituals, ceremonies, and practices has operated as an indigenous conservation scheme for the sustainable use of natural resources (Fernandes 1996).

For millennia a variety of tribal cultures has flourished within and sustained the diversity of natural habitats that surrounded them in an intimate relationship between ecological and human systems. But the fact that tribal communities exist at the very bottom of social, political, and economic power structures and hierarchies has made their lifestyle and habitat vulnerable to exploitation in both colonial and postcolonial India. The languages, belief systems, resource-use practices, species, and ecosystems that make up this complex cultural–ecological diversity is at risk of being lost. Public and private sectors in India, however, are beginning to recognize the symbiotic relationship between tribal culture and habitat and are developing a new approach to resource management that seeks to include local people in the protection and preservation of their natural habitat. Since the degradation of forest ecosystems threatens the livelihood of tribal communities, the conservation of biodiversity and the protection of cultural diversity are now regarded as interdependent projects that have as their common goal the sustainable development of a single habitat—one that encompasses both nature and people (Kothari et al. 1996; Gadgil 1998).

5.3 Mining in the Eastern Ghats of Andhra Pradesh

This case study depicts an episode of maldevelopment that endangered both forest ecology and tribal habitat in the Eastern Ghats of AP. By disregarding fundamental ecological principles and guidelines for land use and land management, maldevelopment degrades ecosystems and undermines the cultures, economies, and resource-management systems of indigenous people who depend on nature for their livelihood (Shiva 1989). In the past decade, the pace of economic development in the tribal forests and hills of the Eastern Ghats has quickened as both state and private industry have increased their efforts to exploit the resource base of this region for economic gain. The significant aspect of this case is not the pattern of maldevelopment it exhibits but its outcome, for, in this instance, maldevelopment was arrested and measures were taken to protect endangered tribal communities and to recover their degraded habitat.

5.3.1 Mining in Tribal Habitats

The case study focuses on the Visakhapatnam district of AP, a forested region rich in mineral deposits and home to numerous tribal communities. Bordering on Orissa, Visakhapatnam district encompasses 10 agencies (*taluks*); each taluk is divided into smaller agency tracts (*mandals*), and each mandal includes numerous councils (*panchayats*). In this case, the proposed mining activity was located near the Nimmalapadu village of the Volasi panchayat, Anantagiri mandal, Paderu taluk, the largest agency tract

(3116 sq km) within the district of Visakhapatnam. A predominately agricultural area, the Anantagiri mandal is a constitutionally protected ("scheduled") tribal habitat with relatively dense forests and a predominately tribal population of almost 40,000 according to a 1991 census. The forests of the Anantagiri mandal possess exceptionally good humus and soil moisture and are classified as "Southern Tropical Semi-Evergreen Forests," a type found in valleys and near perennial streams. Trees reach heights of 21–30 m and girths of 1.2 m and greater. A number of top story trees are deciduous; second story trees are evergreen.

Geographically, the Anantagiri mandal spreads over 153,164 acres, of which 103,466 acres are forested and 29,166 acres are under cultivation. Many streams and tributaries wind their way through this forested region, and the villagers of Nimmalapadu (as well as other villages down stream) derive most of their irrigation and drinking water from the strong flowing Gedda stream in the Volasi panchayat. Leopards are commonly found at higher elevations, and wild jackals and foxes are seen on the fringes of agricultural fields and human settlements. While forests are still found throughout the Paderu agency, the region once possessed a thick forest cover that has suffered gradual environmental degradation. Completely denuded hill tops are now common (Anonymous 1995*b*).

Mining operations in the Anantagiri mandal have increased rapidly in the past few decades. The most serious threat to forest ecosystems and tribal communities posed by this activity has been concentrated in the Borra, Volasi, and Konapurma panchayats, all within relative vicinity of one another. In the Borra panchayat, mining leases were illegally issued by the state to five private companies in 1962 for the extraction of limestone and mica. The main hill ranges of the Borra region are tree covered, and hillsides are utilized by tribals for dry cultivation of such subsistence crops as maize, black and green gram, millet, and ragi. The Borra panchayat encompasses 13 villages and is also home to the Borra caves, whose impressive stalagmite and stalactite formations attract hundreds of tourists a year. The limestone caves of Borra possess geological features not found anywhere else in India; archaeological artifacts have also been recovered from the caves. Since several species of flora and fauna in the region are classified as endangered (e.g., the golden gecko), mining operations in the area are in violation of protective legislation.

Village streams and agricultural fields were affected by high volumes of dust and pollution caused by crushing activities carried out next to villages. In addition to producing water and dust pollution, the horizontal digging of mica below hillside settlements placed these villages at risk of collapse during the monsoon season. One company even conducted mining operations at the very top of the cave's cap-rock, despite the fact that part of the cap-rock had already been dislodged during the development of an adjacent railway line. In general, mining and quarrying activities were undertaken with no regard for environmental and ecological consequences; the geological and archaeological significances of the region were not considered, and the social and economic impact of mining activities on tribal villages were all but ignored. Ecological principles and guidelines were ignored in the interest of profit-taking at the expense of tribal communities and their habitat (Anonymous 1995*a*,*b*).

5.3.2 The Case of Birla Periclase

Since the 1960s, mining activity has continued almost unabated in the region. During the 1980s, a lease, illegally granted to a nontribal individual for mining calcite in the Konapuram panchayat, was taken over by Indian Rayon & Industries. In March 1995, the company began prospecting activities in the middle of agricultural fields belonging to the tribal village of Karaiguda. Standing crops were destroyed; company tents were pitched on deeded tribal land; the village primary school was occupied as a storage building, and a road was built across tribal fields and irrigation canals. Although numerous legal safeguards were violated by this activity, the state failed to intervene (Rebbapragada 1995, 1996). In 1994, Birla Periclase, a subsidiary of Indian Rayon & Industries, initiated twin projects for calcite mining and magnesia production in the Eastern Ghats of AP. The first site of the operation was a seawater magnesia plant, covering over 65 acres, located at Chippada village, a fishing community about 25 km south of the major seaport of Visakhapatnam. The second project site was a calcite mining operation located near the village of Nimmalapadu approximately 85 km north of Visakhapatnam.

Nimmalapadu village belongs to the Volasi panchayat and is part of an extended network of tribal communities. The tribals in Nimmalapadu engage in traditional subsistence farming and also depend on the forest as a source of minor produce for immediate consumption or sale at local markets. To transport the calcite mined at Nimmalapadu village, Birla began construction of a 22-km road through dense forest from the mine site to a connecting road. Intended to service 150 trucks per day, the new road was to be 90 ft wide and cut through 14 tribal settlements and agricultural fields. The calcite was to be used at the Chippada plant for the extraction and purification of magnesia from seawater; magnesia, in turn, is used in the production of steel. Because fresh water is also required for magnesia extraction, the Chippada plant called for the digging of 404 borewells in the Gostani riverbed to withdraw two million gallons of water per day.

In sanctioning the Birla project at Nimmalapadu, the state government of AP was in violation of its own Land Transfer Act of 1959, which prohibits the transfer of tribal land and leases to nontribals. The state government granted a mining lease in 1985 to a village administrative officer, who transferred the lease to Birla in 1993. The state then acquired on behalf of Birla over 220 acres in and around the village of Nimmalapadu, despite the fact that tribals hold valid deeds to 155 acres of this land. For all intents and purposes, the Birla project was a collaborative state-and-private-industry venture. The AP government approved the development of the necessary infrastructure and acted as middleman in helping Birla acquire land and mining leases in legally protected tribal habitat.

5.3.3 Impact of Mining on Tribal Habitat and Livelihood

The construction of the coastal magnesia plant threatened to render as many as 13,000 fishing families jobless since the seawater extraction process endangered many of the species of fish and prawn that coastal communities depend on for their livelihood. Almost all of the land surrounding the village of Nimmalapadu was lost to mining, displacing roughly 400 inhabitants, some of whom had already been displaced by mining activities in neighboring Orissa. As many as 14 villages and 500 acres of agricultural land were threatened by the road construction. Tribals were also distressed by the felling of 275 mango, tamarind, and jackfruit trees crucial to their local economy; hundreds of additional trees were slated to be cut for the completion of the road. Because this area serves as a catchment for the Sarada river and has already been considerably deforested, further depletion of forest cover increased the risk of flooding during the monsoon season. The loss of forests and agricultural lands diminished biodiversity in the region and threatened the livelihood of tribal communities. In disregarding ecological principles and guidelines, the Birla project represents land use and land management at its worst.

An environmental impact study conducted by an independent scientific agency documented the threat of deforestation, topsoil erosion, silting of natural streams and rivers, and air and water pollution posed by the Birla project. According to the report, mining activities were not well planned and no environmental impact study had been conducted prior to the undertaking. The mining project exacerbated soil loss in the region and threatened to silt and damage the area's natural drainage network, which functions as a catchment basin for the Sarada river. Blasting operations at the mine site have polluted air and groundwater, posing a risk to local flora and fauna. Mining wastes leeching from surface excavations have already found their way into streams and rivers (Anonymous 1995*b*). As a relatively soluble mineral, calcite increases water alkalinity, rendering it unfit for aquatic life, irrigation, and human consumption (Rebbapragada 1993). At the coastal Chippada site, the pumping of two million gallons of water a day from the Gostani river for use at the magnesia extraction plant threatened to deplete the water table, which, in turn, would result in seawater intrusion (Mohammed 1995). Also the heavy intake of seawater by the magnesia plant and the resulting discharge of effluents into the ocean would seriously disrupt the marine ecology and threaten rare species of turtles, various species of shrimp, bottlenose dolphins, and sea eagles (Radhakrishna 1995*c*).

Although tribals received vague job promises and were offered Rs. 5000 ($166) per person to compensate for displacement, such measures were no substitute for a degraded environment and the loss of livelihood. As one tribal member put it, "They are talking about payment of compensation. But what will we do with money? If we are relieved of our land, we will perish. They are talking about providing employment in the factory, but as none of us is literate and skilled, we will be given only menial jobs" (Lankesh and Radhakrishna 1995). This assessment is all too accurate. Even when tribals are not displaced by mining activity, the resulting environmental degradation makes the daily struggle for existence all the more difficult. For tribal women who play an important role in maintaining daily household needs, polluted streams and rivers mean much longer walks for fresh water. Also, the loss of forest cover makes it more difficult for women to collect the daily requirement of twigs, firewood, and dry leaves for cooking and grazing (Shiva 1989; Rebbapragada 1993).

The loss of agricultural fields poses a direct threat to tribal survival. While tribals traditionally practiced a sustainable form of shifting cultivation on slopes up to 20° gradient, they are now forced to cultivate ever steeper mountain slopes to replace lost or degraded agricultural fields. In the 1950s, the shifting cultivation cycle was 18–20 years, a fallow period that allowed the forest ample time to regenerate. By the early 1980s, however, the cycle had dropped to 6 years, failing to leave the plot fallow long enough for forest regeneration. What was once a sustainable form of cultivation has become a destructive practice as tribals are increasingly deprived of their habitat. Although deforestation is alien to the conservation-oriented culture of tribals, the erosion of their subsistence-based economy has forced them into such destructive practices as cutting trees for timber contractors and smugglers or for sale as firewood. A vicious cycle is created by the environmental destruction of tribal habitat, undermining what was once ecologically sustainable communities and, in turn, degrading the environment even more (Fernandes 1996).

5.3.4 Decline of Tribal Culture

While tribal men are frequently employed in the many mines, cement factories, and other industries scattered throughout tribal areas, they rarely occupy skilled or even semi-skilled positions. The regional expansion of industry and commerce has disrupted tribal communities and lifestyles. Instead of pursuing traditional cultivation, many tribal men are drawn into the money economy and migrate to the nearby towns to become low-paid daily wage earners. Both tribal men and women are employed in road and building construction or as laborers in brick-manufacturing kilns. The result of this outward migration of tribals (mostly men) from their agriculture- and forest-based communities has been an economic shift away from traditional subsistence economies and a steady erosion of traditional lifestyles and institutions. Although tribal women enjoy equal status within their traditional subsistence communities, their work is devalued and marginalized as tribal men enter the cash economy (Shiva 1989). Tribal women are also vulnerable to sexual exploitation when employed as construction workers outside their local communities. Alcoholism and prostitution disrupt what were once socially cohesive tribal communities and family structures (Rao 1993; Rao 1996).

Indicative of the decline of tribal communities is the fact that between 1983 and 1989 almost 55% of tribal lands in the Eastern Ghats was illegally transferred to nontribals (Radhakrishna 1995*a*). Within AP, the tribal population has been dwindling at a rate of 6% a year, and the death rate among tribals is increasing as more and more of them are displaced from their degraded habitat (Radhakrishna 1995*b*). The Birla case is simply one example of a growing trend within the Eastern Ghats of state governments leasing tribal lands to private industry—ostensibly in the name of national interest and economic development. In addition to granting mining leases, states are also encouraging multinationals to develop large-scale agri-farms in tribal areas. While Indian environmentalists have called for the Eastern Ghats to be declared a "protected zone" and have stressed the importance of afforestation for tribals whose livelihood depends on the forests, the AP government and private industry are intent on pursuing capital ventures with disastrous consequences for the region's ecosystems and indigenous people (Lankesh and Radhakrishna 1995).

5.4 Resisting Maldevelopment

The story of environmental degradation in India, in general, and in the Eastern Ghats of the Anantagiri mandal of AP, in particular, has been one of social, political, and economic domination of indigenous people and the simultaneous exploitation and degradation of their habitat. Rather than carefully applying ecologically based guidelines for land use and management, government and private industry have single-mindedly pursued a policy of short-term economic gain at the expense of both the environment and tribal people. A small, though not insignificant, victory was finally won, however, in the Anantagiri mandal, largely due to the efforts of one environmental and tribal advocacy organization, Samata ("equality"), which has operated in the region for over 10 years. After first filing numerous legal writs on behalf of tribal people, taking their case to the High Court of AP, and eventually appealing a negative judgment of the lower court to the Supreme Court of Delhi, in August 1997 Samata finally won a decision from the highest court in the land banning the Birla operation. The decision was a landmark judgment, for it affected not just the Anantagiri mandal but all of the eight legally protected tribal districts in AP. This victory makes it much more difficult for environmentally damaging industrial and commercial interests to operate with impunity in the region.

5.4.1 Regionalizing the Resistance

As Samata pressed its legal case against Birla, it simultaneously assisted tribals in organizing marches, protest rallies, and demonstrations against the illegal mining. In August 1996, a procession of peaceful demonstrators passed through many of the villages in the area affected by the mining activity to inform people of the consequences of the project. The demonstration culminated in a mass meeting in Paderu attended by some 1500 tribal people from all over the district. Tribals performed street plays, dances, and skits to bring public attention to their situation. Samata notified journalists, politicians, academics, and scientists throughout the region about the case and arranged for an independent scientific agency to conduct an in-depth study of the regional ecological, social, and economic effects of the mining.

While mining officials attempted to buy off tribals, Samata organized exposure visits to mining sites in neighboring Orissa to give tribals from the Volasi Panchayat a firsthand experience of the destructive effects of mining on both community life and the environment (Rebbapragada 1996). By creating public awareness of the Birla case, Samata raised consciousness among tribals about the environmental consequences of mining and the implications of displacement for tribal life. As tribals mobilized throughout the area, they began to see their villages and fields not as separate entities but as extended communities belonging to a commonly endangered habitat. This shift in perspective underscores the significance of guideline 1, which calls for an examination of local decisions and changes in land use in a regional context. By identifying the regional context of the Birla case and observing how adjoining areas were subjected to similar mining operations, tribals were better able to understand and resist the threat to their cultural and ecological habitat.

5.4.2 An Integrated Approach to Ecodevelopment

Samata's efforts to protect tribal habitat and culture focused on an integrated approach to the region that called on a wide range of social, political, economic, and environmental strategies aimed at empowering tribal communities and helping them achieve a self-sufficient and sustainable way of life. Samata's specific objectives were to raise consciousness among tribals to resist external exploitation, to establish locally accountable people's organizations and institutions, to improve the socio-political and economic conditions among tribals, to build strong and self-reliant local economies to meet local needs, to develop a strong leadership base among tribal communities, and to promote the sustainable development of tribal habitat (Rebbapragada 1994). Although only the last goal explicitly mentions the environment, all of Samata's objectives take on ecological significance and contribute to the ecodevelopment of the region. Sustaining the structure and function of ecosystems depends on incorporating an ecological perspective into the land-use and land-management decisions of local communities as well as promoting community practices and institutions dedicated to the implementation of ecological principles and guidelines. This case study illustrates how community-building can become an environmental issue and, in turn, how promoting ecologically sound land-use and land-management decisions reinforces community development and empowerment.

5.4.3 Promoting Ecology and Equity

Samata's integrated approach to ecodevelopment means that the quest for social justice for tribals on the one hand and the protection and restoration of tribal habitat on the other become complementary goals. In considering the exploitation of tribal people and their habitat, Samata's director stated that "time has proved that tribal society cannot be sustained through isolation as the mainstream world neither leaves them alone nor allows them as equal partners in the process of development" (Rebbapragada 1994). Historically, as long as tribals have remained marginalized and disempowered, the protective legislation built around their communities and habitat has been easily circumvented and violated by nontribals, government, and industrial–commercial interests. Samata, therefore, worked to secure social justice for tribals and restore and protect their habitat by helping them build communities that are socially and politically empowered, economically viable, and environmentally sustainable. Both equity and ecology are enhanced by these communities as tribals are now better prepared to resist the external exploitation of their habitat and to meet their subsistence needs without degrading their environment. (Tribals currently engage in an environmentally destructive form of shifting cultivation because they have no other means of meeting their subsistence needs.)

The development of self-reliant, ecologically oriented tribal communities not only offers the best model for the implementation of land-use and land-management practices that are compatible with the natural potential of the area (guideline 8) but also represents the best strategy for avoiding or compensating for effects of development on ecological processes (guideline 7). Given that the region has already been subjected to extreme environmental degradation and can no longer support intensive farming and resource extraction, the ecodevelopment of the region is best promoted by the strengthening of local communities committed to living within the natural limits of their habitats. The sustainable development of tribal habitat can be successful only if land-use and land-management practices in the region are based on ecologically sound principles and guidelines, but the practical implementation of these principles and guidelines requires the support and participation of local tribal people who depend on and most directly relate to the land.

5.5 Building Tribal Communities and Habitat

Tribal communities are especially vulnerable to exploitation because they have been socially, politically, and economically marginalized and disempowered within Indian society. An examination of the strategies employed by Samata to empower tribal communities and restore their habitat reveals how ecological principles and guidelines can be woven into the social, political, and economic fabric of community life to create strong and viable institutions that generate wise land-use and land-management decisions.

5.5.1 Social–Political Empowerment and Ecology

Samata's legal advocacy program was an immediate and effective way of empowering tribal communities by helping them defend their habitat and traditional livelihood. Given their lack of adequate political representation and high illiteracy rates (with woefully underfunded or nonexistent schools and educational facilities), tribals are often unaware of their legal rights and are ill-prepared to defend them in a court of law. Samata educated tribals about their legal rights, helped them seek legal representation, and assisted them in their legal defense of claims and titles. Because of Samata's program, tribals were able to overcome their disadvantage vis-à-vis traditional power hierarchies and to prevail in the highest court of the land in their case against Birla. By successfully pressing their legal case against Birla, Samata and tribals forced the state to examine the impacts of local decisions in a regional context (guideline 1), to preserve rare landscape elements and associated species (guideline 3), to avoid land uses that deplete natural resources (guideline 4), to avoid or compensate for the effects of development on ecological processes (guideline 7), and to implement land-use and land-management practices that are compatible with the natural potential of the area (guideline 8).

Samata also established an independent health program for tribal communities and provided funds for medical supplies and services since state-operated medical facilities in the region were entirely deficient in meeting even the most basic health needs among tribals (Sainath 1996). Further, Samata promoted literary programs and schools to better educate tribals and help them defend their legal rights. Where state-sponsored schools were available for tribals, the curriculum reflected cultural–social values alien to tribal life and culture. By contrast, Samata developed an educational program more finely attuned to tribal life and habitat by offering instruction about local flora and fauna, low-input farming and agricultural techniques, and marketing strategies. Samata's legal, health, and educational programs were important in bolstering the social–political status of tribal communities. As a consequence of these programs, tribals were better able to preserve their lifestyle and habitat and interact with the outside world without fear of exploitation.

5.5.2 Economic Empowerment and Ecology

To break the economic dependency of tribal communities on unscrupulous nontribal moneylenders, Samata assisted tribals in developing a variety of self-reliant farmers' cooperatives and women's groups to strengthen the local economy and meet community needs. Prior to Samata's intervention, nearly every tribal family was heavily indebted to nontribal moneylenders who operated as the only creditors in the region for tribal communities. Tribal members would frequently turn to nontribal moneylenders to borrow at exorbitant rates in order to purchase seed at the beginning of the planting season and to pay for weddings and funerals. If, for example, a tribal family borrowed Rs. 100 ($3.00), they would have to repay the full monetary amount plus a bag of rice. Tribals were therefore compelled to cultivate fewer traditional subsistence crops to provide moneylenders their quota of rice.

As a worst case scenario, many tribals were often forced to give up their land when they were unable to repay even petty loans (Viegas 1991). To sidestep the illegality of this transfer, legal title to the land may be retained by the tribal member while the land itself is actually cultivated by nontribals. Furthermore, when land is removed from tribal cultivation, it is typically converted from subsistence crops and native grains to nonnative cash crops. Finally, as tribals lose more and more cultivable land, they are forced to farm steeper mountain slopes to meet their subsistence needs, and the inevitable result is increased water and soil erosion. The cultivation of these marginal slopes makes it increasingly difficult for tribal communities to satisfy their subsistence needs, and thus they frequently resort to overexploiting natural resources simply to survive.

Working in conjunction with the state tribal marketing agency, the Girijan Credit Cooperative (GCC), Samata ensured that tribals could receive credit at reasonable interest rates. As the economic dependency on external moneylenders was slowly reduced, the effects on tribal life were significant. The outward migration of tribal labor dropped dramatically, and tribals were no longer mortgaging their standing crops to nontribal moneylenders. The traditional subsistence economy of tribal villages was strengthened, and, for the first time, tribals were able to build up a surplus of staples and enjoy relative food security. Land loss to nontribals also fell, and tribals were no longer economically forced to engage in environmentally destructive practices such as shifting cultivation on marginal hillsides or selling firewood and charcoal to generate hard currency for debt servicing. Land uses that depleted natural resources were therefore avoided, in keeping with guideline 4. By strengthening the traditional tribal economy, the credit program also encouraged the planting of native grains and pulses, thereby minimizing the introduction of nonnative species to the region (guideline 6). Further, the less intensive farming practices of tribals are much more compatible with the natural potential of the area than the nontribal production of nonnative cash crops heavily dependent on artificial inputs (Shiva 1991).

5.5.3 Women's Empowerment and Ecology

The success of the tribal credit program served as the catalyst for the formation of women's groups into thrift societies and lending groups, further strengthening tribal economy. Beginning in 1990, tribal women organized community-level meetings and formed four village thrift groups with each member contributing Rs. 5 a month. Instead of turning to nontribal moneylenders, group members were able to borrow money for purchase of clothes, medicine, festivals, marriages, and so on. As the thrift program took root, women from surrounding villages approached Samata for assistance in forming their own groups. By 1994, 74 such groups were in existence with a membership of 2390 women. In continuing to flourish, these saving groups have proved to be a source of security and self-esteem for tribal women. While women used to go begging to moneylenders for paltry sums, they are now using their thrift societies to start up small businesses such as hiring out cookware for festivals and marketing cashews for sale at local towns and markets. Also, in learning how to manage considerable financial resources, women have become more articulate and involved in the decision-making process in their villages; women are more united and courageous than they were in the past in dealing with village disputes and problems.

The most ecologically significant outcome of the thrift societies was the formation of women's banks to disburse loans mainly for agricultural purposes. The establishment of women's banks lessened tribal dependence on the GCC, which often involved cumbersome procedures, delays, and inaccessibility. The women's banks are now handling large sums of money and, with the control of cash flow in their villages, have demonstrated great flexibility in adapting to the needs of local farmers. Both the women's thrift societies and banks have dramatically reduced exploitation by external moneylenders and have gone far in strengthening community ties, building the tribal economy, and preserving traditional farming. As these communities become more self-sufficient in food production and increase their food security, they are no longer forced to degrade their habitat to eke out an existence (guideline 4), but instead can direct more of their energies to land restoration and recovery as suggested by guidelines 3, 7, and 8 (Rebbapragada 1990, 1994).

5.5.4 Environmental Programs

Since beginning its work in the Visakhapatnam district, Samata initiated a number of projects and activities directly aimed at improving environmental conditions within tribal habitats. In addition to helping tribals develop irrigation projects and construct check dams to control soil and water erosion, Samata introduced a much needed seed-collection program among tribals to preserve native grains and cereals, the cultivation of which is slowly dying out. The cash crops that replace tribal grains and millets are dependent on large inputs of artificial pesticides and fertilizers, which over time diminish soil and water quality in the region. By contrast, the mode of subsistence farming as traditionally practiced by tribals, with its reliance upon native seeds (see guideline 6) and collection of minor forest produce, appears to be much more compatible with preserving the ecological integrity and biodiversity of the region (per guideline 8). Furthermore, by forming grain banks in their own villages, tribal farmers are no longer at the mercy of moneylenders to borrow money for the purchase of seed at the beginning of each growing season. Each farmer is now required to deposit one bag of grain at the end of the harvest season. The grain banks provide food security to communities that experience lean months or bad growing seasons.

With the aid of the AP forest department, Samata established numerous tree nurseries throughout the area, stocked with thousands of both horticulture saplings and forest species indigenous to the region. Located in tribal villages, these decentralized nurseries serve a dual purpose. The horticulture saplings (e.g., cashew, papaya, lemon moringa, coconut) are profitably grown for sale at local markets and thus bolster tribal economy. The forest species are used in an afforestation program to regenerate degraded hillsides surrounding tribal villages, thereby preserving local species (guideline 3) and compensating for the effects of past maldevelopment on ecological systems (guideline 7).

By bringing forest officials and tribals together to discuss how best to use and manage forest resources, Samata has been able to mediate what has oftentimes been a conflict-ridden and antagonistic relationship. Samata's mediation role paid important dividends when a government-led initiative on Joint Forest Management (JFM) was launched in the 1990s—a program that involved farmers, peasants, and indigenous people in forest protection and management. (Even prior to the inauguration of JFM, Samata had organized numerous forest-protection committees in tribal communities.) The AP forest department has included Samata and tribal communities in consultations with the World Bank and JFM planning and has employed Samata in identifying tribal villages with varied forest problems for participation in the program. Greater cooperation between tribal communities and the forest department has expanded the institutional scope of land management, making it easier for all ecological shareholders to examine the impact of local land-use decisions and projects in a regional context (guideline 1) and, collectively, to plan better for long-term change and unexpected events (guideline 2).

Finally, Samata organized an eco-tourism program as an alternative environmental development strategy for tribal areas—the first of its kind in the Eastern Ghats—through a tribal youth association. Tribal youth from the Borra panchayat were selected and trained by Samata to conduct eco-friendly treks and camps for tourists and educational groups (Rebbapragada 1997). This land use in an area that was slated for mining is not only compatible with the natural potential of the region, as dictated by guideline 8, but also enhances the economic standing of traditional tribal communities.

5.5.5 Technology, Community, and Habitat

Samata has functioned as a catalyst and facilitator for the empowerment of tribal communities, not by attempting to protect tribals from the market economy but by helping them interact with the market economy on their own terms. According to Indian ecofeminist Vandana Shiva, however, the encroachment of the market economy onto the traditional survival economies of women, peasants, and indigenous people inevitably results in the destruction of nature and the disempowerment of women. Nature, subsumed under the patriarchal power structure of the market economy, is no longer viewed as a wellspring of subsistence but as a resource to be siphoned off and transformed into capital. The central place of women's work and women's knowledge in the maintenance of survival economies is simultaneously devalued and marginalized as market forces draw men into the cash economy (Shiva 1989). (Samata's organization of women's savings and lending groups reinforced women's egalitarian status in traditional tribal economies threatened by outside exploitation.)

Shiva's analysis is an accurate description of what happens when powerful industrial and commercial interests encroach upon *disempowered* traditional communities and their habitats. It is a mistake, however, to draw a sharp line between the market economy and the survival economies of rural women, peasants, and tribals by insisting that nature and traditional communities can be protected only by holding market forces at bay. The experience of tribal communities within AP suggests several ways in which indigenous people can preserve their habitat and traditional lifestyles through market interactions.

The establishment of women's thrift and lending institutions, a project that has granted tribal members economic autonomy from the exploitation of outside moneylenders, was an empowering move that allowed tribal communities to interact with the market on a more equal footing. As a next step, tribals were taught simple principles of weights and measures so that they would not be taken advantage of when selling their goods at local markets. Also, tribals were made aware of current market rates and prices so that they can more profitably market their commodities. Samata has also assisted tribals in purchasing weighing machines, in obtaining and renewing business permits, and in linking with larger urban markets where they can sell their goods at higher prices. Tribals are now marketing organic produce to urban consumers interested in purchasing pesticide-free food items. In

1995, tribal farmers from the Dumbriguda mandal of Visakhapatnam district of AP approached Samata for help in marketing their vegetables. These tribal farmers, called Malis, are traditional vegetable growers who earn their living in an extremely fertile but remote valley on the Andhra–Orissa border. Although they are successful in cultivating a wide variety of vegetables, the remoteness of their region makes it difficult to transport and market their produce. With Samata's help, a society of vegetable growers consisting of 180 farmers was formed and registered; an assessment study of production costs and estimated profits was conducted; cultivation practices were evaluated; water and soil conservation measures were improved, and an afforestation program was initiated. The society eventually received a loan from the state-run Intertribal Development Agency for the purchase of a truck to transport their vegetables to outside markets.

A final problem remained. To market their vegetables profitably, the farmers needed to supply in bulk, and, although their harvests were bountiful, high losses were incurred due to the lack of storage facilities. While villages required electricity for setting up storage units, the remoteness of the area made it likely that it would be years before the state would provide conventional electricity to the region. Samata, therefore, pursued an innovative micro-hydel power project that utilized the natural streams and waterfalls of the valley. Through Samata's efforts, a water turbine was provided by the Dutch government, and in 1997 the project was completed, bringing electricity to 5 villages and 420 families in the valley. The introduction of electricity into this remote area allowed tribals to store and, therefore, market their crops in bulk. The locally generated electricity is also used for such domestic enterprises as rice mills and oil extraction, mitigating labor-intensive activities typically performed by women. Samata recognizes that, while no uniform technology is suitable for all tribal communities and all areas of the Eastern Ghats, whenever technology is contemplated, it must draw upon local resources and skills, contribute to the self-reliance of the tribal economy, harmonize with tribal traditions and institutions, minimize capital investment, and not degrade the environment (Rebbapragada 1997). The micro-hydel project represents a model of an environmentally sustainable form of technology that takes advantage of the natural potential for productivity and water cycling in the region (guideline 8), while avoiding the effects of development on ecological processes (guideline 7).

5.6 Conclusion

The successful application of ecological principles and guidelines in land-use and land-management decisions depends on a proper understanding of the causes of change in land cover and land use. While the current study does not assert that all instances of the domination and degradation of nature can be traced to social domination, it makes the case that the erosion of the ecological and cultural diversity of tribal habitats and communities can be explained in terms of the low standing tribals occupy in the social, political, and economic hierarchies and power structures of Indian society. The problem posed by the study is how wise land-use and land-management decisions incorporating ecological principles and guidelines can be implemented in the context of these harmful power structures and hierarchies. The land-use model most appropriate for addressing this difficulty is one that integrates social, economic, political, and ecological considerations; the development of such a model, as Dale et al. note in Chapter 1, is still in its nascent stages. The findings of this study suggest that a dynamic relationship exists between tribal communities and their habitat. The promotion of socially, politically, and economically self-reliant and viable tribal communities is essential for the protection and ecological regeneration of degraded tribal habitat. At the same time, the very development of such communities depends on reinforcing practices and institutions that generate land-use and land-management decisions grounded in ecological principles and guidelines.

A broad-based conservation model that grants local communities a greater voice in land-use and land-management decisions is emerging within India, challenging the traditional top-down, state-controlled model of resource management. This new model envisions strong local communities actively involved in the conservation and management of their own resources and supported by a coalition of outside agencies (e.g., forest department, environmental nongovernment organizations, research and educational institutions) to monitor proper resource use and management, to coordinate decision-making with neighboring communities, and, when necessary, to enforce protective legislation aimed at preserving natural resources. These agencies could also serve as an important source of information for local communities regarding inputs, marketing strategies, and value-added processes that would strengthen local economies (Kothari et al. 1996; Gadgil 1998). While many questions remain regarding its practical implementation, this alternative program captures many of the insights of the current case study and merits further consideration as a prototype of a workable land-use model for developing countries such as India seeking to repair the severed link between culture and nature and between community and habitat.

References

Anonymous. 1995*a*. A preliminary study on the impact of mining in and around Borra caves, Anantagiri mandal, Visakhapatnam, Andhra Pradesh. [Environmental impact statement prepared by independent scientific agency.] Located at: Academy for Mountain Environics, Dehradun UP, India.

Anonymous. 1995*b*. Impacts of calcite mining in Nimmalapadu village, Visakhapatnam district, Andhra Pradesh. [Environmental impact statement prepared by

independent scientific agency.] Located at: Academy for Mountain Environics, Dehradun UP, India.
Bookchin, M. 1991. The ecology of freedom: the emergence and dissolution of hierarchy. Black Rose Books, Montreal, Canada.
Des Jardins, J. 1997. Environment ethics. Wadsworth Publishing Company, Belmont, California, USA.
D'Souza, N. 1997. Tenurial rights and displacement of indigenous people [Laya NGO report]. Located at: Visakhapatnam, AP, India.
Environmental Protection Training and Research Institute. 1995. The Eastern Ghats 1(1):1–8.
Environmental Protection Training and Research Institute. 1996. The Eastern Ghats 1(2):1–12.
Fernandes, W. 1992. National development and tribal deprivation. Indian Social Institute, Delhi, India.
Fernandes, W. 1996. Land reforms, ownership pattern and alienation of tribal livelihood. Social Action: A Quarterly Review of Social Trends 46:428–453.
Fox, W. 1989. The deep ecology-ecofeminism debate and its parallels: a defense of deep ecology's concern with anthropocentrism. Environmental Ethics (11):5–25.
Gadgil, M. 1998. Conservation: where are the people? Survey of the Environment 1998. The Hindu 107–137.
Gadgil, M., and R. Guha. 1992. This fissured land: an ecological history of India. University of California Press, Berkeley, California, USA.
Gadgil, M., and R. Guha. 1995. Ecology and equity: the use and abuse of nature in contemporary India. Penguin Books, New Delhi, India.
Kothari, A., S. Neena, and S. Saloni. 1996. People and protected areas: towards participatory conservation in India. Sage Publications, Delhi, India.
Lankesh G., and G.S. Radhakrishna. 1995. July 30–Aug 5. Assault on the hills. Sunday:44–51.
Mohammed, S.H. 1995. Birla mining project unviable: study. Economic Times 24 (col 1–4).
Radhakrishna, G.S. 1995*a*. Desam, Cong made a killing of tribal land: Andhra's receding forests, IV. Telegraph, 27 June.
Radhakrishna, G.S. 1995*b*. Disease plays havoc with tribals: Andhra's receding forests, III. *Telegraph,* 26 June.
Radhakrishna, G.S. 1995*c*. Birla's magnesia plant butt of tribal anger: Andhra's Receding Forests, II. Telegraph, 25 June.
Rao, B. 1996. Dry wells and deserted women: gender, ecology and agency in rural India. Indian Social Institute, Delhi, India.
Rao, K.M. 1993. Socio-cultural profile of tribes of Andhra Pradesh. Report of the Tribal Cultural Research and Training Institute. Available from: Tribal Welfare Department, Hyderabad, AP, India.
Rebbapragada, R. 1993. Yearly report. [Samata NGO document.] Located at: Paderu mandal, Visakhapatnam, AP, India.
Rebbapragada, R. 1994. A review: 1990 July–1994 Dec. [Samata NGO report.] Located at: Paderu mandal, Visakhapatnam, AP, India.
Rebbapragada, R. 1995. Half yearly report. [Samata NGO document.] Located at: Paderu mandal, Visakhapatnam, AP, India.
Rebbapragada, R. 1996. Half yearly report. [Samata NGO document.] Located at: Paderu mandal, Visakhapatnam, AP, India.

Rebbapragada, R. 1997. Final report of Samata for the national fund proposal. [Samata NGO document.] Located at: Paderu mandal, Visakhapatnam, AP, India.

Sainath, P. 1996. Everybody loves a good drought: stories from India's poorest districts. Delhi: Penguin Books, Delhi, India.

Shiva, V. 1989. Staying alive: women, ecology and development. Zed Books, Atlantic Highlands, New Jersey, USA.

Shiva, V. 1991. The violence of the green revolution: third world agriculture, ecology and politics. Zed Books, Atlantic Highlands, New Jersey, USA.

Viegas, P. 1991. Encroached and enslaved: alienation of tribal lands and its dynamics. Indian Social Institute, Delhi, India.

6
Applications of Historical Ecology to Land-Management Decisions in the Northeastern United States

EMILY W.B. RUSSELL

"Change over time is fundamental to analyzing the effects of land use" (Dale et al., Chapter 1, this volume). Changes in the past, which set the scene for the present, often constrain decisions and potential for future change (Russell 1997; Foster and Motzkin 1999). Former events, conditions, and processes may also provide a yardstick for the goals of planners and managers. Historical ecology is the study of the past to understand the range of variation in natural ecosystems and the processes that have led to current conditions. For example, historic conditions may form a template for planning, when land use aims to reproduce ecological features or processes that existed at some time in the past or to commemorate a historic event. I refer to this as using the past as an analogue for the present.

Even when former conditions or processes are not a goal, the past always forms a prologue to the present and future (Russell 1997, pp. 238–239). The influence of the past on the present may be apparent, for example, when past overgrazing has produced an eroded landscape overgrown by invasive weeds, or subtle, for example, when selective logging in the past has altered the species composition of a forest. Because of the effects of prior human activities on ecosystems, the range of natural variability in the system may be difficult to ascertain and even more difficult to incorporate into management (Landres et al. 1999; Swetnam et al. 1999).

Two case studies illustrate the importance of considering the past as part of the temporal framework for land-use planning. In the first, Saratoga National Historical Park, the aim of management is to reconstruct and manage a landscape using as a template the battlefield at the time of the battle of Saratoga, 7 October 1777. This cultural historical template was a varied landscape that supported a variety of ecological communities, some of which had been essentially unmodified by the European settlers. Its reconstruction would recreate an eighteenth-century cultural landscape, and protect open space in a region subject to suburbanization. An ecological approach to managing and interpreting the landscape has had several consequences. Planners and managers have realized that the natural landscape is dynamic and cannot be reconstructed as a stable museum piece, even if the details of that scene could be precisely known. They have also rethought their goals in the light of historical ecological processes, such as the extinction of species that were present in the precolonial forest and the importance of the last several centuries of human activities in modifying the potential for forest regeneration. This ecological point of view has led to management that takes ecological processes into account.

The second example is the Rome Sand Plains, a region defined by a distinctive geological substrate that supports several rare plant communities. Here it is obvious that the ecology of the ecosystems is critical for land-use planning. The history of the system, however, provides vital clues to the factors and processes that have produced the current communities. The ecological past of the system also brings into question the "naturalness" of the rare communities: Are they artifacts of the last two centuries of human-caused disturbances or are they under the control of nonhuman factors such as soils and hydrology? Regardless, historical processes, whether caused by people or not, must be incorporated into plans for the future if the rare communities are to be preserved.

In the last few centuries, both landscapes have undergone major human alterations, which both constrain decision-making and, at the Rome Sand Plains, possibly have contributed to the very features for which the landscape is being managed. While both are being managed for public goals, the processes of deciding what is the ideal goal and working within temporal constraints and potentials are generally applicable to ecologically informed land management. Planning at both sites must consider modifications of the sites, the species, and especially the disturbance processes over the last several centuries, as recommended by ecological land-use planning (Chapter 1).

6.1 Saratoga National Historical Park

This park includes about 1400 ha of land located 40 km north of Albany, New York, on the west side of the Hudson River (Fig. 6.1). It is the site of the battle of Saratoga, a critical early battle of the American Revolution, fought in the fall of 1777. At the time of the battle, the land was mostly forested, with a few cleared farm fields (Snell 1949; Russell 1994). The National Park Service decided in the mid-twentieth century to attempt to reconstruct the field and forest configuration as it existed at the time of the critical battle on 7 October 1777 (Hamilton 1944). By the 1990s these efforts had led to a landscape that consisted of fields interspersed in regenerating forest. There remained several ecological issues related to land use and management. First, what features of the 1777 species composition, structure, and ecological processes could be known and reconstructed? Second, how were postcolonial disturbance regimes affecting the reestablishment of target communities and processes?

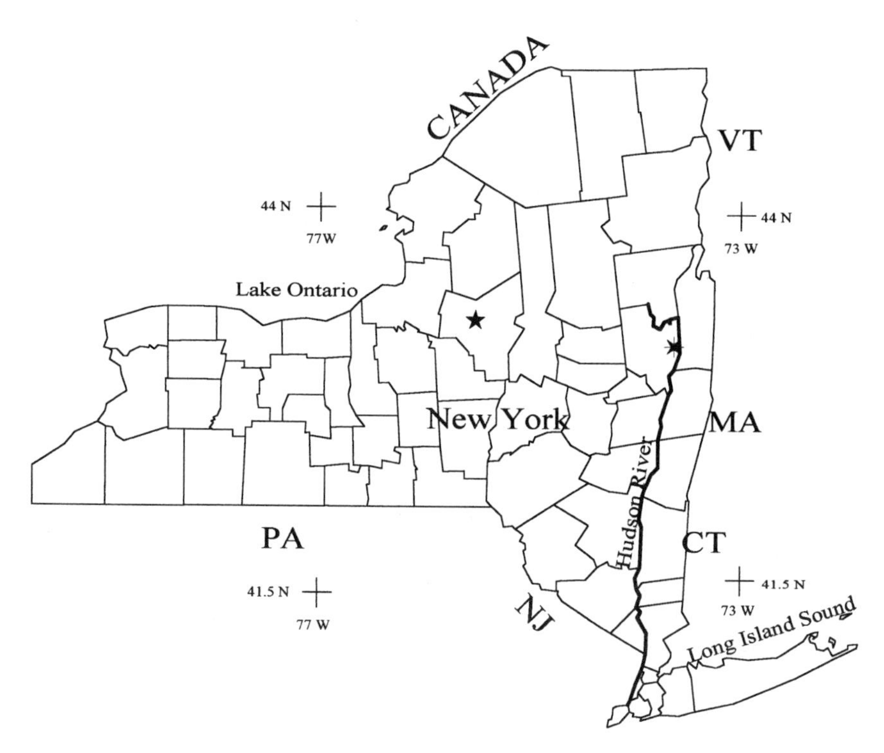

FIGURE 6.1. Location of sites: ✶, Saratoga National Historical Park; ★, Rome Sand Plains.

To answer these questions required historical research into the landscape of the eighteenth century as well as changes in the intervening centuries that affect the current and future conditions—in other words, studying history both as analogue and as prologue. The research could then be translated into ecologically sound management options.

6.1.1 Site Characteristics

Site characteristics form the basis for ecologically informed land management (Dale et al., Chapter 1). At Saratoga, they also determined the configuration of the battlefield and subsequent land use.

6.1.1.1 Site

Elevation ranges from 28 m on the floodplain of the river on the east boundary west to hills about 110 m high. Shale bedrock is at the surface in a few places, but in most areas thick surficial deposits of Wisconsinan till, glacial lake clays and sands, and river alluvium cover the bedrock. The surficial deposits exert primary control on the land; streams have cut deeply into the lake sediments, producing dissected topography in the eastern third of the park, with steep-sided ravines leading to the Hudson River floodplain. The river in this vicinity is deeply incised, leaving a narrow floodplain bounded by 30-m-tall bluffs. Soils on the till to the west are rocky and fairly well drained, while those in other areas are poorly drained or only fairly well drained, with slow permeability. Some soils on dune sands are very well drained, with high permeability. The few shale outcrops have very shallow soils (USGS Soil Conservation Service 1990).

6.1.1.2 Species

The species pool consists of native species of the temperate deciduous forest biome, agricultural weeds and crops, and other nonnative species. Hardwood species, including oaks (*Quercus* spp.), hickories (*Carya* spp.), American beech (*Fagus grandifolia*), basswood (*Tilia cordata*), and red and sugar maples (*Acer rubrum* and *A. saccharum*), dominate the uplands and south-facing slopes. Hemlock (*Tsuga canadensis*) is common in the steepest ravines on north-facing slopes. White pine (*Pinus strobus*), American elm (*Ulmus americana*), and clones of big-leaved aspen (*Populus grandidentata*) are the first trees to colonize abandoned old fields, pine and aspen on better-drained sites and elm on poorly drained ones.

Several species have major impacts that affect park management and potential. The native white-tailed deer have benefited from recent changes in land use. There are about 30 deer per square kilometer, three times the maximum population density recommended to prevent intense browsing damage (Underwood et al. 1989; Porter 1991; Austin 1992). Another very successful native species, which was not common a century ago, is gray dogwood (*Cornus racemosa*), a shrub that forms thickets in some abandoned fields and encroaches on fields that are slated to remain open. Research, however, suggests that neither the deer nor the thick shrubs are responsible for the slow pace of reforestation on the abandoned fields (Underwood et al. 1989; Austin 1992).

Nonnative diseases and insect pests have a long history in this region, and continue to affect tree growth. The introduced chestnut blight of the early twentieth century eliminated this species from the forests, and continuing infections make reestablishment of the species unlikely in the absence of disease-resistant cultivars. The Dutch elm disease, also a twentieth-century introduction, prevents native elm trees from flourishing in the successional forests. Because elm is the most common species of the wetter parts of abandoned farmland, this disease is inhibiting reforestation there. White pine blister rust slows white pine growth, on dryer microsites, but does not appear to be eliminating this species from the forest. An introduced herbaceous plant, knapweed (*Centaurea jacea*), has spread over the still open fields, producing a field species composition that differs from that present in the eighteenth century.

6.1.1.3 Disturbances

In the absence of people, several natural disturbance regimes have shaped this landscape, and continue to influence it. The most important of these is erosion in the ravines and river terraces, where slumping on steep slopes continues to widen and lengthen ravines. Some deposition of alluvium occurs in the river floodplain. Erosion of shallow soils on outcrops of bedrock maintains these shallow soils. Occasional wind damage and droughts have affected the forests. Naturally ignited fires have most likely been very rare in the last few millennia. It is possible that some of the droughtier soils in the northern part of the park supported pitch pine forest in precolonial times. Over the last 300 years people have imposed new disturbance regimes, which continue to influence ecosystem dynamics and landscape configuration.

6.1.1.4 Landscape

The site conditions exerted a major influence on patterns of land settlement, choice of this location for the battle itself, the disposition of the battle, and potential forest regeneration. It is likely that the Amerindians had farms on the floodplain, and used the uplands for hunting and gathering, but there is little record of their impacts. The Dutch also chose the floodplain of the Hudson for their farms, in the early eighteenth century, and hunting pressure no doubt increased in the uplands as their populations increased along the river. Later in the colonial period, immigrants from New England first cleared farms along the northern part of the site and in areas underlain by glacial till (Russell 1994).

The American army chose this location to stop the advance of the British down the Hudson River in large part because of the narrowness of the river valley here and steep bluffs commanding a strategic overview of the river. The rough ground away from the river made passage of troops through the uplands difficult. The Americans enlarged farm fields to create open space for their encampment and defenses. The British also set up an encampment and cleared trees to enlarge open areas. The actual battle was fought in the open areas of the northern part of the current park.

In the century after the battle, farmers cleared almost all of the rest of the forest for agricultural fields. After less than a century, however, they began to abandon some of the fields, especially on the steepest and the most poorly drained land. When first the state, then the federal government began to acquire the property in the early part of the twentieth century, farming was declining and more fields had been abandoned. What had been mixed farming had shifted to mainly sheep grazing. Fields were separated by hedgerows, where elms were a predominant tree species. White pines were common on the steep slopes down to the river.

Early park development included continued grazing of sheep to maintain open fields, but by the mid-twentieth-century mowing had replaced the sheep as a means of keeping the fields open and white-tailed deer had replaced sheep as the primary herbivores. Park management also included logging some forests that were growing on areas that had been fields at the time of the battle. In the late-twentieth century, controlled burns were introduced to maintain the fields, to keep gray dogwood out of the fields, and to try to control knapweed.

6.1.2 Guidelines for Land Use

This park has a specific mission: to commemorate this important battle of the American Revolution. Thus, historical cultural values have predominated over ecological values in planning the park. When the decision was made to allow forest regrowth to reconstruct the battlefield as it had existed on the day of the famous battle, however, there was an implicit decision to allow much of the landscape to revert to a "natural" condition by letting natural regeneration produce the forest. It is still, however, the cultural, not the natural, landscape that is the goal of management. I will thus consider the ecological guidelines that have been important in managing this cultural landscape. The historical ecological approach to understanding this landscape has led to several suggestions for management.

6.1.2.1 Forest Regeneration

Forest regeneration in the fields abandoned in the twentieth century has been very slow. The extent of clearing, the pattern of field abandonment in the twentieth century, limited historic seed sources, and disease are possible causes, while the importance of the thick shrub cover and deer browsing are problematical (Fig. 6.2). Given this slow development of forests on most of the park, it is important to maintain and protect the existing mature forests, even if they are located where there were open fields at the time of the battle in 1777. This is important for ecological reasons, to perpetuate the forests, but also can be defended culturally as these are the only stands at present where visitors can see what the forests of the eighteenth century may have looked like.

6.1.2.2 Forest Species Composition

Details of the forest species composition in the eighteenth century are not available. Extrapolating from other New York state forested areas, however, it seems likely that the forests were mixed oak/northern hardwoods. Chestnut trees were present but probably not abundant in the eigh-

FIGURE 6.2. Regenerating forest at Saratoga National Historical Park. The tall tree at the right marks an old field line. It stands at the edge of an area intended for forest regeneration, which, after 40 years, is still a gray dogwood/viburnum shrub thicket with some small trees.

teenth-century forest. Paleoecology and current study of forest dynamics confirm that forest vegetation is dynamic. It thus seems both ecologically and culturally appropriate to set as a goal a mature hardwood forest on the uplands, with hemlock in ravines, with little precision about the species that should be present in the forests. The current mature forests stands also support a diverse native herbaceous flora, which should be protected. Pitch pine was probably eliminated, if it had been present at all, by farming and mining sand. Though it has populations 15 km or so away, it is unlikely that this site will become a pitch pine forest in the future without active management.

6.1.2.3 Disturbances

Occasional disturbances due to wind storms, ice, landslides, and disease are a natural part of the dynamics of these forests. These will change the structure and species composition in ways that cannot be predicted. It is not appropriate, therefore, to think of the forest as a museum piece that precisely replicates the forest that was on the site in October 1777. This is neither possible nor probably desirable.

6.1.2.4 Fields

Open fields were most likely not part of the prehistoric landscape. It is possible that the very shallow soils on rock outcrops and the frequently disturbed sites in the river floodplain were not forested, and small landslides produced temporarily denuded patches within the forest. However, maintaining open fields to recreate the historic scene provides habitat for currently declining open-field species of birds, and is an appropriate management aim in terms of maintaining biodiversity, especially where the management techniques can eliminate nonnative species from the fields.

6.1.2.5 Other Ecological Guidelines

The regional landscape is one of forest regeneration, some farming, and suburbanization. Thus, many of the "natural" processes and plant communities, those that predated Euroamerican settlement of the area, are not likely to be reestablished. However, the management of the park with a mix of cultural and ecological goals will best maintain a diversity of native species in this region of accelerating regional suburbanization. The historical perspective allows the management to understand the features of the cultural landscape that are and are not appropriate to try to reconstruct. Goals for the reconstruction have been modified by an understanding of the role of natural disturbances, both past and present, in regulating the dynamics of a natural landscape. The ecological approach reinforces the historical goals of the park to maintain open space, and to work to maintain open space in the general region.

Other ecological guidelines for land-use decisions, such as maintaining large open areas and avoiding land uses that deplete natural resources (Dale et al., Chapter 1), are captured in the effort to regenerate the historic scene, especially the forest vegetation. Some, however, deserve special note. The open fields, which can be maintained only by active management, support the major rare elements of the landscape. It is likely that the existence of these elements is dependent on past active management of fields, which prevents the encroachment of the natural native forest vegetation, and that continued active management will be necessary to maintain it. Nonnative species were most likely not major elements of the eighteenth-century scene, so efforts are made to eliminate them, though for historical and cultural rather than for ecological reasons.

6.1.2.6 Lessons

The past is gone, and can only be recaptured in its broad outlines. Landscape configuration is one of these broad categories that can be reconstructed, but the individual elements of the landscape will differ in significant ways from what existed several hundred years ago. However, trying to recapture a historic landscape, especially one of the eighteenth or nineteenth century, provides ideal habitat for a variety of species that are today becoming rare because the altered habitats of those times that fostered them are themselves disappearing. Ecological principles of land-use management—working with native species and natural processes, maintaining open space, and using a landscape perspective—complement and enhance efforts to reach cultural goals.

Managing this small portion of the landscape is still dependent on what is happening around it, such as the proliferation of nonnative species, deer hunting, and air, water, noise, and sight pollution. A regional scheme is necessary to control the overall landscape and its values. Change, however, will be continuous. The environment is favorable to forest, and it is likely that a forest will eventually regenerate, but the historical research indicates that the composition and structure of this forest, either of the trees or the other species, cannot be predicted. Past activities will continue to exert an influence on the patterns of the forest regeneration, with differences across old field lines, and old field line trees contributing seeds. Less effort will be necessary to maintain the naturally occurring forest communities than the human-constructed open fields.

6.2 Rome Sand Plains

Botanists have long recognized the Rome Sand Plains (Fig. 6.1) as a unique feature of the north-central New York landscape. For example, in 1840, John Torrey referred to them as the "pine plains near Rome" (Torrey, Flora of Oneida County, 1840, seen in Bonnano (1994)). More recently, surveys by The Nature Conservancy (TNC) have described three rare plant communities there: pitch pine–heath barrens, pine barrens vernal ponds, and pitch pine–blueberry peat swamp. The state of New York has declared the area a top open-space priority in its Open–Space Plan (Reschke 1990; Bonnano 1994). The overall goals of management of this landscape are to preserve its unique ecosystems and the open space surrounding them.

Reaching these goals has presented some major questions, however. In 1994 land ownership was fragmented. The state of New York owned 240 ha and the city of Rome and TNC each about 50 ha, while private individuals and corporations held the remaining 860 ha. The state had designated its lands as unique but had no management plan for them. The city of Rome had acquired its acreage by tax foreclosure and had also no designated management. Some of the private land had been used for sand mines, dumping, and some logging, but these uses were mostly declining or discontinued. The land was forested and available for conservation mainly because of its historical undesirability for conventional farming. Much of it was, however, suitable for suburban development. The threat of development of these rare ecosystems and open spaces led to current concerns about managing this landscape for its rare communities and species and for recreation.

When TNC, the City of Rome, the New York State Department of Environmental Conservation, and other partners joined to formulate appropriate conservation initiatives to preserve this open space in the face of development pressures, questions arose about the best way to maintain the rare plant communities. The major questions were (1) What role have historic disturbances, especially fires, played in the establishment and maintenance of these ecosystems? (2) How long have these systems existed here, and have human disturbances been instrumental in their formation? (3) Given answers to (1) and (2), what management is appropriate and necessary to manage this area as a natural landscape? A historical ecological approach has yielded critical information for answering these questions.

6.2.1 Site Characteristics

Here as at Saratoga battlefield, site characteristics are critical for ecologically sound management, especially as they relate to the occurrence of the rare communities.

6.2.1.1 Site and Species

The unique nature of The Rome Sand Plains depends on the unusual surficial geology of about 1200 ha of north-central New York State. During the retreat of the Wisconsinan glaciers, there was a large lake in this area. In the vicinity of the lake, retreating glaciers left deposits of glacial till, but in the lake basin itself, clays and sands dominated the surficial, soil-forming deposits. In the early Holocene, about 10,000 years ago, after the lake had drained, winds reworked much of the sand to form parabolic dunes. These fossilized dunes now provide topographic relief in the otherwise flat regional landscape. The water table intersects the surface in the lower areas between the dunes, while the soils on the dunes themselves are dry. Deep (at least 3 m thick) deposits of peat have formed in some of the depressions between the dunes.

This landscape today forms a stark contrast with the surrounding flat, fertile farmland, which is almost all cleared of trees and planted in crops. Uplands, including some dunes, support forest vegetation of American beech, white oak (*Quercus alba*), red and sugar maples, white and pitch pine (*Pinus strobus* and *P. rigida*), gray birch (*Betula populifolia*), hemlock, aspen (*Populus* spp.), American elm, and other northern hardwood species. Some uplands are also characterized as pitch pine heaths, dominated by pitch pines with an understory of blueberries (*Vaccinium* spp.) and other related (ericaceous) shrubs (Bernard and Seischab 1995; Leimanis 1993). Pitch pine is the characteristic tree of the wetlands, along with aspen, gray birch, and red maple, with an ericaceous shrub layer.

Pitch pines dominate the rare plant communities, as they do many rare plant communities in northeastern North America. Some of these, such as the Pine Barrens of southern New Jersey and the Pine Plains of Long Island are extensive, but many others are scattered and small, such as the stands in Waterboro Barrens in Maine and the Montague Pine Plains in Massachusetts (Patterson 1993; Motzkin et al. 1996).

6.2.1.2 Disturbance

The most conspicuous disturbances here are those caused by people in the last two centuries. Historic records suggest that all the merchantable timber was removed in the nineteenth century (Bonnano 1994). This harvest would have included all of the pines as well as the hardwoods and hemlock. Consistent with this logging history is the young age of the pitch pines; the average age of trees sampled in two typical stands was 34 and 52 years (Bernard and Seischab 1995). Sand and peat have provided resources for sale, and both have been mined, most likely with consequences not only for the topography and vegetation, but also for the hydrology of the area. Sand mining continues to some extent today. Other human-caused disturbances include agriculture, blueberry harvesting, and possibly mining of bog iron. Although the soils are not very good for agriculture, the surrounding land is heavily agricultural, and, even if the sand plains were never plowed, it is likely that they were grazed.

Because the site supports the highly inflammable pitch pine and blueberries, recent fires have been frequent and large, as documented by written descriptions and aerial photographs from 1938 to 1955. Railroads ignited many of the fires, which burned not only dry pine sites, but even burned through peat deposits during droughts. Many of the fires were very intense crown and ground fires. People started almost all of them, either directly or indirectly, though lightning may have started some. Because blueberries fruit best if they are burned often, over the last century blueberry pickers frequently set fires to improve the harvests (Fried 1995). Only a small part of the area has not burned since 1915, but the frequency and intensity of fires has decreased greatly in the last decades, because of postwar fire suppression and the change from steam to diesel engines on the trains (Leimanis 1993).

6.2.1.3 History of the Vegetation and Fires

This disturbance history and the lack of many characteristic pine barrens species raises major questions about the factors that have led to the development of the pitch pine communities. How long have they been there? Have they resulted from the last century of human-caused disturbances? Is it necessary to continue disturbances, especially intense fires, that characterized the early part of this century to maintain the pitch pine communities?

Fire is a major land-management technique used to maintain many pitch pine preserves. Is it appropriate here? More historical research was required to answer these questions, taking the history back to before there were written documents.

Unmined peat deposits of the Rome Sand Plains form an excellent stratigraphic record of the last several thousand years. Pollen and charcoal preserved in these sediments record past vegetation and fires in the vicinity of the deposit. To study this history, we extracted a 230-cm-long core of peat from a depression bounded by fossil sand dunes. The peat in the core had been accumulating for about 4700 years, according to a radiocarbon date from the bottom sediment. Over that time, beech, hemlock, sugar maple, white pine, and birch dominated the vegetation, with some oak (Fig. 6.3). There was very little pollen of herbaceous plants, suggesting that the area has been heavily forested for millennia. The small amounts of charcoal suggest a low fire frequency for most of the time recorded in the sediment. About 500 years ago, however, before the increase of ragweed pollen indicating post-Revolutionary land clearing, there was an increase in the amount of pitch pine and charcoal in the sediment, along with a decrease in the amounts of beech and hemlock. After farmers cleared the surrounding land, the amount of charcoal in the sediment increased greatly and beech and hemlock became even less common, being replaced by birch. At least one big fire burned the peat itself, leaving a distinct layer of charcoal and a centimeter of lake-deposited sediment.

The overall outlines of this scenario fit well with the broader regional picture of vegetational change over the last several millennia (Davis 1983; Webb 1988; Russell et al. 1993). For example, the amounts of beech and hemlock have decreased throughout the northeastern United States in the last several centuries, as the amount of birch, a genus that responds favorably to disturbances, has increased. The changes do not directly reflect global cooling of the "Little Ice Age," about 1450 to 1850 AD, since more local vegetational changes mask this signal (Bradley and Jones 1995).

What is unusual in the local picture, as suggested by this sediment core, is that there was an increase in pitch pine and fire well before the advent of European-style agriculture and clearing. The amount of charcoal remained very small, however, in comparison with many other areas in the northeast where pitch pine dominates the vegetation (Patterson et al. 1987). Only the two peaks, one reflecting the fire that burned the peat and the other large recent fires described in historical documents (Leimanis 1993), come even near the amounts reported from pitch pine areas in Maine (Patterson 1993) and the Shawangunk Ridge in New York (Laing 1994).

This record still leaves the question of what factors led to the increase in pitch pine, the species that dominates the rare plant communities, before European clearing of the area. The sedimentary record combined with historical documents provides several clues. First, the hydrology of the site was changing constantly over the time that the peat accumulated. Only in

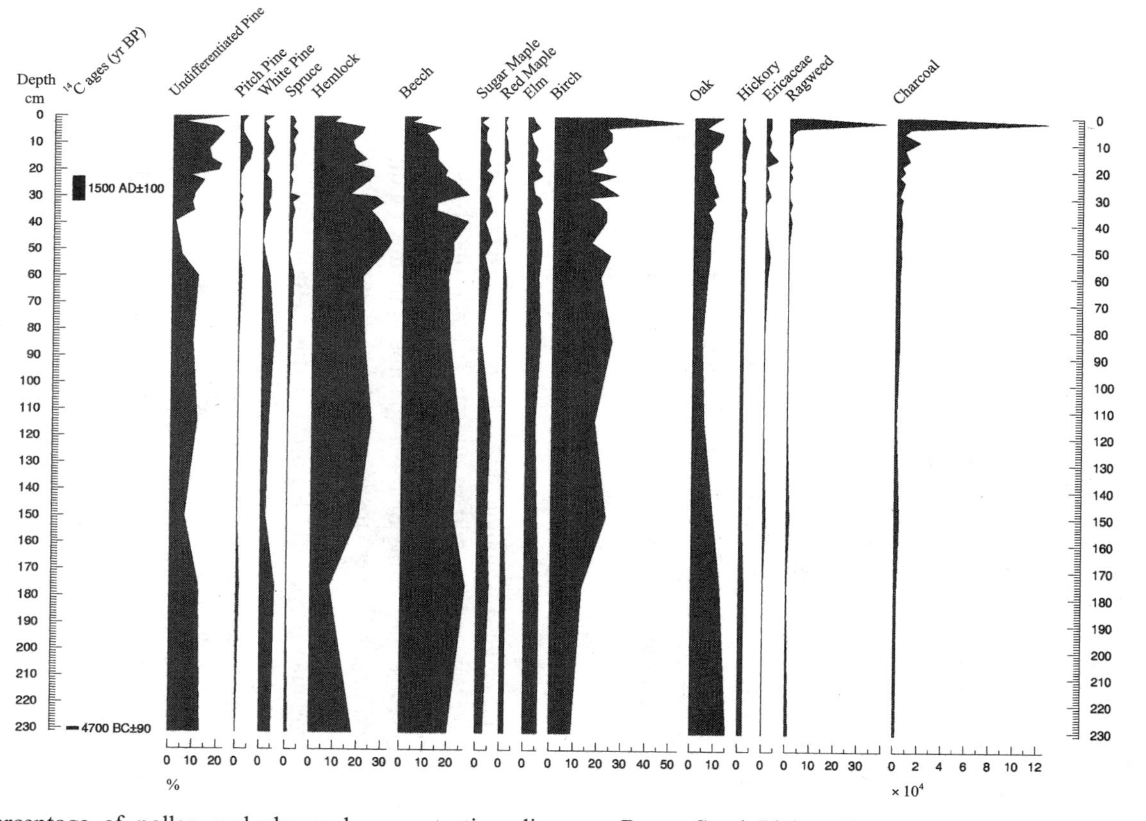

FIGURE 6.3. Percentage of pollen and charcoal concentration diagram, Rome Sand Plains. Tree taxa as percentage of tree pollen; herbaceous taxa as percentage of total pollen; charcoal as μ^2 charcoal/pollen grain.

the layers just above the charcoal bed caused by the peat fire is there any evidence that this site was ever a pond. The peat did not grow here filling a body of water; rather, the water table rose as the peat grew, more than 2.3 m. At some point, this locally elevated water table, possibly perched on an impermeable clay layer at the bottom of the depression, may have altered the hydrology of the surrounding dunes. If the same thing was happening in other depressions, this may have led to a regional change in the vegetation.

Second, there is some historical evidence of cultural change in this region about AD 1300 as the Iroquois arrived locally (Lake 1970). The Rome Sand Plains were part of a major trading route between the Mohawk River and St. Lawrence River drainages. In addition, long before extensive clearing for agriculture by European settlers, trade would have been increased by European demand for furs. This increased activity may have led to more fire ignitions, which favored pitch pine. Increased pitch pine and increased fires were apparently contemporaneous events, but cause and effect cannot be disentangled.

It appears that the large and frequent fires of the late nineteenth and early twentieth centuries were a historical anomaly. Thus, the recent control of them is part of a return to pre-Revolutionary conditions. The prehistoric increase in pitch pine occurred in the absence of these large fires.

6.3 Management Implications

This landscape still stands in contrast to its surroundings in terms of substrate, topography, and vegetation. It has probably done so in terms of vegetation for centuries at least, and in terms of substrate and topography for millennia. The value of the landscape is in this uniqueness; mining of sand and peat have perhaps altered the landscape, but have not eliminated its special character. However, current development pressures threaten the plant communities if not also other features, for example, if structures are built on the fossil dunes.

How do these findings relate to land-use planning and management? In contrast with Saratoga, it is the present configuration of communities that is of value here. The questions of planning relate more to the impacts of regional change, nonnative species, and other human-caused disturbances on these rare ecosystems. Historical research has indicated that the rare ecosystems may have originated only in the last few centuries, but that they predate major clearing and increases in fires in the region. The most important question is the role of these past human disturbances on the continued existence of the rare communities. If these disturbances are critical for maintaining, or expanding, the communities, should they be continued? And in what form? Some past fires have been intense enough to burn into peat bogs, but whether this has been important for establishing the rare wetland communities is not known. These may gain value if they are prehistoric, but more important is the influence of the last 200 years, or more, or human activities.

The past is valued more as prologue than as an analogue. The hope is to be able to perpetuate the present conditions, which include rare communities. However, if these communities are relics of nineteenth- and twentieth-century human activities, their perpetuation may require very active management, which thwarts natural tendencies. Whether this is desirable or not will continue to be a value judgement.

6.4 Conclusions

History provides a template for land use and management at Saratoga National Historical Park. Even here, however, with a plethora of historical documents to reconstruct the historic scene, there are questions about details to be reconstructed. The issue of a template, or goal, is even more difficult to resolve at Rome Sand Plains, where the goal is the preservation of rare plant and animal communities. There is no straightforward documentation of conditions that might have favored these communities in the past, or of how (or even if) current communities resemble those that existed in the past. History can provide an ideal to strive for, but this ideal is constructed both from the perceptions of those making the plans and the very sketchy evidence of the past. It is never a simple past that can be recaptured in specific features or processes.

Even more important, changes on a landscape over the last several centuries have altered its potential. In almost all landscapes, these changes have affected species, site conditions, disturbance regimes, and landscape configuration, the ecological principles identified by the Ecological Society of America land-use committee (Dale et al., Chapter 1). Some native species, such as American chestnut and American elm, cannot regain their place in the ecosystem because of diseases. Disturbances in the last centuries, such as grazing and frequent, intense fires, have altered the species and age structure of stands in ways that cannot be easily reversed (Mendelson 1998). The pattern of different communities on the landscape has been greatly modified over the last few centuries, and, again, will continue to exert an influence for many decades, at least. Sometimes these new disturbances and patterns favor rare species, as in open fields at Saratoga and the rare pitch pine communities in the Rome Sand Plains. The importance of changed conditions over the last few centuries, as well as the last millennia, and projected changes in the future must be integrated into decisions about the processes and patterns to favor in management decisions.

Acknowledgments. The work at Saratoga National Historical Park was funded by The National Park Service, Northeast Regional Office, Cultural

Landscape Program, directed by Nora J. Mitchell. I especially thank her and the staff at the Park for their help. The work at the Rome Sand Plains was funded by the Central and Western Chapter of The Nature Conservancy. I thank Kris Agard in that office for initiating the project and help in the field.

References

Austin, K.A. 1992. Gray dogwood (*Cornus racemosa* Lam.) as a refuge from herbivory in old fields of Saratoga National Historical Park, New York. Ph.D Thesis, State University of New York, Syracuse, New York, USA.

Bernard, J.M., and F.K. Seischab. 1995. Pitch pine (*Pinus rigida* Mill.) communities in northeastern New York State. American Midland Naturalist **134**:294–306.

Bonnano, S. 1994. Site Basic Record. Rome Sand Plains. The Nature Conservancy, Central and Western New York Chapter.

Bradley, R.W., and P.D. Jones. 1995. Recent developments in studies of climate since A.D. 1500. Pages 666–679 *in* R.W. Bradley and P.D. Jones, editors. Climate since 1500, 2nd edition. Routledge, London, England.

Davis, M.B. 1983. Holocene vegetational history of the eastern United States. Pages 166–180 *in* H.E. Wright, editor. Late-quaternary environments of the United States. University of Minnesota Press, Minneapolis, Minnesota, USA.

Foster, D.R., and G. Motzkin. 1999. Historical Influences on the landscape of Martha's Vineyard: perspectives on the management of the Manuel F. Correllus State Forest. Harvard Forest Paper No. 23, Harvard Forest, Harvard University, Petersham, Massachusetts, USA.

Fried, M.B. 1995. The huckleberry pickers. Black Dome Press, Hensonville, New York, USA.

Hamilton, W.F. 1944. Proposed archeological program for Saratoga NHP, Saratoga NHP, Administrative History Files; R.J. Koke, 1947. A report on the reforestation program for Saratoga National Historical Park. Saratoga NHP, Administrative History Files, Saratoga, New York, USA.

Laing, C. 1994. Vegetation and fire history of the dwarf pine ridges, Shawangunk Mts., New York. Report to The Nature Conservancy, Eastern New York Regional Office.

Lake, R.M. 1970. Prehistoric Rome and Romans. Rome Historical Society, Rome, New York, USA.

Landres, P.B., P. Morgan, and F.J. Swanson. 1999. Overview of the use of natural variability concepts in managing ecological systems. Ecological Applications **9**:1179–1188.

Leimanis, A. 1993. Vegetation and fire history of the Rome Sand Plains: report to the Nature Conservancy Central and Western New York Chapters.

Mendelson, J. 1998. Restoration from the perspective of recent forest history. Transactions of the Wisconsin Academy of Sciences, Arts and Letters **86**: 137–148.

Motzkin, G., D.R. Foster, A. Allen, J. Harrod, and R. Boone. 1996. Controlling site to evaluate history: vegetation patterns of a New England sand plain. Ecological Monographs **66**:345–365.

Patterson, W.A. 1993. The Waterboro Barrens: fire and vegetation history as a basis for ecological management of Maine's unique scrub oak–pitch pine barrens ecosystem. Report to the Maine Chapter of The Nature Conservancy.

Patterson, W.A., K.J. Edwards, and D.J. Maguire. 1987. Microscopic charcoal as a fossil indicator of fire. Quaternary Science Reviews **6**:3–23.

Porter, W.F. 1991. White-tailed Deer in Eastern Ecosystems: Implications for Management and Research in National Parks. Natural Resources Report NPS/NRSUNY/NRR-91/05, United States Department of the Interior, National Park Service.

Reschke, C. 1990. Ecological Communities of New York State. New York Natural Heritage Program and NYS Department of Environmental Conservation, Latham, New York, USA.

Russell, E.W.B., R.B. Davis, R.S. Anderson, T.E. Rhodes, and D.S. Anderson. 1993. Recent centuries of vegetational change in the glaciated north-eastern United States. Journal of Ecology **81**:647–664.

Russell, E.W.B. 1994. Cultural landscape history: Saratoga National Historical Park. Saratoga NHP, Administrative History Files.

Russell, E.W.B. 1997. People and the land through time. Yale University Press, New Haven, Connecticut, USA.

Snell, C.W. 1949. A report on the ground cover at Saratoga National Historical Park, October 8, 1777. Saratoga NHP, Administrative History Files.

Swetnam, T.W., C.D. Allen, and J.L. Betancourt. 1999. Applied historical ecology: using the past to manage for the future. Ecological Applications **9**:1189–1206.

Underwood, H.B., K.A. Austin, W.F. Porter, R.L. Burgess, and R.W. Sage, Jr. 1989. Deer and vegetation interactions on Saratoga National Historical Park. Report to NPS. Saratoga NHP files.

USGS Soil Conservation Service. 1990. Draft soil map, Saratoga Battlefield area.

Webb, T. III. 1988. Eastern North America. Pages 385–414 *in* B. Huntley and T. Webb III, editors. Vegetation history. Kluwer Academic, Dordrecht, The Netherlands.

7
India's "Project Tiger" Reserves: The Interplay Between Ecological Knowledge and the Human Dimensions of Policymaking for Protected Habitats

CYNTHIA A. BOTTERON

This chapter poses a challenge to the utility and value of the Ecological Society of America's guidelines for land use and management (Dale et al., Chapter 1) by conducting a hypothetical, post hoc analysis of India's utilization of the inviolate "national park" model as a tool for saving the Bengal tiger. Both the utility and value of the ESA's guidelines are affirmed, but not without revision. For protected area/species preservation efforts to work, the principles guiding planning decisions must be informed not only by ecological science, but at all points, by the social, cultural, and political environment of the human communities directly impacted by those decisions.

The most reliable method for testing the utility of a set of policy guidelines is to use them and then compare the outcome achieved to the outcome desired. The Ecological Society of America's (ESA's) guidelines for land use and management (Dale et al., Chapter 1) are a relatively recent development and it is unlikely that a statistically significant number of projects that used the guidelines during the planning phase currently exist. The few that do are likely to have been too recently implemented to provide a full illustration of the guidelines' strengths and weaknesses. In this light, another method for testing is to conduct a hypothetical, post-hoc analysis that answers the following, Was there an opportunity for direct application of the guidelines? An affirmative answer can demonstrate utility. Second, if the project would have incorporated the guidelines, is it defensible to conclude that ecosystem integrity and functioning would have been better assured than it is at present? An affirmative answer can demonstrate positive value.

National park planning and management in India is used as this hypothetical case study to investigate the utility and value of ESA's guidelines. Most relevant to national park planning and management are the ecological principles related to the evolution of natural systems, the role of disturbance in promoting change, and the interdependency of species and landscapes (Dale et al., Chapter 1). As an extension of these principles, planners and managers are advised to follow the guidelines recommending the preservation of rare landscape elements and associated species, to plan for long-term change and unexpected events, and to examine the impact of their decisions in a regional context (Dale et al., Chapter 1).

There are few better case studies to challenge the utility of the guidelines than India's adoption of the inviolate "national park" concept as a means of providing sanctuary to the beleaguered Bengal tiger (*Pantheris tigris*). A comparison between planning decisions made by the Government of India in devising Project Tiger and ESA's guidelines will serve two purposes. The first will be to illustrate the guidelines' utility, highlighting the points where they would have easily been incorporated in the planning phase. Second, the analysis will answer, in the affirmative, the second question posed above, that ecosystem integrity and function probably would have been more assured than it is at present. Project Tiger is now facing a myriad of problems threatening to undo the gains made over the past 30 years. Some of the most endemic problems are directly linked to one weakness that plagued park planning in the past and may diminish the value of ESA's guidelines if not addressed. In the context of protected area policy in densely populated, economically developing agrarian nations, land-use and land-management decisions must raise to a prominent position considerations of the social, cultural, and political environment of the human communities impacted by a protected area scheme. More problematic is that for this change to occur, a truly interdisciplinary effort is required—essentially a breaking down, on some issues of mutual interest, of the distinction between "natural" and "human" science.

7.1 National Parks and the ESA's Guidelines for Land Use and Management

National parks were originally conceived as a form of land use intended to preserve a "pristine nature" at a time when the United States was relatively sparsely populated and wilderness "was something that only the cultivated ... could truly understand and value" (Budiansky 1995). Wilderness had value because of what it could do to elevate us:

> While our forefathers were carving a civilization from the wilderness, the land in turn made enduring impressions on their minds and thoughts. ... [T]he qualities of perseverance, independence, and initiative being developed and refined, as the American character was shaped on vast stretches of virgin prairies, beside rolling rivers and in lonely mountain passes. It is in the national parks that these influences on the United States can be maintained and kept pure, so that this and future generations may know and feel—and benefit from—the same wondrous exposure that our forefathers experienced.
>
> (Wirth 1962)

National parks later came to be valued for more pragmatic reasons: They are the last and best sites for preserving wild species (Badshah and Bhadran 1962), ecosystem processes, and biodiversity. "We must make every effort to preserve, conserve, and manage biodiversity. Protected areas, from large wilderness reserves to small sites for particular species, and reserves for controlled uses, will all be part of this process" (WRI, IUCN, and UNEP 1992).

What, exactly, is the entity "national park?" This is seemingly an innocuous question, but it is one that has taken decades of research by scientific, governmental, and transgovernmental organizations to answer. The criteria that differentiate a national park from other protected area types have changed over time and been refined and fitted for a broad set of conditions. The list of protected area categories has also been expanded in the effort to respond to a myriad of habitat preservation needs and contexts (for an elaboration of protected area categories, see IUCN (1994)). Adopted by India as a model for Project Tiger is the definition agreed upon in 1969 by member states of the International Union for the Conservation of Nature (IUCN, whose name has recently been changed to The World Conservation Union).

IUCN and the United Nations Educational, Scientific, and Cultural Organization (UNESCO) have been ardent promoters of the national park concept throughout the world. UNESCO and IUCN, together, create and maintain the globally recognized set of standards and categories of protected areas that delineate the U.N. list of National Parks and Equivalent Reserves. The list serves as an honor roll for nations that have met IUCN and UNESCO's rather exacting standards. IUCN had struggled for nearly a decade to devise a national park definition acceptable to its member nations. At its 10th General Assembly Meeting in New Delhi, 1969, IUCN's long-standing definition for a national park was adopted. Because of its importance to this discussion, the entire resolution is reproduced here.

A National Park is a relatively large area:

1. where one of several ecosystems are not materially altered by human exploitation and occupation, where plant and animal species, geomorphological sites, and habitats are of special scientific, educative, and recreative interest of which contains a natural landscape of great beauty;
2. where the highest competent authority of the country has taken steps to prevent or to eliminate as soon as possible exploitation or occupation in the whole area and to enforce effectively the respect of ecological, geomorphological or aesthetic features which have led to its establishment; and
3. where visitors are allowed to enter, under special conditions, for inspirational, educative, cultural and recreative purposes (IUCN 1970).

Closely read, the above definition of a national park establishes the types of interactions between humans and nature that are allowed: We can visit, appreciate, and recreate; we cannot exploit or occupy. Prohibiting the use of resources found within national park boundaries has proven to be the most contentious aspect of the scheme. Park advocates, in the early part of the movement, were strident in their belief that occupying communities had to be removed and the use of resources prohibited. Their belief was not entirely without justification:

> Experience has shown that some settlers have been extremely unscrupulous, and their presence in the sanctuaries has been fraught with danger to wildlife. Their bows, arrows, and traps do not require any separate mention, let alone the modern versions of this equipment of destruction.... Grazing of livestock in national parks should be strictly prohibited, as cattle resorting to grazing in the area become serious competitors to the herbivores in the matter of fodder, besides communicating cattle diseases to the wildlife.
>
> (Badshah and Bhadran 1962)

> It is impossible to determine how far man intervenes naturally, how far he can still be looked upon as making up part of the natural biotic totality. Every attempt in the national parks to maintain so-called primitive societies in proper balance with the environment has proved itself a failure.
>
> (Verschuren 1962)

For densely populated, agrarian nations the requirements to prohibit access to resources and to move occupying communities have been difficult to fulfill. The ESA's guidelines to consider the implications of decisions at a regional scale and to plan for long-term change would have drawn the attention of planners to the question of how protected areas were to be situated in a landscape dominated by human need.

7.2 India and National Parks: A Tool for Wildlife Preservation

7.2.1 Background

India has a long, established history of land use by hunting–gathering communities, shifting and settled agriculturalists, nomadic populations, and livestock herders. Along with the history of use is an established legal and philosophical tradition of land-use regulation. Taught through religion and law, or experienced through the hand and plow, India touts a centuries-old reverence for some animals and some forested areas (GOI 1996). The tradition in the emergent scholarship on India is to argue that British imperial rule fundamentally disrupted this time-honored understanding of nature by the imposition of a commercially oriented forest policy, the centralization of power, and the profligate exploitation of forests and wildlife (e.g., Arnold and Guha (1995), Guha (1989), Gadgil and Guha (1992), and a classic piece—Ribbentrop (1900)). The stories the British told about their own hunting exploits lent credence to this interpretation of history. F.B. Simon, the author of *The Sport in Eastern Bengal,* claimed to have shot 600 tigers over his 21-year career with the British Administration in the mid-1800s. In Kathiawar, a cavalry officer was reported to have shot approximately 80 lions. Colonel Pollock, a Military Engineer working in Assam, claimed to have shot a rhino or guar (Indian buffalo) nearly everyday for breakfast. India's ruling princes and "leisured gentry" were also guilty of contributing to this slaughter. The Cooch-Bihar organized shoots that, over a period of 35 years, claimed 370 tigers and 208 rhinos. The lifetime record for the number of tigers' shot is 1116, held by one Maharaja of Surguja in Madhya Pradesh (Raghavan 1967).

However, the British also established a legal foundation that provided broad-based protection to favored species and forest areas. Region-specific legislation was adopted that protected commercially important species (e.g., the elephant), regulated hunting, and restricted the use of forests by tribals and villagers. In 1912 the Wild Birds and Animals Protection Act No. VIII was adopted. In it, fines were assigned for the killing, selling, or possession of listed species (Panjwani 1994). Species listed in the Act were those deemed to be important to tribals, villagers, or the environment: antelopes, asses, bison, buffaloes, deer, gazelles, goats, hares, oxen, rhinoceroses, sheep, and 16 species of birds. Not listed were animals hunted for trophy by the Maharajas and British: lion, tiger, leopard, bear, crocodile, and a variety of birds. Subsistence and ceremonial hunting by tribals was effectively banned (Gadgil and Guha 1992). Hunting by the leisured and ruling class was not. The resentment by tribals and villagers over the government's choice of privileging an alien class over their needs will be fueled once again when these same groups come to see Project Tiger as another instance of the government's antipathy and ambivalence toward their needs and poverty.

In 1935, the Wild Birds and Animals Protection Act was revised and adopted as the Wild Birds and Animals Protection Act No. XXVII. In this Act, provincial governments were given the authority to declare any area a sanctuary for birds and animals (Panjwani 1994). Uttar Pradesh, in 1936, passed the Hailey National Park Act creating Hailey National Park, whose name was changed to Corbett National Park in 1957 (Singh 1988).

In 1947, India won its independence from Britain. Two problems that pressed upon the new government were the lack of self-sufficiency in food production and poverty in the countryside. To that end, the central government launched its "Grow More Food" campaigns. Permission was given for the diversion of large tracts of forest for cultivation. As habitat for wildlife was transformed into habitat for humans, increasingly wild animals and humans came into conflict. To assist the rural and tribal communities, the government relaxed restrictions on crop-protection gun licenses. Wild animals in the near vicinity of agricultural plots were summarily wiped out.

George Schaller (1967) likens India's post-Independence period of destruction to that of our own "slaughter on the American prairies in the 1880's." This same story is told repeatedly when answering the difficult question of what initiated India's most recent wildlife crisis (Stracey 1960; Putnam 1976; Dhyani 1994).

In 1952, the central government convened the Indian Board for Wildlife (IBWL) as a mechanism to check the deteriorating conditions of India's forests and wildlife. The Board was empowered to recommend and approve the declaration of national parks and sanctuaries, to oversee the trade in wild species, and to promote nature education among both rural and urban communities. One of the Board's first acts was to devise a set of guidelines for India's national parks. The Board was cognizant of the conditions in rural areas and made allowances in the national park standard for the use of forests by tribal and other forest-dwelling communities. The Board restricted commercial forestry operations, recommending that "islands" in the park be designated as "Sanctums Sanctorum" or "Abhayaranyas"— areas not to be developed or mined for timber. However, over time, the list of prohibited activities in sanctums sanctorum gradually evolved to include grazing, gathering forest produce, lopping tree limbs, grass cutting, and cultivation. The bulk of a protected area continued to be available to meet these needs. In effect, through "zoning" the IBWL sought to balance protection with the established land-use practices and resource needs of the communities living in and near the protected areas.

In 1970, the IBWL received a report from its Expert Committee on National Parks, Wildlife Sanctuaries, and Policy for Wildlife Conservation (IBWL–Expert Committee 1970). Some Committee members attended the 1969 General Assembly meeting and had several of IUCN's recommendations incorporated in the report. They were included because Committee members deemed them relevant to India's policy goals and thought they should be brought forward for consideration by the nation's leadership. Most critical was the adoption of IUCN's definition for national parks (IBWL–Expert Committee 1970) as a guideline for India. In accordance with the IUCN definition, the Committee concluded that the islands of inviolability (sanctums sanctorum) ought to be abandoned and the entire park should be free from "human exploitation or occupation" (IBWL–Expert Committee 1970). The feasibility of this recommendation was questionable because park managers were having a difficult time enforcing the no-use zones (sanctums sanctorum). Expanding the no-use policy to the entire park area would require a significant increase in the number of enforcement staff and fortitude to withstand protest against the policy by resource-dependent communities. In other words, it would take a high profile project to instill sustained commitment to the new approach to national parks. Project Tiger proved to be that scheme.

7.2.2 Project Tiger

At IUCN's 10th General Assembly in 1969, the "precarious situation of the Indian tiger" was considered (IBWL–Task Force 1972). The Assembly adopted a resolution calling for a moratorium on all forms and reasons for tiger hunting until the species' demographic and viability status was known. IUCN recommended to all tiger-range nations that it be given complete protection. The Indian Board for Wildlife, in turn, recommended to all states in the Indian Union that a 5-year moratorium on tiger hunting be adopted. At the behest of Prime Minister Indira Gandhi, there was a near unanimous adoption of the ban by July 1970.

Despite the ban on hunting and efforts to enforce the prohibition against the commercial export of tiger skins, the smuggling, poaching, and destruction of the tiger and its habitat continued. Guy Mountfort, a trustee of the World Wildlife Fund–International (now the World Wide Fund for Nature), "apprised her [Indira Gandhi] of the grave situation of the tiger and concern of the conservationists of the world for saving the species from extinction." One million dollars was pledged by Mountfort for tiger preservation (IBWL–Task Force 1972). Catalyzed by this interest, Indira Gandhi constituted the Tiger Task Force to devise a project for India. The Task Force, in consultation with scientists and preservation advocates from IUCN and WWF–International, created Project Tiger. The project was defined by the following:

- Saving relevant patches of habitat to provide the tiger with a place to live. These reserves would serve as "a breeding nucleus from which surplus animals could migrate to surrounding forests" (IBWL–Task Force 1972).
- Creating and maintaining a viable population of tigers through the improvement of the "biotope and stimulation of its diversity, according to sound principles of conservation" (IBWL–Task Force 1972). To answer the pragmatic question of the number of tigers required to maintain a viable population, Paul Leyhausen, Ph.D., advisor to IUCN, theorized that 300 tigers were needed in a contiguous population. Additionally, Leyhausen speculated that this would require protected areas of approximately 2000 km^2 (IBWL–Task Force 1972).
- Removing all villages from core areas of the reserves.
- Allowing tourism and management-oriented research in the reserves. All "other forms of human disturbance, such as commercial fellings, collection of minor forest produce, mining, excessive traffic, heavy grazing by domestic livestock... must be phased out for complete elimination" (IBWL–Task Force 1972).

The Tiger Task Force found few areas of 2000 km^2 that could realistically be set aside for tiger preservation. Deviating from Leyhausen's recommendations, the Task Force decided it would be "better to concentrate efforts to administer small units as model parks to preserve the tiger" (IBWL–Task Force 1972).

On 1 April 1973 Project Tiger was christened. In the Prime Minister's inaugural address, the philosophy and intent of the project was broadcast:

> [T]he tiger cannot be preserved in isolation. It is at the apex of a large and complex biotope. Its habitats, threatened by human intrusion, commercial forestry, cattle grazing must first be made inviolate. Forestry practices designed to squeeze the last rupee out of our jungles, must be radically re-oriented at least within our National Parks and Sanctuaries, and pre-eminently in our tiger reserves. The narrow outlook of the accountant must give way to a wider vision of the recreational, educational, and ecological values of totally undisturbed areas of wilderness.

Leading up to this moment had been a series of decisions on reserve siting, management regimes, and the type of relationship that the reserves would have with adjacent communities. It is at this stage of project planning that the guidelines would have proved most useful.

7.3 ESA's Guidelines and Project Tiger: A Difference That Would Have Made a Difference?

The wilderness ideal that drove and sustained the national park movement did not provide guidance for management or a convincing argument for the virtue of preservation to those who had not had a conversion experience (Budiansky 1995). Because the wilderness ideal has its roots so deeply embedded in an antipathy toward people, where the "world of non-human life" becomes a "sanctuary" (Budiansky 1995), the ideal proved "increasingly hard to reconcile not only with an ever less romantic and more crowded world, but with the realistic tasks of park acquisition and park management" (Nicholson 1974). Operating at the level of the subconscious is the belief that humans are intruders in nature, and it is this belief that undergirds ecology, and national park policy and management (Worster 1994; Budiansky 1995).

ESA's guidelines are not entirely neutral on this issue (Dale et al., Chapter 1). The guidelines recommend a set of constraints on human activity to maintain the integrity of an ecosystem. Certainly, to ensure sustainability this is necessary. The guidelines are also clear about the value of human activity. Human use of the environment is akin to an avalanche or fire—creators of disturbance, degraded habitat, and perpetrators of extinction. Humans are rarely praised for being the creators of diverse habitat, and as a consequence, sustainers of a diversity of species and the creators of landscapes we now call "natural" (Worster 1994). The following discussion demonstrates that even when we believe we are talking about nature or managing nature without humans, such subtle and unstated beliefs about what is natural and nonnatural impact management and planning decisions with consequences that range more broadly than typically anticipated. ESA's guideline to place local decisions into a regional context holds the most promise for focusing planning deliberation on considerations of the impact on human communities. There is little disagreement in India that hope for maintaining a "wild nature" now rests primarily on the shoulders of those who had previously been victimized by the policy to preserve it (Kothari et al. 1996).

7.3.1 Species and Landscape Principles: Preserve Rare Landscape Elements and Associated Species

Dale et al. (Chapter 1) recommend that to preserve indicator species and ecological processes that encompass both organic and inorganic components, large contiguous or connected areas should be preserved; national parks are a highly utilized means to that end. Given that the goal was tiger preservation, planners needed to identify critical tiger habitat and suggest sufficiently large areas for protection. Natural science is fully qualified to advise and create a plan to accomplish this goal. However, Project Tiger illustrates that choosing reserve sites, delineating reserve boundaries, and devising management regimes that will ensure the tiger's long-term viability are a complex mix of natural science, economics, and global and national politics. Arguably, an inclusion of these considerations into what would be a relatively straightforward application of the guideline is justified.

Tigers are well adapted to a wide range of habitat types and occur throughout the Indian subcontinent. Given tremendous fiscal constraints with a human population of over 500 million with more than one-third living in rural and forested areas (Census of India 1998), the Government of India, in 1972, could not feasibly protect all of the tiger's range. In that light, the Task Force chose to protect a limited number of sites that had a high probability of containing a near viable number of tigers in a contiguous population. To identify these sites and populations, tigers needed to be located and counted. As wildlife biologists are fully aware, finding and enumerating solitary mammals is extremely difficult and this has become one of the most politicized and contentious aspects of Project Tiger.

In the early 1970s, wildlife experts and policymakers in India and IUCN deliberated on the best method for counting tigers. India's leadership decided to use the "pug-mark" and "tiger tracer" method against the recommendation of IUCN's experts. Inadvertently, the pug-mark method became a showcase of the competency of India's experts. Taken from personal correspondence between Field Directors in India: "I'm so happy that we are at last giving it [tracer method] a trial. Entirely Indian methods with no foreign aid or so called foreign expertise."

The pug-mark method rests on the assumption that during the hot season ("pinch period") a tiger will need to visit one of the relatively scarce water sources at least once in a 48-hour period. At the water sources, census takers erase all animal tracks and then return within 24 hours. Upon finding tiger tracks, a framed glass platform is placed over the pug-mark and its outline is traced onto the glass—the tiger tracer. The outline is transferred onto a sheet of paper with the following information: the path of the tiger, its size (determined by the distance between pug marks), its gender (determined by the pug-mark shape), and whether the dam was with cubs. Census takers also traverse the forest looking for pug-marks that might denote the presence of tigers that had not visited one of the water holes. The tracings and the identifying information are consolidated and analyzed for duplication at the field office. They are then sent for further analysis to the Chief Wildlife Warden of the state who sends the final tally to the Project Tiger Directorate in New Delhi (GOI n.d.).

There are numerous objections to this method (Kumat 1992). Receiving the most criticism is the additional assumption that pug-marks are unique to each tiger and that the characteristics that make the track unique to the tiger can be captured by the tiger-tracer method. Those not fully enamored with the method have argued that pug-marks will differ in shape and size, depending on the ground's surface-type (i.e., hard ground vs. sand vs. mud), and, as a result, tigers may be missed or double counted. Another concern is that there may be a characteristic in the tiger's gait that obliterates the distinguishing features of the pug-mark (i.e., dragging sand over the print or double-stepping into the track). Alternative methods, such as the camera trap or transect methods, too, have their detractors both in and outside of India (Tilson and Seal 1987; Seidensticker et al. 1999).

Despite the controversy, the first all-India tiger census was conducted and concluded in 1972. As expected, tigers were found throughout India and the Task Force was faced with the difficult position of having to choose where to site the reserves. To assist in this decision, the Task Force relied on Resolution 25, adopted at the 1969 IUCN General Assembly meeting. Resolution 25 recommended that nations design protected area networks in such a manner that "representative types of natural vegetation" were preserved. This narrowed the options, because the Task Force made every effort to site reserves in a variety of ecosystems, not duplicating any.

To distribute the financial burden of Project Tiger among tiger-range states, the Task Force additionally decided that no state should have more than one reserve. States were also required by the Task Force to make a convincing case that they wanted to be included in the scheme. Enthusiasm for the project became a criterion because IUCN and the World Wide Fund for Nature–International had, by this time, drawn global attention to India's efforts and a failure or poorly implemented project would create a negative image for India. In keeping with these criteria, nine reserves were eventually chosen (Fig. 7.1).

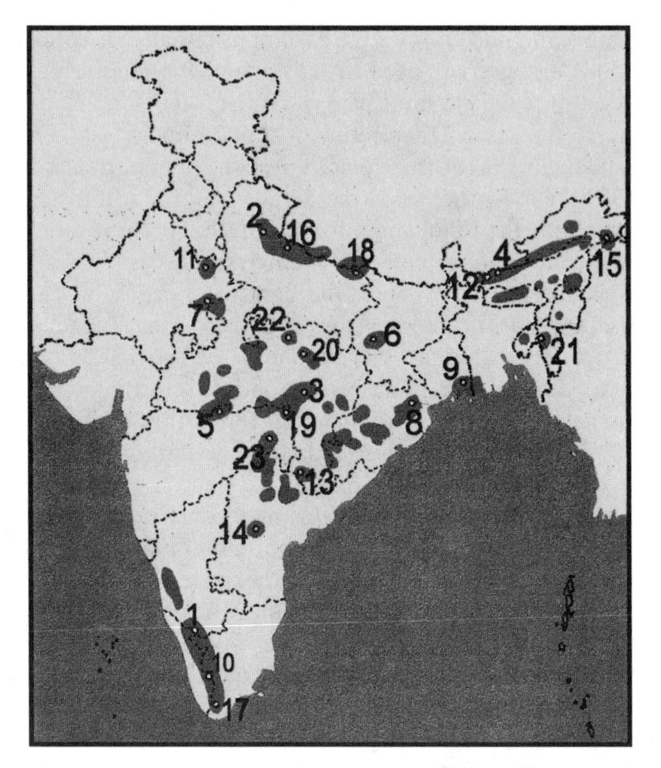

FIGURE 7.1. Map of India showing tiger ranges and Project Tiger reserves as of 1997 (WWF–India 1996); reserves in ALL CAPS are the 9 original declared in 1972:

Declared in 1972
1. BANDIPUR, Mysore
2. CORBETT, Uttar Pradesh
3. KANHA, Madhya Pradesh
4. MANAS, Assam
5. MELGHAT, Maharashtra
6. PALAMAU, Bihar
7. RANTHAMBHORE, Rajasthan
8. SIMLIPAL, Orissa
9. SUNDERBANS

Declared in 1978
10. Periyar, Kerala
11. Sariska, Rajasthan

Declared in 1982
12. Buxa, West Bengal
13. Indravati, Madhya Pradesh
14. Nagarjunasagar, Andhra Pradesh
15. Namdapha, Arunachal Pradesh

Declared in 1988
16. Dudhwa, Uttar Pradesh
17. Kalakad-Mudanthurai, Tamil Nadu
18. Valmiki, Bihar

Declared in early to mid 1990s
19. Pench, Madhya Pradesh
20. Bandhavgarh, Madhya Pradesh
21. Dampha, Mizoram
22. Panna, Madhya Pradesh
23. Tadoba, Maharashtra

Dispersed reserve siting served another purpose. India is a country divided by religion, language, and caste; Prime Minister Indira Gandhi saw unifying potential in Project Tiger:

> [T]he Tiger Reserves are situated in eight different States, in different climates, in all the four corners of the country from Assam to Rajasthan and Uttar Pradesh to Mysore. When developed they will function as focal points of home and foreign tourism, thus contributing towards the emotional integration of the country.
>
> (IBWL–Task Force 1972)

The guideline to preserve large and contiguous areas was not feasible in this context. Nor could the guideline authors have anticipated the political forces brought to bear on India's leadership. Given these constraints, did the planning process, as it stood, produce a successful project?

The 1972 census showed that there were 258 tigers residing in Project Tiger reserves (Table 7.1). The all-India census logged 1856 tigers (Table 7.2). Project Tiger, in its initial phase, was providing upgraded protection to approximately 14% of India's tigers. As the protected area network expanded in size and range (see Fig. 7.1), a greater percentage of tigers were provided upgraded protection. In 1994, approximately 33,000 km^2 of habitat or 5% of India's total forest area were designated as Project Tiger reserves. The 1997 census found 1458 tigers in the reserves out of 3235 tigers—approximately 45% of the total population in India.

Interpreting these figures as a success or failure is much like the proverbial dilemma of choosing whether the glass is half empty or half full. The fear expressed by tiger advocates is that tigers outside specialized reserves are extremely vulnerable to poachers and greater variance in habitat management. However, India does have an extensive system of national parks and sanctuaries—as of 1995 there were 80 national parks and 441 wildlife sanctuaries covering over 4% of India's total land area (GOI 1996). It is highly probable that a good number of tigers not living in tiger reserves are scattered throughout other protected area categories, but precise figures are not readily available. The degree of protection and habitat management for wildlife varies widely among reserves and tiger preservation efforts will vary. Because of their international stature, tigers living in Project Tiger areas are arguably the best protected and cared for in India.

ESA's guideline to preserve "rare landscape elements" was certainly relevant and applied, albeit, in a restricted form. Revealed by this case study is what many ecologists already know, that science alone is not equipped to answer the political and fiscal questions faced by a nation's leadership in assigning priority to a project and ascertaining that project's scope and chances for success. Planning guidelines must consider the inevitable forces brought to bear upon decision makers in the face of competing interests, institutional limitations, and scientific uncertainty.

TABLE 7.1. Number of tigers in Project Tiger reserves.

Reserve	State	Number of tigers								
		1972	1975	1977	1979	1984	1989	1993	1995	1997
Bandipur	Karnataka	10	19	26	39	53	50	66	74	75
Corbett	Uttar Pradesh	44	55	73	84	90	91	123	128	138
Kanha	Madhya Pradesh	43	48	55	71	109	97	100	97	114
Manas	Assam	31	41	105	69	123	92	81	94	125
Melghat	Maharashtra	27	32	57	63	80	77	72	71	73
Palamau	Bihar	22	30	33	37	62	55	44	47	44
Ranthambhore	Rajasthan	14	20	22	25	38	44	36	38	32
Similipal	Orissa	17	50	60	65	71	93	95	97	98
Sunderbans	West Bengal	50	—	181	205	264	269	251	242	263
Periyar	Kerala	—	—	—	34	44	45	30	39	—
Sariska	Rajasthan	—	—	—	19	26	19	24	25	24
Buxa	West Bengal	—	—	—	—	15	33	29	31	32
Indravati	Madhya Pradesh	—	—	—	—	38	28	18	15	15
Nagarjunasagar	Andhra pradesh	—	—	—	—	65	94	44	34	39
Namdapha	Arunachal Pradesh	—	—	—	—	43	47	47	52	57
Dudhwa	Uttar Pradesh	—	—	—	—	80	90	94	98	104
Kalakad-Mundanthurai	Tamil Nadu	—	—	—	—	20	22	17	16	28
Valmiki	Bihar	—	—	—	—	—	81	49	—	53
Pench	Madhya Pradesh	—	—	—	—	—	—	39	27	29
Tadoba-Andheri	Maharashtra	—	—	—	—	—	—	34	36	42
Bandhavgarh	Madhya Pradesh	—	—	—	—	—	—	41	46	46
Dampha	Mizoram	—	—	—	—	—	—	7	4	5
Panna	Madhya Pradesh	—	—	—	—	—	—	—	22	22
Total		**258**	**295**	**612**	**711**	**1221**	**1327**	**1341**	**1333**	**1458**

Source: Project Tiger Directorate.

TABLE 7.2. Number of tigers in India broken down state wise.

State	Number of tigers			
	1972	1989	1993	1997
Andhra Pradesh	35	235	197	171
Arunachal Pradesh	69	135	180	—
Assam	147	376	325	458
Bihar	85	157	137	103
Gujarat	8	9	5	1
Karnataka	102	257	305	350
Kerala	60	45	57	—
Manipur	1	31	—	—
Madhya Pradesh	457	985	912	927
Meghalaya	32	34	53	—
Maharashtra	160	417	276	257
Mizoram	—	18	28	12
Mysore	102	—	—	—
Nagaland	80	104	83	—
Orissa	142	243	226	—
Rajasthan	74	99	64	58
Tamil Nadu	33	95	97	62
Tripura	7	—	—	—
Uttar Pradesh	262	735	465	475
West Bengal	73	353	335	361
Total	**1856**	**4328**	**3745**	**3235**

Source: Project Tiger Directorate.

7.3.2 Disturbance and Time Principles: Plan for Long-Term Change and Unexpected Events

Disturbance and time principles draw attention to the long-held finding that natural systems are not static, nor do the components making up a system change at a uniform pace, which makes management an incremental process of perturbation, observation, mitigation. National park managers largely have incorporated the principles of change and disturbance into their thinking about landscape management. However, as previously discussed, ESA's guidelines derive from ecological principles that relate primarily to a nature that does not incorporate a view of human activity as anything other than a disturbance to which natural systems respond and overcome. Consequently, management's task is to ensure or reestablish "natural" ecosystem functioning, which requires that all human activity be prohibited or severely curtailed. Presented here is a challenge to the assumption that all human activity is antithetical to the goal of ecosystem integrity and functioning.

Ecology, although more equipped now than in the early stages of Project Tiger, could not advise or project the precise effects of tribal and rural villager's subsistence use of the land. Authors of the Project Tiger planning document admitted this fact, made a brief assessment of the benefits and costs of allowing human occupation and continued use of the resources, and, in the face of uncertainty, decided to continue on course with the inviolate national park model (IBWL–Task Force 1972). Project Tiger Field Directors were advised to make every effort to enforce the prohibitions and to remove all human inhabitants. The Director of Project Tiger, Kailash Sankhala, believed that once this disturbance variable was removed, nature would be free to return to its course. However, India's Field Directors would soon share with Africa's managers the revelation that "strict protection has, paradoxically, ended in an extensive modification of the very environment which it was desired to preserve intact" (Bourliere 1962).

> There is scarcely any doubt that the survival of certain animal species is connected indirectly, or even directly, with human intervention. ... Biologists are not wanting who believe that certain open botanical formations in Africa, as well as in other continents, were established by and, chiefly, owe their continuance to their being regularly overrun by agricultural fires. ... The fact remains that these 'open' areas, which are due partly to fires, correspond to a "T" with the areas inhabited by the larger ungulates in Africa.
>
> (Verschuren 1962)

The case of the barasingha in Kanha Tiger Reserve provides an excellent illustration of this issue. Taken from the Project Tiger pamphlet, Kanha is described as having "breakthtaking views" that "present large, sparsely wooded grassy plateaus, sprawling slopes covered with a dense miscellany of tall trees and bamboo, and finally the lush green sal extending in stately groves from the lower slopes onto the rolling meadows in the valleys" (GOI 1990). The meadows of Kanha, mentioned twice in this brief description, are its unique contribution to the Project Tiger network. These meadows are home to a remnant population of hard-ground barasingha, a stately antlered ungulate. The barasingha is a grazer, dependent on tall-grass meadows. Medium- to light-intensity human use is largely responsible for these habitats: shifting cultivation, cyclic desertion of village sites, annual burning, and rotational large-scale cattle grazing.

Upon the declaration of Kanha as a Project Tiger reserve in 1972, the relocation of villages and the prohibition against subsistence use of the area commenced. The tall-grass meadows reemerged in the deserted sites. However, because burning and grazing were also prohibited, the meadows began to succumb to a rapid colonization of woody species. Park managers faced a difficult decision; if they allowed encroachment to continue, the sal forest would rapidly overrun the meadows, adding to the "vastness of domineering sal expanse at the cost of near total loss of habitat diversity and a sure extinction of the barasingha" (Kotwal and Parihar n.d.). The decision was made in favor of the barasingha and diversity; managers began experimenting with burn regimes to maintain the tall-grass meadows.

More unexpected than this complex ecosystem response was the reaction of blackbuck to these efforts. The blackbuck is a gloriously colored long-horn deer that prefers short-grass meadows. Short-grass meadows are the result of relatively intensive human activity: clear-fallow cultivation and heavy livestock grazing. Prohibiting use and promoting the growth of long-grass meadows diminished the amount of habitat favorable to the blackbuck. The decline in their population was rapid. A choice was required.

Kanha's Field Director, H.S. Panwar, employed a criterion of "relative endangerment" to determine which species to favor. Kanha's long-grass meadows were the last and only site in India that featured the hard-ground barasingha; the blackbuck, however, was relatively abundant in other areas (e.g., Rajasthan) despite its rarity in Kanha. The choice was made to maintain the central meadows for the barasingha while a smaller expanse of short-grass meadow on an upper plateau is managed for the blackbuck. The blackbuck's population in Kanha has remained relatively stable, hovering in the 30s, whereas the barasingha now number well into the 500s (Kotwal and Parihar n.d.).

It is generally agreed that judicious manipulation of habitat and species may be required to maintain the specific natural value for which a protected area was declared. Because of what we now know to be complex interrelations and dynamics operating at the ecosystem level, the *precise* nature of change upon the enforcement of an inviolate protected area scheme could not have been anticipated. Contributing to that uncertainty was the belief that human activity was always a negative force in the environment. This belief blinded park mangers to the established complex dependencies between human activity, wild species, and ecosystem creation and maintenance. Ecologists have only recently begun interrogating these assumptions and are devising research projects aimed at investigating ecosystem dynamics reliant on human intervention and activity. All to say that the merits of the edict to plan for long-term change are crucial to park management. In addition, a challenge is made to reassess the effects of subsistence use on ecosystems with an eye to the possibility that it has become an integral aspect of the landscape.

7.3.3 Landscape Principle: Examine the Impacts of Decisions in a Regional Context

As the previous section illustrated, nature comprises complex interdependencies and the assortment of species inhabiting a specific area is likely the result of larger-scale processes and attributes of the landscape. When making a land-use decision that will affect a local area, planners are advised by the ESA guidelines to expand considerations to a regional scale (Dale et al., Chapter 1). Here I argue that they must also simultaneously integrate with ecological considerations, considerations of the social, economic, and political context. In effect, decisions should note the presence of two types of relationships: physical and human (Garrat 1983). Park managers concur:

> They [protected areas] cannot be seen as islands which exist in isolation from their surroundings. They are important parts of the regions in which they are situated, and the mutual relationships and linkages between them and adjacent land must be understood and applied to management.
>
> (Garrat 1983)

It was only after failed efforts to implement a protected area scheme that neglected to consider the regional impact of park management and planning decisions did the human component of land-use planning for protected areas become a concern equivalent to that of saving habitat and species. Project Tiger is an apt example of what has become a global trend.

When India's Tiger Task Force, in 1972, were conducting surveys of potential reserve sites, they took an inventory of the services provided by each reserve to occupying and adjacent communities. They also noted the existence of potential threats to the reserves, such as industrial development, roads, chances for disease transfer from the domestic to wild ungulate populations, and cultural practices (Table 7.3). With the exception of Bandipur Reserve, every site was found to have a high probability of conflict between stakeholders upon the Project's implementation. However, the commitment to an inviolate reserve model was so strong that these indicators were ignored. Had the valuation of human activity been determined as anything other than damaging by definition, had the commitment to inviolability not been so strong, and had the regional implications of the project been considered, it is likely that at least the following questions would have arisen during the planning process. Better still would have been a planning phase that reconfigured the project to mitigate the problems being signaled.

- *Village relocation:* How resistant are the communities to relocating? Are there equivalent sites for relocation? Who would be carrying out the relocation efforts? Are adequate finances promised and fixed in both state and central government budgets?
- *Resource use prohibition:* Where else can villagers go to meet their basic resource needs? Would those alternatives be at an equivalent cost? Would there be equivalent access? If not, are compensation measures planned? Are adequate finances promised and fixed in both state and central government budgets?

Because these questions were not addressed in a systematic manner, the crucial links between the Forestry Department, District Revenue Office, Project Tiger staff, and tribal development offices were not established with any regularity. There is, however, historical precedent for the systematic exclusion of considerations for tribals and rural villagers' resource limitations and liabilities to ensure favor for state-sanctioned schemes.

TABLE 7.3. 1972 Tiger Task Force inventory of land-use pressure on potential Project Tiger reserves and the surrounding "buffer zone."

Reserve, State	Number of villages	Number of villagers	Livestock population	Other forms of land-use
Manas, Assam	154	6100	50,000	26,000 hectares of cultivated land next to reserve.
Palamau, Bihar	136	30,000	50,000	4-year cycle of bamboo harvest in reserve.
Simlipal, Orissa	70	Not Given	Not a Problem	Hunting by tribals.
Corbett, Uttar Pradesh	None	None	Yes. No number provided.	Forests worked on selective felling system. Reserve experiences poaching pressure.
Ranthambore, Rajasthan	Not Given	Not Given	1400 buffaloes, 2600 cows, 700 bulls, 8000 goats and sheep, 200 camels	Reserve experiences poaching pressure.
Kanha, Madhya Pradesh	67	12,300	19,900	High land use in adjoining areas. Surrounded by tribals interested in hunting and poaching is a problem.
Melghat, Maharashtra	58	12,000	12,000	Shifting villages out of reserve would be particularly difficult. Teak forests in and near reserve are intensively worked.
Bandipur, Nagarhole	Not given	Not given	Grazing prohibited.	Migratory cattle pass diseases to ungulate population. Reserve experiences poaching pressure.

Source: IBWL–Task Force (1972).

Over the greater part of India's forest policy history, forests were valued primarily for their commercial utility: ship building, railroads and rail cars, fuel, buildings, and foreign revenue. This approach was dominant when Project Tiger was first implemented and was additionally strengthened in the 1976 report of the National Commission on Agriculture, which laid out its plan for forest development. The Commission concluded that the *raison d'etre* of India's forests was the production of industrial wood. As for the rural villagers' need of fodder, housing timber, fruit, medicial herbs, and other forest products,

> no special programme were taken which could directly contribute to the upliftment [sic] of the tribal economy. The programmes executed were essentially forest development programmes which benefited the tribals only indirectly ... through wage earning opportunities.
>
> (Saxena 1995)

The recommended "social forestry projects" were developed, primarily, to "lighten the burden on production forestry" (Saxena 1995). In other words, the projects were intended to develop nonforested areas and shift villagers' resource demands to them. Initially, these programs received very little budgetary support, and, subsequently, were not a great success.

This same strategy is found in the Project Tiger planning document. Approximately 90% of the budget was slated for job creation; many of the jobs were intended as incentives for tribals to adhere to the new prohibitions. In reality, Project Tiger Field Directors were not given this type of discretionary funding nor did they have control over the reserve's buffer zones where most villages would be moved or where social forestry projects would be developed. As a result, the success of enforcing the inviolate national park scheme varied dramatically and most reserves were not successful for the previously mentioned reasons, because acceptable alternative sites for villages did not exist, or because villagers could not be convinced to move. Consequently, most sanctuaries and national parks in India are overrun by villagers who have few, if any, options for meeting their basic needs (Fig. 7.2). As of 1996, 63% of Project Tiger reserves had villages in the core area (GOI 1996).

As might be expected, given this tremendous use pressure, "conflicts over the use and control of natural resources become law-and-order problems

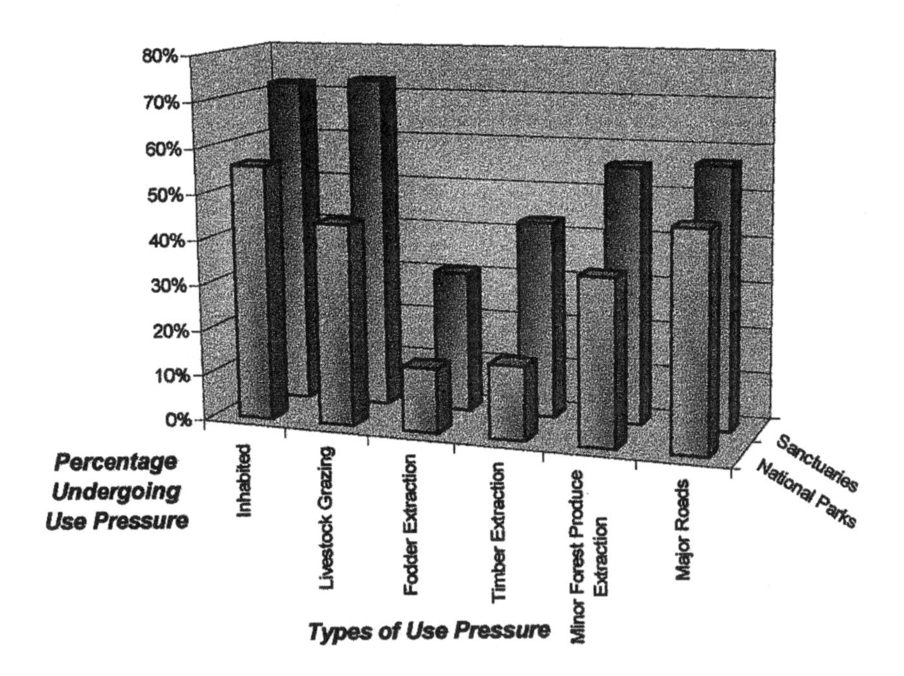

FIGURE 7.2. Percentage of national parks and sanctuaries undergoing designated use pressure for the years 1976–1984 (Kothari et al. 1989).

and result in physical confrontations between the people and the authorities" (Kothari et al. 1989). In a survey conducted in the early 1980s by the Indian Institute of Public Administration, 37% of national parks and 18% of sanctuaries reported physical confrontation between local people and park officials (Kothari et al. 1989). All indicators point to an increase in the number and ferocity of these conflicts. In Uttar Pradesh alone, over the past decade 53 Forest Department staff have been killed by what are euphemistically called the "forest mafia," those involved with illegal timber felling. In one small region of Uttar Pradesh, the Hardiwar area, over a 3-year period (approximately 1993–1996) there have been 15 separate attacks on forest staff (*Tiger Link News,* December 1999). These clashes occur most often when forest staff attempt to enforce the prohibitions against the felling of trees, poaching, and grazing. The number and types of infractions recorded from 1976 through 1984 for national parks and sanctuaries are provided in Figure 7.3.

Project Tiger staff and the government of India hope that new plans for developing alternative resource sites and devising sustainable income-

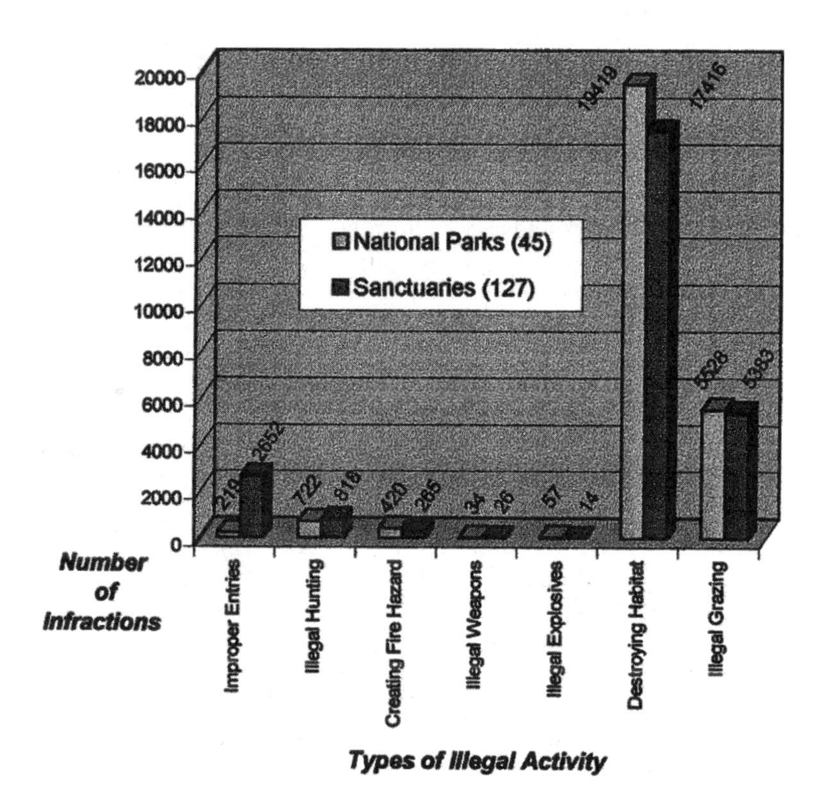

FIGURE 7.3. Number of infractions filed by Forest Officers for the years 1976–1984 by type of illegal activity (Kothari et al. 1989).

generating projects in reserve buffer zones will bring about two positive trends. First, they hope that the strained or severed relations between reserve staff and villagers can begin to mend as both communities work together on these projects. Second, if villagers witness reserves playing a positive role in their daily life, staff are hoping that villagers will take an interest in protecting the area and will identify known poachers, will not poach, or, minimally, will not harbor poachers, as has been the case in the past. Currently, approximately 81% of the Project Tiger reserves dedicate funds to these types of eco-development projects (GOI 1996). It is generally agreed by Field Directors that it is too soon "to comment on tangible effects" (GOI 1996). The ESA guideline to consider the regional implications of a local land-use decision is clearly relevant provided that it includes systematic consideration of the economic, social, and political context in which decisions are being rendered.

7.3.4 Landscape and Time Principles: What Do You Do If You Are Successful?

The primary goal of Project Tiger is the recovery of the tiger population. Over and again, Project Tiger reserves were described as "nurseries" that would protect the vulnerable population until the time that "excess" tigers could spill out of them and into the surrounding areas, not unlike the experience in the United States with bears, moose, elk, wolves, and mountain lions. Combining the guidelines to plan for long-term change and view local decisions in a landscape context, planners should have asked whether tigers spilling out into the surrounds is a politically, socially, and culturally viable goal. If the project were successful, tiger populations would be expanding, as would the human population. India's population grows at about 2% a year and in the year 2000 is approximately 1 billion. More problematic is that only 29% of the population is living in urban areas (Census of India 1999). Where will the tigers go?

It can be expected that as both tiger and human populations expand, they will increasingly come into conflict. Unlike the time when villagers were allowed to kill tigers in defense of their property and lives, tigers are now protected. India adopted, in 1972, broad-sweeping wildlife-protection legislation. In the Wildlife (Protection) Act, 1972 (GOI 1972), a person is allowed to kill an "endangered" animal in self-defense provided the person is not in violation of any other law, such as illegally entering a reserve, grazing cattle, or gathering forest produce (Wildlife (Protection) Act, 1972: Chapter III, Para. 11.2). Animals on less restricted lists (i.e., those deemed threatened or rare) can be killed if they pose a threat to life or property (Wildlife (Protection) Act, 1972: Chapter III, Para. 11.1b). The penalties for killing an endangered species are relatively steep if one is a tribal; they are infinitesimal if one is a well-connected poacher. The penalty for killing a Schedule I species (i.e., an endangered animal) is imprisonment for a term not less than

1 year up to 6 years, and a fine not less than Rs. 5000 with no upper limit (Wildlife (Protection) Act, 1972 (1991): Chapter VI, Para. 51.1).

How critical has the human–tiger conflict become? As imagined, reliable statistics of tiger attacks on domestic livestock and humans are difficult to obtain, not because they are not available, but because several dynamics work to either inflate or deflate them. First, unless compensation by the government is offered there is no incentive to report the incident. Second, in the case where compensation is offered, it is provided only when the villager is not in violation of any other law. As might be expected, some attacks go unreported (Kothari et al. 1989). Third, it is to the benefit of policymakers to underreport the negative aspects of the project and, concurrently, to the benefit of tiger preservation advocates to not ferret out any discrepancies between the actual position as opposed to what is reported to avoid provoking challenges to the project. Finally, when compensation is available, there is an incentive to claim that a death was the result of a tiger attack when in fact it was not. Although park level data differentiates between the total number of claims and those claims that were rejected after an investigation, unless specified, it is impossible to tell whether quoted figures of attacks are pre- or postinvestigation. With that said, the following figures are illustrative of the relationship between humans and tigers in and around protected areas.

In the Sunderbans, West Bengal, although a unique circumstance because tigers are noted man-eaters there, from 1975 to 1993 approximately 45 men per year were killed (Jordan 1998). In the survey of national parks conducted by Kothari et al. (1989), they found that out of the 39 national parks and 167 sanctuaries responding to the survey, between 1976 and 1984 there were 629 attacks on humans and tigers were responsible for 221 of those (190 in the Sunderbans alone) (Kothari et al. 1989). Of the 629 attacks, 485 (77%) were fatal. Not all states provide compensation for attacks, but for those that do, amounts vary from Rs. 10,000 to Rs. 200 for fatalities (The exchange rate is approximately Rs. 38 to U.S. $1 as of 1998). Because of the unique situation in the Sunderbans, the West Bengal government pays family members Rs. 20,000 for a death caused by a tiger (Jordan 1998).

Attacks on domestic livestock are even more problematic to track. Project Tiger does pay compensation for cattle depredation. However, statistics reported at the national level are provided at rupees paid, not the number of livestock killed. The value assigned to an animal will vary dramatically from incident to incident, and precise figures as to the number of livestock killed are almost impossible to discern. However, some Field Directors do keep these records and a general picture of the problem can be inferred (Table 7.4).

Villagers are in a double bind: Animal populations are increasing, the number of clashes between wildlife and people appear to be on the increase as well, and laws restricting villagers' ability to defend themselves and their property are becoming more stringent, which "have tilted the odds against

TABLE 7.4. Attacks on cattle and compensation paid for Corbett Tiger Reserve, Uttar Pradesh, and Kanha Tiger Reserve, Madhya Pradesh.

	Corbett National Park, Uttar Pradesh		Kanha National Park, Madhya Pradesh	
	Cattle attacks reported	Rupees paid	Cattle attacks reported	Rupees paid
1983–1984	83	63,350	49	33,475
1984–1985	45	41,050	100	68,100
1987–1988			112	71,162
1998–1999	1050[a]			

[a]This is the number of lost cattle reported to the WWF–Tiger Conservation Programme operating in Corbett Tiger Reserve, between January 1998 and September 1999. WWF paid Rs. 1,218,019 (approximately US $29,000) for 1225 kills around 5 reserves. 1050 of those occurred around Corbett (*Tiger Link,* December 1999).

the villagers" (Ashraf 1993). In Corbett Tiger Reserve, Uttar Pradesh, "Unable to protect their crops from the marauding animals, the villagers decided to exercise their next best option: They demanded that the government settle them elsewhere" (Ashraf 1993). For these villagers, despite their centuries old attachment to the land, "today it is the question of our survival. Between hunger and emotion, the choice is clear" (Ashraf 1993).

The question to ask is whether the integrity of India's protected areas and the legitimacy of efforts to save species that directly challenge human livelihood can be maintained over time. Ullas Karanth, a research scientist focusing on the Bengal tiger, concludes: "We must learn to live with the fact that the park's neighbors will always be hostile," and maintaining protected areas will be "primarily a policing job" (Ward 1992). At the other extreme, a growing movement in India is catalyzed around the view that local people must be primarily responsible for the management of protected areas if they are to remain viable sanctuaries for wildlife (Kaushal and Tandon 1993). Regardless of where one stands on the issue, a government cannot lend the impression that it is consistently favoring one population over another and hope to maintain its legitimacy. This must be the concern of tiger advocates, because the voices of those living in and near tiger reserves are speaking loudly: Mahichandra, one of the villagers who asked to be moved out of Corbett:

> Tell me, when we do not even have proper drinking facilities, is it fair that the government should be spending money to build water holes for animals? Why, they even keep salt bricks in the park. Are animals more important than human beings? (Ashraf 1993)

Could attention to the combined issues of long-term change and regional impact have structured relationships and institutions in such a manner that this critical and increasing challenge to Project Tiger been avoided?

The answer is not one that any environmentalist wants to hear. Unless the problems of landscape utilization favoring wild species is not placed within a larger context of encouraging humans to move to cities, increasing the standard of living there, and increasing the productivity of land already used for natural resource exploitation so as to provide space for wild animals, the time may have already come and gone for species like the tiger, elephant, lion, and rhino.

7.4 Conclusion

Perhaps as the result of our creation and efficient utilization of technology, humans are impacting natural systems far beyond what our biomass alone would predict. Reducing our impact is necessary if we hope to sustain not only our own species but those we value for aesthetic, religious, and pragmatic reasons, as well as other reasons not easily categorized. This is the goal of the Ecological Society of America's guidelines for land use and management. What we cannot do, as the case of India's Project Tiger demonstrates, is assume that explanations and the resulting recommendations shaped to fit the conditions of one context are exported, without revision, to another. Human activity, social structure, political institutions, and understandings of nature are widely divergent and one model or approach cannot fit all.

The utility and contributing value of the ESA's guidelines were affirmed, but not without revision. The guidelines were derived from and fully consistent with principles ascribed to nature by ecological science. Ecology has principally been the science of natural systems—"natural" as distinct from "nonnatural," human-impacted nature (Worster 1994). Policy guidelines informed by a science that excludes the human component will have limited applicability and success; having less chance of success if applied to projects in regions or nations with widely divergent human institutions. To increase applicability of the guidelines, increasing the odds of maintaining ecosystem integrity, questions of ideology, politics, economics, and the cultural context of potentially impacted communities must be highlighted at all available points. Second, the guidelines should be broadly construed so as to provide sufficient flexibility to address a diversity of human approaches to nature, institutions, and needs. For this to occur, on some issues of mutual interest, there must be a breaking down of the distinction between "natural" and "human" science.

Acknowledgments. This research was funded by a Doctoral Dissertation Research Improvement Grant from the National Science Foundation's Ethics and Values Studies Program, Grant SBR-9521562. I thank the anonymous reviewers for their comments and suggestions.

References

Arnold, D., and R. Guha. 1995. Nature, culture, imperialism: essays on the environmental history of South Asia. Oxford University Press, Delhi, India.

Ashraf, A. 1993. Corbett wildlife, villagers fight over habitat. The Pioneer, New Delhi 4 March.

Badshah, M.A., and C.A.R. Bhadran. 1962. National Parks: their principles and purposes. Pages 24–33 *in* A. Adams, editor. Proceedings of the First World Conference on National Parks. National Park Service, U.S. Department of Interior, Washington, D.C., USA.

Bourliere, F. 1962. Science and parks in the Tropics. Pages 64–68 *in* A. Adams, editor. Proceedings of the First World Conference on National Parks. National Park Service, U.S. Department of Interior. Washington, D.C., USA.

Budiansky, S. 1995. Nature keepers: the new science of nature management. Free Press, New York, New York, USA.

Dhyani, S.N. 1994. Wildlife management. Rawat Publications, New Delhi, India.

Gadgil, M., and R. Guha. 1992. This fissured land: an ecological history of India. Oxford University Press, Delhi, India.

Garratt, K. 1983. The relationship between adjacent lands and protected areas: issues of concern for the protected area manager. *In* J.A. McNeely and K.R. Miller, editors. National parks, conservation, and development: the role of protected areas in sustaining society. Proceedings of the World Congress on National Parks 11–22 October 1982. Smithsonian Institution Press, Washington, D.C., USA.

Government of India. 1996. Recommendations of the committee appointed by the Honorable High Court of Delhi on wildlife preservation, protection and laws.

Guha, R. 1989. The unquiet woods: ecological change and peasant resistance in the Himalaya. Oxford University Press, Delhi, India.

India, Government of (GOI). n.d. Draft scheme for conducting census of tigers in India. Doc. No. F.25/90/70/FD.

India, Government of (GOI). 1972. Wild life (protection) bill, 1972. Bill No. 78, 17 August 1972.

India, Government of (GOI). 1990. Project Tiger 1990. Ministry of Environment and Forests, New Delhi, India.

India, Government of (GOI). 1993. A review of Project Tiger. Ministry of Environment and Forests, New Delhi, India.

India, Government of (GOI). 1996. Recommendations of the Committee Appointed by the Honorable High Court of Delhi on Wildlife Preservation, Protection and Laws. New Delhi, India.

Indian Board for Wildlife–Expert Committee (IBWL). 1970. Wildlife conservation in India: report of the Expert Committee. Forest Research Institute, Dehra Dun, India.

Indian Board for Wildlife–Task Force (IBWL). 1972. Project Tiger: a planning proposal for preservation of tiger (Panthera Tigris tigris). Forest Research Institute, Dehra Dun, India.

International Union for the Conservation of Nature. 1970. IUCN Supplementary Paper No. 27. IUCN, Gland, Switzerland.

International Union for the Conservation of Nature. 1994. Guidelines for protected area management categories. IUCN, Gland, Switzerland.

Jordan, M. 1998. Safety from man-eaters. Christian Science Monitor **90**(62):10.
Kaushal, A., and R. Tandon. 1993. Doon declaration on people and parks. Pages 35–36 *in* Report on National Workshop on declining access to and control over natural resources in national parks and sanctuaries. Forest Research Institute, Dehra Dun, India.
Kothari, A., P. Pande, S. Singh, and D. Variava. 1989. Management of national parks and sanctuaries in India: a status report. Indian Institute of Public Administration, New Delhi, India.
Kothari, A., N. Singh, and S. Suri. 1996. People and protected areas: towards participatory conservation in India. Sage Publications, New Delhi, India.
Kotwal, P.C., and A.S. Parihar. n.d. Management plan of Kanha National Park and Project Tiger Kanha for period 1989–1990 to 1998–1999. Officer of the Conservator and Field Director, Mandla, Madhya Pradesh, India.
Kumat, R.S. 1992. Enquiry into missing tigers. Government of Rajasthan, Jaipur, India.
Nicholson, E.M. 1974. What is wrong with the national park movement? Pages 32–38 *in* Sir Hugh Elliott, Bt. editor. Second World Conference on National Parks. IUCN, Morges, Switzerland.
Panjwani, R. 1994. "Evolution of wildlife laws in India." *In* The Wildlife (Protection) Act, 1972 (as amended up to 1991). Natraj Publishers, Dehra Dun, India.
Putnam, J.J. 1976. India struggles to save her wildlife. Pages 299–343 *in* National Geographic, September 1976.
Raghavan, S. 1967. Wild life management trends in India. Government of India Press, New Delhi, India.
Ribbentrop, B. 1900. Forestry in British India. Indus Publishing Company, New Delhi, India.
Saxena, N.C. 1995. Forests, people & profit: new equations for sustainability. Natraj Publishers, Dehra Dun, India.
Schaller, G.B. 1967. The deer and the tiger: a study of wildlife in India. University of Chicago Press, Chicago, Illinois.
Seidensticker, J., S. Christie, and P. Jackson, editors. 1999. Riding the tiger: tiger conservation in human-dominated landscapes. Cambridge University Press, Cambridge, England.
Singh, B. 1988. Corbett National Park. *In* Indian wildlife: Sri Lanka, Nepal. S. Israel and T. Sinclair, editors. APA Publications, Singapore.
Stracey, P.D. 1960. Wild life management in India. Leaflet No. 3 issued by the Indian Board for Wildlife, Ministry of Food and Agriculture. Government of India Press, New Delhi, India (reprinted in 1966).
Tiger Link. 1999. Cattle compensation. Page 23 in TigerLink News **5**(3) December.
Tilson, R.L. and U.S. Seal. 1987. Tigers of the world: the biology, biopolitics, management, and conservation of an endangered species. Noyes Publications, Park Ridge, New Jersey, USA.
Verschuren, J. 1962. Science and nature reserves. Pages 270–276 *in* A. Adams, editor. Proceedings of the First World Conference on National Parks. National Park Service, U.S. Department of the Interior, Washington, D.C., USA.
Ward, G.C. 1992. India's wildlife dilemma. National Geographic **181**(5):2–28.
Wirth, C. 1962. National parks. Pages 13–21 *in* A. Adams, editor. Proceedings of the First World Conference on National Parks. National Park Service, U.S. Department of the Interior. Washington, D.C., USA.

World Wide Fund for Nature–India. 1996. The tiger call. Dhriti Printers, New Delhi, India.

WRI, IUCN, and UNEP (World Resources Institute, the World Conservation Union, and the United Nations Environment Programme). 1992. Global biodiversity strategy: guidelines for action to save, study, and use earth's biotic wealth sustainability and equitably. IUCN, Gland, Switzerland.

Part III
Alternative Futures

8
Alternative Futures for Monroe County, Pennsylvania: A Case Study in Applying Ecological Principles

CARL STEINITZ and SUSAN MCDOWELL

Several years before the Ecological Society of America's principles and guidelines were published, Monroe County, Pennsylvania, was the subject of a study conducted by researchers from the Harvard University Graduate School of Design in collaboration with representatives of the U.S. Environmental Protection Agency and the county government. That study analyzed the trends of growth in Monroe County, determined the possible effects of that growth, and provided some insight into how that growth might best be managed. The study was conducted from a planning perspective and an ecological perspective, but it reflected a number of ecological concerns. It identified six key issues (geologic landscape, biologic landscape, visual landscape, demographics, economics, and politics) as necessary points of discussion, decision, and action. The research derived six alternative futures for a time 25 years hence. These alternative futures were determined by modeling the results of (1) following the county's comprehensive plan, (2) allowing development to be market-driven, (3) pursuing the strategic development of each township, (4) adopting a policy of land conservation with an emphasis on outdoor recreational opportunities, (5) concentrating new development in a corridor served by public transportation, and (6) conserving all existing undeveloped land. These six possible patterns of future land use reflected a spectrum of natural resource uses over a broad area. Models of the six selected processes of growth and development produced maps of expected development outcomes, allowing the citizens to visualize the consequences of such changes and to progress through a series of decisions in an informed manner. This process allowed decision makers to consider how changes to the environment would affect the future of their county.

Monroe County in northeastern Pennsylvania lies in the "heart of the Poconos." Its beautiful scenery and year-round recreational opportunities have made it an ideal destination for tourists for the past 100 years. Recently, these valuable landscape resources and improved transportation have attracted new residential development, making Monroe County the second-fastest-growing county in Pennsylvania. An estimated 90,000

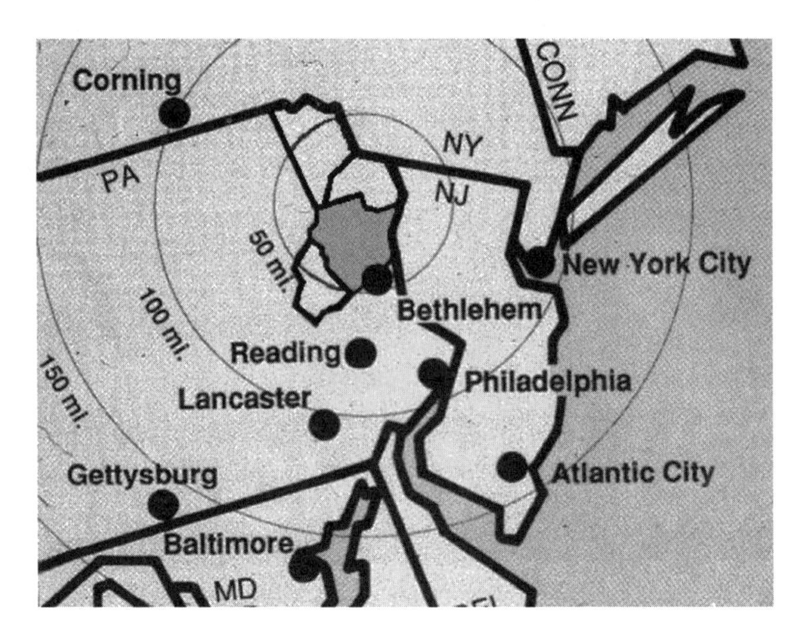

FIGURE 8.1. The Pocono Mountains area.

additional people are expected to locate there by the year 2020, doubling the current population. As a result, Monroe County faces a crisis, the classic dilemma of conservation versus urban development. In addition, New York City and Philadelphia are only 90 miles away, putting 60 million people within a 4-hour drive of the recreational attractions of the area (Fig. 8.1). The question arises, How will historical, current, and future land-use practices influence tomorrow's landscape? The county needed to take a hard look at its future.

Monroe County and the U.S. Environmental Protection Agency (EPA) asked Professor Carl Steinitz of the Harvard University Graduate School of Design to conduct a study that would enable the county and its citizens to visualize the ramifications of its land-use policies, development patterns, and transportation strategies and their subsequent impact on the conservation of its natural diversity and landscape features. The study was also to consider the realities of Pennsylvania's Home Rule, state municipal-planning codes, and a system of local (municipal) government that is entrenched in county and local politics. This political environment will affect future actions.

8.1 Background

The study was based on a framework offered by Steinitz (1990, 1994) that allowed Monroe County to evaluate alternatives to current land-use planning. The subsequent analyses provided a set of options that would help remedy existing policies and practices that are leading to a fragmented landscape, degraded ecological processes, and the loss of important species and networks. The study framework, when combined with ecological principles and guidelines such as those described in Dale et al. (Chapter 1, this volume), can be used by county or municipal planners and by land managers overseeing public or private lands, especially if direct or adjacent development is foreseen. In this chapter, we consider the consistency between the Monroe County analysis and the Ecological Society of America (ESA) principles and guidelines. Although the work by the ESA Land-Use Committee was completed after the Monroe County Study, the two approaches are similar, especially with regard to the landscape context and the focus on preserving rare elements of the ecological system. Indeed, casting the Monroe County case study in the light of the ESA principles and guidelines demonstrates the relevance and applicability of those principles while showing how alternative-future approaches can be used to apply or expand the ESA principles and guidelines. The case study presented here also reviews subsequent actions taken by Monroe County and its municipalities.

8.2 The Study

In October 1993, the study team, comprising graduate students and faculty advisors with interests that spanned landscape architecture, urban planning, urban design, ecology, and law, visited Monroe County to gain firsthand knowledge of the issues surrounding the future development of the area. Six key sets of decision-related issues were identified: geologic landscape, biologic landscape, visual landscape, demographics, economics, and politics. These issues defined the content and methods of the study and formed the basis for evaluating the existing conditions of Monroe County and comparing alternative futures.

8.2.1 The Study Framework

The framework within which this study was organized (shown schematically in Fig. 8.2; Steinitz 1990, 1994) identifies six questions:

1. *How should the state of the landscape be described in terms of content, boundaries, space, and time?* This level of inquiry leads to representation models.
2. *How does the landscape operate, and what are the functional and structural relationships among its elements?* This level of inquiry leads to process models.
3. *Is the current landscape functioning well?* The metrics of judgment (whether ecological health, beauty, cost, nutrient flow, or user satisfaction) lead to evaluation models.

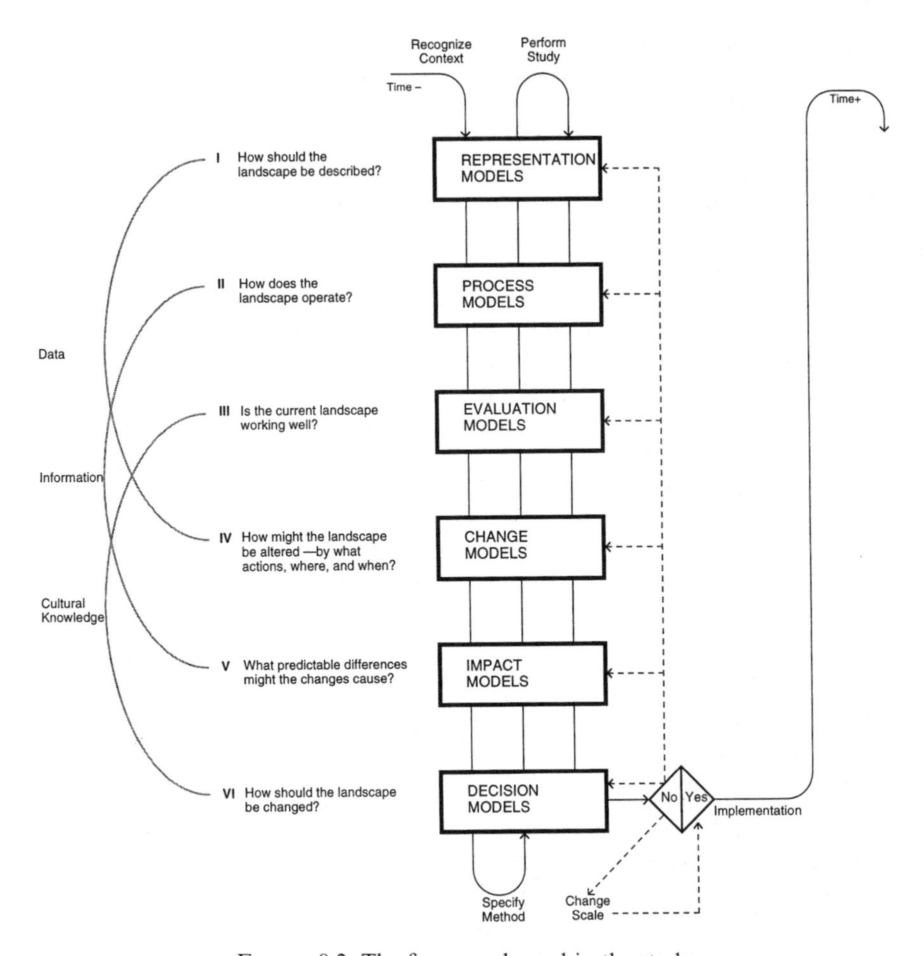

FIGURE 8.2. The framework used in the study.

4. *How might the landscape be altered; by what actions, where, and when might it be altered?* This fourth level of inquiry leads to change models. These changes must be described in the same terms as used in the representation models that evolve from question 1. At least two important types of change should be considered: change by current projected trends and change by implementable design, such as plans, investments, regulations, and construction.
5. *What predictable differences might the changes cause?* This fifth level of inquiry shapes impact models, in which the process models from question 2 are used to simulate change.
6. *Should the landscape be changed, and if so, how is a comparative evaluation among the impacts of alternative changes to be made?* This sixth level of

inquiry leads to decision models and is similar to question 3 in that both are based on knowledge and on cultural values.

Each of these questions is related to a modeling type, which must be based in usable and valid (or presumed to be valid) theory. The six questions are applicable to, but not limited to, the issues encompassed by the ESA's ecological principles. The framework is "passed through" at least three times in any project. The first step is downward in defining the context and scope of a project (defining the questions) by passing through each level of inquiry:

- Representation
- Process
- Evaluation
- Change
- Impact
- Decision

The second step is upward in specifying the project methodology (how to answer the questions). And the third step is downward again in carrying the project forward to its conclusion (getting and providing the answers). Implementation could be considered another level, but this framework considers it as a forward-in-time feedback to the first question, the creation of a changed representation model.

The six levels are presented in the order in which they are normally recognized. However, considering them in reverse order is a more effective way of organizing a landscape-planning study and specifying its method. A project should be organized and specified upward through the levels of inquiry, with the output of each model being defined by the input requirements of the next model above it in the framework:

- To decide about proposing or making a change (or not), one needs to know how to compare alternatives.
- To compare alternatives, one needs to predict their impacts from having simulated changes.
- To simulate change, one needs to specify (or design) the changes to be simulated.
- To specify potential changes (if any), one needs to know the current conditions.
- To evaluate the landscape, one needs to understand how that landscape works.
- To understand how a landscape works, one needs a representational schema to describe it.

At the extreme, two decision choices present themselves: "no" and "yes." A "no" implies a backward feedback loop and the need to alter a prior level. All six levels can be the focus of feedback; "more research" and "redesign of the proposed changes" are frequently applied feedback strategies.

A "contingent yes" decision (still a "no") may also trigger a shift in the scale or size or timing of the study. In a scale shift, the study will again proceed through the six levels of the framework and continue until it achieves a positive ("yes") decision. (A "do not build" conclusion can be a positive decision). A "yes" decision implies implementation, and (one assumes) a forward-in-time change to new representation models.

When repeated and linked over scale and time, the framework may be the organizing basis of a very complex study. Regardless of complexity, the same questions are posed again and again. However, the models, their methods, and their answers vary according to the context in which they are used. While the framework looks orderly and sequential, it frequently is not so in application. The passage through any design project is not a smooth path; it has false starts, dead ends, and serendipitous discoveries. But it does go through all the questions and models of the framework before a "yes" can be achieved.

8.2.2 Representation

To describe Monroe County, a geographic information system (GIS) was prepared. The available data resources included a digital terrain model, interpreted satellite images, the national (digital) highway map, the national (digital) wetlands inventory, infrastructure plans, and maps with field notes showing areas of ecological sensitivity. These uncoordinated data sets were registered to a raster with a spatial resolution of 25 m, and each map of the county has about 5 million sampling points. Monroe County is roughly 30 miles long from north to south and 35 miles across from east to west. From the air, and on the Vegetation Cover Map, which was derived at Cornell University from interpreted satellite imagery, it appears that much of Monroe County is covered by trees. But, a closer look based on other data reveals that the area under much of that tree canopy is actually developed and that little of the undeveloped land is protected from future development.

8.2.3 Key Issues

Six key issues (geologic landscape, biologic landscape, visual landscape, demographics, economics, and politics) were identified as necessary points of discussion, decision, and action. These issues overlap with the ecological principles of time, space, species, landscape, and disturbance; and dealing with these issues required taking many of the principles into account. However, several of the issues considered go beyond the ecological principles and deal with social concerns, such as landscape aesthetics, population, economics, and politics.

8.2.3.1 The Geologic Landscape

Monroe County has been shaped by glaciers. Evidence of this natural process is seen in the county's landforms, soils, wetlands, and unique bogs. These landforms have afforded Monroe County both an environment capable of supporting a wide array of plant and animal species and a high quality of life: good drinking water, diverse scenic quality, and substantial outdoor recreational opportunities. The quality of surface water is dependent on stream-bank vegetation, which absorbs pollutants that would otherwise be carried into the water systems through runoff and errosion. A portion of this buffer zone is regulated by the state, but its management can become ineffective in the face of overuse, development, or unregulated activities.

Because most of the county's drinking water is supplied by underground wells, it is important to actively protect the quality of the groundwater supply. Although the county's groundwater quality is currently rated high, it is extremely vulnerable to contamination in areas of development that are neither constructed on suitable soils nor sewered. Much of this development already exists in Monroe County. The southwestern zone of Monroe County has sizable areas of active agricultural soil with very high productivity. These agricultural areas contribute to the local economy in two ways: directly by the crops they produce and indirectly by attracting tourists to their landscapes. An evaluation of the geologic landscape reveals the most sensitive geologic sites are in the agricultural regions of the southwest.

8.2.3.2 The Biologic Landscape

The special biologic quality of Monroe County is widely recognized. The EPA has identified Monroe County as one of the areas of highest biodiversity within the Middle Atlantic Region of the United States. Biodiversity is defined by the EPA as "the variety and variability among living organisms and the ecological complexes in which they occur." One strategy to keep plants and animals off the Endangered Species List is to preserve areas of existing species richness as a part of landscape planning rather than to engage in crisis management after a species has been identified as threatened. This strategy meshes with the ESA principles and guidelines to preserve rare landscape elements and associated species. The biodiversity evaluation for Monroe County is based on estimates of species richness and was derived from interpreted satellite data of ground cover and a model of habitat capability (Smith and Richmond 1994).

Of special importance to Monroe County is the black bear. The county's bear population is among the most prolific in the nation and boasts the largest average size of the species's individuals in the country. From school mascots, to postcards, to regional advertising campaigns, the bear is Monroe County's image of choice. Because of this, bear habitat, which consists of wetlands and low shrub areas, was included as a special concern in the biologic-landscape analysis. Through working with local experts and using the GIS, all known bear habitats were combined with all similar habitat areas in Monroe County. This analysis resulted in a patchwork habitat pattern, which must be connected in a network of movement corridors. Most of the wetland habitat is regulated, but the linking corridors, which are essential for bear survival, are seriously threatened by development pressure. The resulting pattern identifies key areas needed to maintain the bear's existence in Monroe County. They are typically stream corridors, wetlands, large and diverse habitat areas, and connection corridors for wildlife movement.

Another source of biologic analysis was the Natural Areas Inventory, which was previously prepared by the state and county. For example, the pine woodland on the Pocono Plateau in the northwestern part of the County has been identified as the only mesic pine barren in the world that is partially dry and partially wet. The unique environment that allows this ecosystem to work was also the result of glaciation 10,000 years ago.

8.2.3.3 The Visual Landscape

The county's visual character has, for many years, served as a destination for the tourist trade, and in recent years as an attraction for new residents. And, because most people's understanding of the landscape is through what they see, the visual landscape of Monroe County is an issue of great concern. As part of our study, several areas were identified as being "visually sensitive" and should be protected to maintain the existing and highly valued aesthetic character of the county. Views of scenic landscape elements include lakes, streams, wetlands, steep slopes, and open landscapes, including farmland and grassland.

For most people, the landscape is seen from roads and major public viewing points. Scenic roads in the county, as listed in the many official tourist guides of the area, and view sheds from accessible high elevations and roads along ridge tops were considered especially important.

8.2.3.4 The Demographic Landscape

Accelerating demographic change has occurred in Monroe County since 1890. The population has doubled in the past 20 years and will double again in the next 20. Within the county itself, the population is shifting from the older town centers to a more diffuse and sprawling pattern.

8.2.3.5 The Economic Landscape

Economic factors must be assessed when considering Monroe County's future, for current land value that will change as a function of future policies and investments. The highest-valued land is that near interstate highway access points, adjacent to state and county roads, having sewer service or septic capability, and having important landscape amenities, such as water access or scenic views. These locations are also the areas of greatest recent development.

8.2.3.6 The Political Landscape

William Penn organized Pennsylvania politically around the idea of the township. Today, the townships provide the most immediate level of government, responsible for education, public safety, planning, zoning, water management, sewage treatment, and some roadway construction. While there are obvious benefits to a localized decision-making process, there are also serious limitations and potentially dangerous disadvantages. Township borders are discrete, but the landscape is continuous; ecological and visual processes do not normally recognize political boundaries. Coordination among political entities is required to take into account issues larger than the township. Thus, the ESA guideline of examining local decisions on a broad context should be heeded.

8.3 The Use of Ecological Principles in Alternative-Future Analysis

In retrospectively considering the Monroe County study in terms of the ESA principles and guidelines, several of those principles can be observed coming into play. Paradoxically, the time principle is at once critical to the study and irrelevant to it. This strange situation arises because the study was designed and conducted from a planning perspective rather the ecological perspective adopted by the ESA principles. The study was called for because social processes like population growth, transportation advances, and the economy of land use and ownership were changing at a rapid rate, far outstripping the rates of change of geologic or biologic processes. From a planning perspective, 25 years is a mid- to long-term horizon for decisions and investments; instead of the hundreds of years involved in the successional change of a forest, the study group was faced with large changes in the built environment occurring in a decade or two. It was clear that, to preserve the landscape elements that change at geologic or long biologic time scales, action must focus on the forces acting at the greatest rate of change. It was also clear that the major changes to the ecosystem would result not from ecological processes but from social processes. Although not taking sides in the debate between development and preservation, the study sought to aid the local decision makers in striking a balance that would reflect the long-term best interests of the county (i.e., for the next 25 or so years) and its citizens. In dealing with changes in ecological processes, the study had to focus on those that occur at short time scales, such as groundwater recharge and surface-water cleansing. Conservation, degradation, or loss were the endpoints for ecological processes that occur at long time scales, such as land cover and wildlife diversity.

The species principle was a major concept in the study in that much concern was directed at the preservation of plant and animal species, both wild and domesticated. Obviously, the crops and herds of local farms are important to the local population. But the study noted that the game fish, large mammals, low-lying vegetation, and forests also play an important role in the economic and social life of Monroe County. For example, the Pocono black bear population, an umbrella species, has one of the highest reproductive rates in the country (Gary Alt, personal communication). However, to support this black bear population, the county and surrounding areas must retain a mosaic of wetland and upland habitats and preserve the wildlife corridors that the black bear uses. These same ecological features contribute to good water quality, flood retention, and recreational opportunities. The public and private lands, stream corridors, and wetlands that currently serve to provide the necessary habitat and these ecological services are threatened by future development. The case study shows how local land-use decisions will influence the county and the Pocono region's ability to maintain the habitat requirements of key wildlife species and other natural infrastructure that is necessary for a high-quality environment.

The place principle was central to the framework adopted by the study. To different degrees, it examined the local climatic, hydrologic, edaphic, and geomorphologic factors as well as the biotic interactions that affect ecological processes and the abundance and distribution of species in Monroe County. These assessments underlaid an effort to understand what makes Monroe County be Monroe County. The unique geography, geology, history, soil types, stream locations, and habitat patterns were analyzed to determine their historical development and contribution to the current conditions. Moreover, those conditions were modeled and projected into several versions of the future to allow the citizenry to understand the effects of their actions on the setting of their community.

In terms of the disturbance principle, the study was, by necessity, again focused on changes brought about by social and economic pressures rather than ecological disturbances like wind and fire damage. Disturbances created by 100-year floods paled in comparison to the changes that could be brought about in a decade by unconsidered development.

Conversely, the study was acutely aware of and intensely interested in the landscape principle. The projections of potential change within the county were framed largely in terms of the size, shape, and spatial relationships of land-cover types, considering the ecological, agricultural, and built landscapes alike. In analyzing the effects of changes in these landscapes, the study took particular note of (1) the influences visited upon populations of flora and fauna, communities of organisms, and ecosystems and (2) the resultant feedbacks on the quality of life of the county's residents.

In summary, the alternative futures derived and described for Monroe County strove to examine the impacts of local decisions in a regional context as well as on a local or site-by-site basis. They did not plan for long-term change and unexpected events; rather, they were concerned solely with the short term, with all contingencies expected to be known and expected. A central motivation for conducting the study and deriving the alternative futures in the first place was to preserve rare landscape elements, critical habitats, and associated species. The alternative futures represented a variety of land uses that reflected a spectrum of depletion of natural resources over a broad area, allowing the citizens to visualize the consequences of such changes. Several of these alternative futures retained large contiguous or connected areas that contain critical habitats. However, no consideration was given to the introduction and spread of nonnative species. The alternative futures depicted a variety of landscapes that sometimes avoided or compensated for the effects of development on ecological processes and implemented land-use and land-management practices that are compatible with the natural potential of the area, but sometimes did not; again, the purpose was to show the decision makers and residents the different outcomes from different actions. Thus, although this study predates the ESA ecological principles, it can be seen to be generally consistent with those principles and other published conservation-planning objectives and landscape-ecology principles (e.g., Forman and Godron 1986).

8.4 Change: The Alternative Futures

Any alternative future is the product of the interaction of private and public actions over time. There are substantial public interventions in landscape and infrastructure policy that can influence the urban pattern. The public actions listed in Table 8.1 were considered and selectively combined to produce each alternative.

Starting from existing conditions circa 1993 (Fig. 8.3), six possible alternative futures for Monroe County in the year 2020 were prepared. Two were derived by extending and extrapolating current development practices. Four were derived from alternative policy/plan actions. For each alternative, the 2020 land-use pattern and a computer-produced aerial view of the future landscape are shown. The aerial perspective of existing conditions (Fig. 8.4a; see color insert) is a view facing north toward I-80 between Bartonsville and Stroudsburg. Interstate-80 and the large cranberry bog can be seen in the background. For reference, the view spans approximately 2 miles from left to right.

The Plan-Trend Alternative continues the development practices currently used in the county and is guided by the Monroe County Comprehensive Plan (Fig. 8.5). It assumes that only what is currently publicly owned

TABLE 8.1. Public interventions in landscape and infrastructure policy.

Existing conservation	The enforcement of laws and regulations, especially on public lands
Proposed conservation	A proposed pattern for the acquisition of conservation land or rights
Billboard policy	The prevention of, and removal of, "unrooted" signs
Recreation	The expansion of commercial recreation and "ecotourism"
Pocono Raceway	A plan that expands this facility while not harming nearby Long Pond
Road improvements	Three priority projects
Railroad service	Restoration of service or a new route
Railroad stations	Six new station possibilities
Sewer service	Several alternative proposals
Zoning	New low-density and mixed-use types
Development guidelines	Several innovative suggestions

FIGURE 8.3. Existing land uses, land covers, and landscape features.

FIGURE 8.5. Land uses, land covers, and landscape features projected for the Plan-Trend Alternative.

or regulated will be conserved. Its infrastructure assumes that (1) only townships that have already proposed sewer plans will implement wastewater-management improvements, (2) the existing railroad alignment will be used, and (3) no major road improvements will be undertaken. Residential development in this alternative follows the same patterns of broadly dispersed low-density housing that have been characteristic of the past decade. In this alternative, all land that can be developed will be developed, primarily at low density, as proposed by the Plan. Higher densities are allocated only to land within prior township sewer plans. Almost all trees are lost in the almost fully developed landscape. This alternative does not take into consideration the impact of increased development (and population growth) on the health and integrity of those existing conservation areas, nor does it seek to maintain or protect existing wildlife corridors (unless in otherwise protective status). For the black bear, planned development will continue to erode the matrix of upland and wetland habitats it uses unless those areas are part of the current inventory of conservation lands (public or private). The loss of wetlands and associated upland habitats will contribute to lower water quality and a decline in biological diversity and will threaten the existence of rare animal and plant associations, such as bogs, tannic streams, and scrub oak/pitch pine barrens. This alternative will not prevent future degradation of ecological processes *without the intervention of new policies, subdivision regulations, and ordinances that specifically target these issues.*

The Build-Out Alternative is the most pessimistic view of Monroe County's future by assuming that market-driven development will overwhelm the planning and investment abilities of the townships and the county (Fig. 8.6). Only existing conservation is maintained. The infrastructure projection

FIGURE 8.6. Land uses, land covers, and landscape features projected for the Build-Out Alternative.

assumes that the townships will not fulfill their transportation or sewer plans and that extensive low-density development, which is economically and technically feasible, will result. The land-use pattern of this alternative will look like any other metropolitan suburb, with low-density housing everywhere. As a result, the aerial perspective view could be mistaken for many areas near New York or Philadelphia.

The Township Alternative maintains local political control and proposes strategic development of each township, while minimally threatening its most valued landscape features (Fig. 8.7). It assumes the implementation of the proposed acquisition of conservation land or rights. For developable lands, exurban density rezoning is proposed near existing or proposed conservation lands. Innovative wastewater treatment and sewer technology is proposed for each town in areas with higher-density development. Because most existing residential developments are not connected to a central sewage

FIGURE 8.7. Land uses, land covers, and landscape features projected for the Township Alternative.

treatment plant or occur on lands that do not percolate well enough for standard in-ground septic systems (hence the high use of sand mounds), using new technologies for wastewater treatment would help to reduce the impact of development on water quality. These new technologies would serve small groups of houses or residential developments and thus would not induce the additional sprawl development that often follows the pipeline leading to the sewage treatment plant.

This alternative assumes the implementation of the new I-80 railway alignment and road improvements. The proposal envisions higher-density development in existing subdivisions and new higher-density residential and urban mixed-use development near existing town centers. This strategy promotes compact development while allowing more residents access to large natural areas. The aerial perspective (Fig. 8.7) shows a new exurban residential area in the foreground. Near the new railroad station along I-80, new high-density mixed-use development can be seen.

The Township Alternative, however, will not allow the ESA ecological principles to be fully implemented. Township government, as practiced in Pennsylvania, provides few opportunities or incentives to coordinate with neighboring townships and municipalities, coordination that is necessary to "retain large contiguous or connected areas that contain critical habitats." Nor does this political system allow planners and decision makers to easily view their individual land-use decisions and associated impacts (positive or negative) within a regional context. Unless townships work together, any attempts toward achieving conservation measures that are meaningful and that address problems at the landscape ecological scale would be piecemeal and ineffective.

The Southern Alternative addresses the two distinct characters of Monroe County, and offers each area the benefits of maintaining its uniqueness (Fig. 8.8). The northern part of the county is typified by areas of high scenic quality, high biodiversity, and other sensitive geologic and biologic areas. Despite this, development there continues. The less fragile southern part of the county has remained underdeveloped because of inadequate infrastructure. The proposed alternative assumes implementation of the new proposed conservation plus an additional area for extensive outdoor recreational opportunities in the north. The planned infrastructure upgrades southern roads and has alternative-technology sewer systems, especially in developable locations adjacent to conserved agricultural areas. This alternative results in extensive conservation in the north and, by introducing road improvements and sewer service to selected locations in the south, promotes compact and environmentally responsible growth. The aerial perspective (Fig. 8.4b; see color insert) shows new development directed toward already developed areas, in which some existing residential areas have been rezoned for increased density and mixed-use development.

The Southern Alternative goes far in recognizing the special landscape and geologic features that result in areas of high biodiversity and/or areas

FIGURE 8.8. Land uses, land covers, and landscape features projected for the Southern Alternative.

of unique habitat and species associations. By encouraging growth in the southern part of the county through various techniques (such as the transfer of development rights, conservation easements, and conservation design of residential development), the county can minimize development impact on its more sensitive ecosystems and retain the natural character of the county. However, this alternative would require that new ordinances be adopted that would allow for increased density in existing residential areas or the use of conservation designs that promote open space within the developments; see, for example, Arendt (1994, 1996).

This alternative does spark controversy. The southern part of the county is where most of the farms operate. Although not a major economic source, these farms have some of the richest soils in the county. It will be necessary to have policies in place that will allow the southern part of the county to accommodate new development while ensuring that the active farms themselves are not sacrificed in the process.

The Spine Alternative recognizes the interdependencies of Monroe County with the metropolitan New York–Philadelphia areas and, to an increasing degree, the new employment centers in the suburbs of New Jersey (Fig. 8.9). The most efficient strategy to accommodate the county's anticipated growth, promote tourism, and protect the natural landscape is to concentrate new development in the corridor between Mount Pocono and Stroudsburg and

FIGURE 8.9. Land uses, land covers, and landscape features projected for the Spine Alternative.

maintain the rural character of the rest of the county. A key decision is increased dependence on high-speed rail transportation. The design for this alternative assumes the implementation of the proposed new conservation plan and also hopes to draw development away from the sensitive areas by making it much more attractive to locate in the central corridor between Stroudsburg and Mt. Pocono. The infrastructure is based on the new I-80 rail alignment with four new station stops. The roads serving this more intensely developed urban corridor would be improved. Sewered areas in Pocono and Stroud townships would be upgraded. The central urban spine would be the focus of future urban growth and zoned as high-density residential, with mixed use allowed along the principal state and county roads. The perspective view (Fig. 8.4c; see color insert) shows the railroad station and concentrated development located east of the I-80/railroad corridor with new conservation and park lands on the western side.

This alternative provides the greatest opportunity for the conservation of developable lands that have not succumbed to the pressure to be the next residential or resort development. This alternative builds upon the existing infrastructure and, in the process, minimizes opportunities for sprawl and its associated outcomes: traffic congestion and increased vehicle miles travelled, "green field" development, and loss of habitat and biodiversity. These conserved lands can help preserve the natural ecological systems and corridors that link the more valuable habitats in other parts of the county. By concentrating growth and development along the "spine" corridor, the county can focus its conservation efforts on areas that will enhance stream quality, protect its underground aquifers, restore the riparian buffers and forested watersheds, and better preserve irreplaceable landscape features.

The Park Alternative (Fig. 8.10) advocates conserving all of the existing undeveloped land in Monroe County. It envisions the county following an updated vision of Warren Manning's National Landscape Plan of 1923 (Manning 1923) and the creation of a major landscape reserve for the northeastern United States. All undeveloped land not already in conservation would be bought for further conservation and for recreational development. The infrastructure adopts the new I-80 railroad alignment with one station stop. Sewers would be extended to high-density residential and mixed-use commercial and residential zones. The proposal enables preservation of the last unprotected northeast-metropolitan corridor of such a large size and high quality. The management of this reserve would be directed both to the protection of Monroe County's natural habitats and to the development of its recreational possibilities without neglecting the needs of the present and future county residents. The perspective (Fig. 8.4d; see color insert) shows a landscape similar to that of today. Given the rate of change, now may be the last chance for such a bold approach.

Not all of these alternatives can be considered to fully include the ecological principles described by Dale et al. (Chapter 1). Indeed, several of them clearly violate those principles because, in the real world, there are

FIGURE 8.10. Land uses, land covers, and landscape features projected for the Park Alternative.

times when any set of objectives (including ecological principles and guidelines) will have internal incompatibilities. As a result, some of the alternative futures intentionally sacrificed some concerns to derive other, preferred benefits.

The alternative futures were devised to show the different outcomes that may occur when different approaches, policies, and values are employed in making decisions about the growth, development, and occupation of Monroe County. The results of the study were used to graphically portray to the citizens and decision makers of the county what would likely happen if ecological and other concerns were or were not thoroughly considered in decisions regarding the county's future. The adherence to the ecological

TABLE 8.2. The ecological priniciples and alternative futures for specific issues in Monroe County.

Concerns	Alternative futures					
	Plan-Trend	Build-Out	Township	Southern	Spine	Park
Principles						
Time				Y	Y	Y
Species				Y	Y	Y
Place	Y		Y	Y	Y	Y
Disturbance						Y
Landscape	Y		Y	Y	Y	Y
Issues						
Geologic	Y	Y	Y	Y	Y	Y
Biologic				Y	Y	Y
Visual	Y		Y	Y	Y	Y
Demographic	Y	Y	Y	Y	Y	Y
Economic	Y	Y	Y	Y	Y	Y
Political	Y	Y	Y	Y	Y	Y

principles and the concern with specific issues in the different alternative futures are roughly sketched out in Table 8.2.

8.5 Impacts of the Alternative Futures

The future alternatives presented here are complex. To highlight their relative impacts, the alternatives were compared with the six criteria sets used in the earlier evaluations: the geologic landscape, the biologic landscape, the visual landscape, demographics, economics, and politics (Fig. 8.11). Some impact highlights:

- The Plan-Trend Alternative and Build-Out Alternative have the most negative landscape impacts because both cause widespread development. Simply stated, existing conservation policy does not sufficiently protect the sensitive landscape areas of Monroe County.
- The Southern Alternative strategy of dividing the county into two areas for planning purposes may successfully protect the landscape of the area.
- By concentrating development in the central section of the county, the Spine Alternative protects much of the existing undeveloped land. However, the preservation of this open space comes at the expense of severe impacts in the expanded Route 611 corridor, which will be irrevocably urbanized.
- The Park Alternative provides the greatest protection for the landscape because it proposes conservation of the largest amount of land and uses the strategy of densification in several areas.

At the time of the study, the population of Monroe County was about 100,000 people. None of the alternatives was designed with the intention of

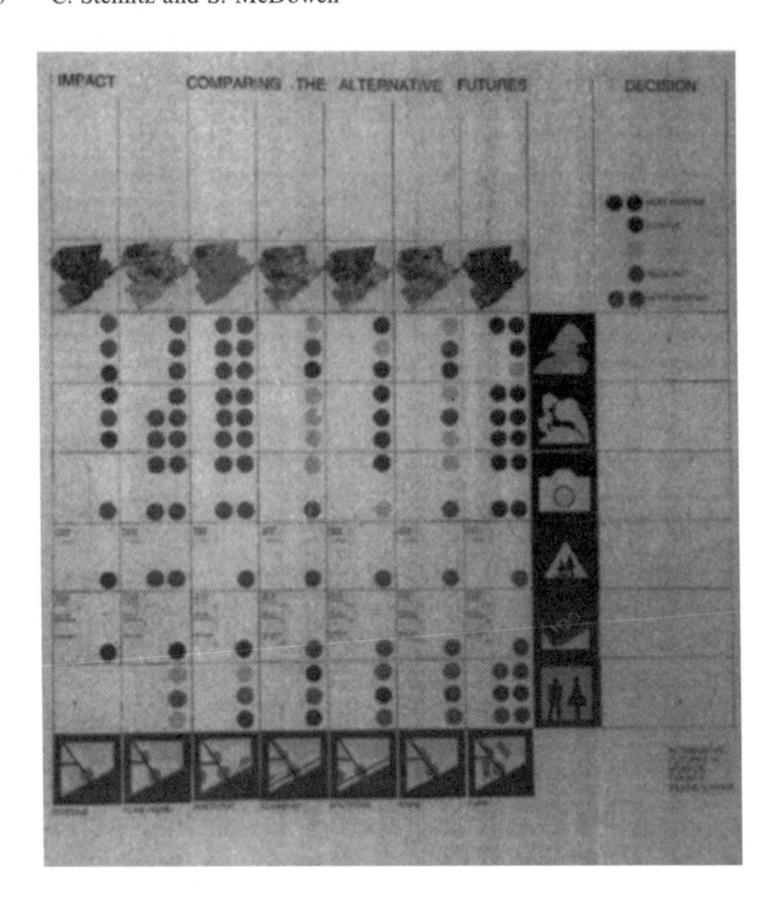
IMPACT
COMPARING THE ALTERNATIVE FUTURES
DECISION

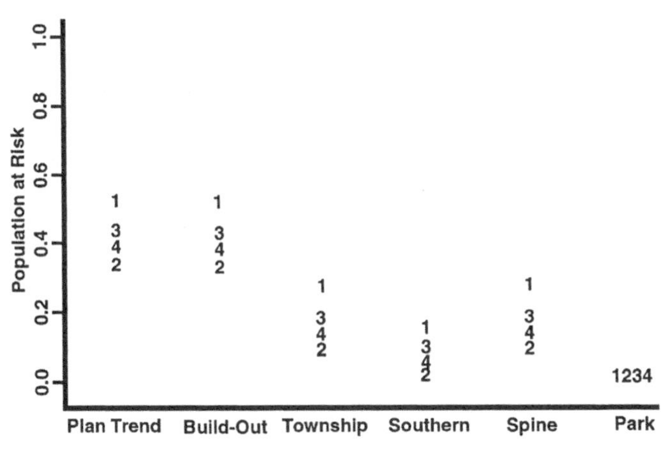
Population at Risk
0.0
0.2
0.4
0.6
0.8
1.0
Plan Trend
Build-Out
Township
Southern
Spine
Park
1234
1 Herps
2 Birds
3 Mammels
4 All Vertebrates

TABLE 8.3. Annual costs of the alternative plans.

Alternative	Population capacity	Annual cost (millions of dollars)
Plan-Trend	600,000	1
Build-Out	400,000	1
Township	300,000	104
Southern	350,000	106
Spine	350,000	120
Park	300,000	At least 135

limiting Monroe County's growth or arriving at any particular population capacity. However, each alternative would sustain a different optimum population and would come at a different cost (Table 8.3). Most of the capital expenditures for roads, railroad, and the cost of land conservation would be shared among the county and other sources. Annual costs associated with each alternative were calculated under the assumption that the county issued a 20-year bond for its share.

8.6 Decision

There is a "bottom line." In the long term, and in full awareness of the costs, any of the four designed alternatives is a better long-term decision than continuing the trend toward build-out. This is especially true of the ecologic considerations. However, this study does not present a recommendation. Its purpose was to present alternatives, not a single answer. The many decisions needed for any of the alternatives are clearly the responsibility of the people of Monroe County. Given the enthusiastic initial responses to this study, it was clear that we enabled the people of Monroe County to see more clearly some of the important issues they will face during the next 25 years.

Three of the ecological principles are exemplified by this case study. First, the study clearly encourages land managers to examine the ecological impacts of local decisions within a regional context. The case study also illustrates the difficulty in implementing this principle. The public and private decisions that accumulate to form the 2020 alternatives are almost all local, mainly because the level of government responsible for land-use planning in Pennsylvania is the township or borough. The allocation of future land use

FIGURE 8.11. A decision-aiding matrix that compares the existing conditions and the six altenatives in terms of the six key decision-related issues that were identified: geologic landscape, biologic landscape, visual landscape, demographics, economics, and politics.

in the six alternatives was guided by local zoning, sewer, road, and conservation proposals. The Plan-Trend Alternative and Township Alternative were based on the allocation of each individual township's forecast population. In this way, the Monroe County study exemplifies the guideline regarding local decisions, or, as faculty advisor Michael Binford would describe it, "the tyranny of small decisions."

However, there are also decisions taken at a regional or larger context that must be examined at the local level. When long-term changes are being considered, these decisions may be of even greater impact. Perhaps of greatest importance when assessing long-term ecological change are national and international trends in demographics. State-level activities, such as the promulgation and enforcement of various environmental regulations and the acquisition of state parks and reserves, come into play. Actions supported at the federal- and state-government levels, such as major infrastructure investments, also must be considered. In this study, and with the exception of the Plan Alternative and Township Alternative, all of the alternatives were produced on a regional basis, in this case at a county level. All of their impacts were assessed both on county and township levels.

The reader should also consider that this particular case study was designed within the parameters of several technical decisions that required that the "region" be defined as Monroe County, with a politically defined boundary. These decisions were made, in part, because of how data were made available and also because of the tradition of reporting politically sensitive studies within the spatial limits of the relevant decision-making levels of government. Neither of these reasons is particularly compelling. Indeed, they lead to a spatial limitation that is inconsistent with the more appropriate geographical bases of ecologically oriented studies, such as river basins or arbitrary boundaries that more than encompass both ecologically and politically relevant areas. However appropriate these geographically based analyses may be, the ability to connect "data and analysis" to local land-management decisions will be constrained by the reality of political boundaries. At best, the results of these ecologically based studies must still be delivered as products that are relevant to the local land-use decision maker.

The second guideline that this study exemplifies calls for the preservation of rare landscape elements, critical habitats, and associated species. Indeed, an underlying theme of the alternative futures was the conservation of biological diversity. In an associated study, White et al. (1997) examined the impacts of the possible future land-development patterns of each alternative on the biodiversity to the Monroe County landscape. The species data included lists of all bird, mammal, reptile, and amphibian species in the study area; their habitat associations; and area requirements for each. The measures of biodiversity were species richness and habitat abundance (Fig. 8.12). Species richness was based on the presence of diverse natural habitat, and it changed little from present to future. There were distinctly

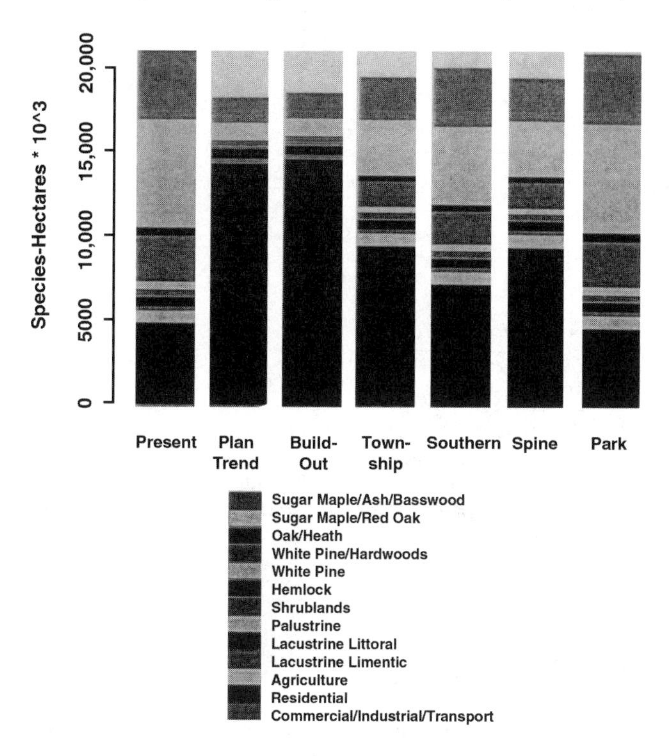

FIGURE 8.12. Area in each of four aggregated habitat classes for the present and future landscapes.

greater risks to habitat abundance in landscapes that extrapolated from present trends or zoning patterns, as opposed to landscapes in which land-development activities followed more constrained patterns (Fig. 8.13). These results were stable when tested with Monte Carlo simulations and sensitivity tests on the area requirements.

Finally, the ecological guidelines encourage land-use and land-management practices that are compatible with the natural potential of the area. The citizens, decision makers, developers, businesses, and local, state, and federal agencies are attempting to do just that. In the next section, we describe how the study has influenced policies and actions that will move Monroe County toward a more desirable and sustainable future. The reader should note that the changes described below *are significant* when considered within the political, economic, and social context of Monroe County and the Commonwealth of Pennsylvania. Furthermore, it was this study that prompted, explicitly or implicitly, the actions that have occurred.

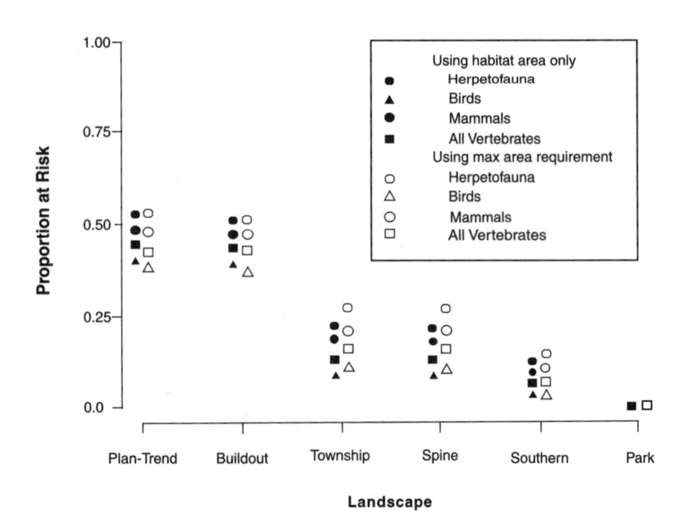

FIGURE 8.13. Risk to terrestrial vertebrate habitat, by future landscape and by taxonomic group, estimated from (1) total habitat area only and (2) maximum area requirements.

8.7 A Catalyst for Change

When the Alternative Futures Study for Monroe County was undertaken, the time was ripe for a vast awakening by residents and local officials. The U.S. Environmental Protection Agency, Region III, had been working with Monroe County on a demonstration project aimed at habitat protection. The opportunity arose to bring Harvard University Graduate School of Design into the project. The key players involved in the overall EPA project believed that a more comprehensive analysis of the county was needed to help frame subsequent conservation and land-use decisions. The nexus of opportunity and need was timely, and the educational value derived from this study far exceeded any expectation.

Once the public meetings were held and the Harvard University report was published, the scenario-based, data-driven study served as a catalyst, a call to action, as many would come to view it. In 1993, Harvard University presented stark, yet believable, visions of the future for Monroe County. Faced with the possibility of a future that would hasten environmental degradation, increase traffic congestion, further fragment wildlife habitat, and burden local governments with increased infrastructure costs, the residents of Monroe County under the leadership of the county commissioners and planning commission, heeded the call to action. The alternative scenarios that the study designed provided a backdrop and reference for county leaders and citizens as they began to take ownership of the tasks that faced them. Several significant outcomes can be attributed to this catalytic project:

- The initiation of Monroe 2020, a stakeholder-driven process that helped to formulate the basis for a new County Comprehensive Plan
- A commitment to use a GIS and the data derived from the study
- The successful passage of a $25 million Open Space Referendum in 1998
- The initiation of municipal community audits
- The addition of two new county planning positions

The legacy of the Alternative Futures Study is impressive. Several earlier attempts at broad-based planning and design had failed to inspire change. The county and its local governments continued along the path they had embarked upon 30 to 50 years ago, developing residential and commercial properties with little consideration to the constraints of the landscape and the environmental services it provided. As Alan Price Young, Chairman of the Executive Committee for Monroe 2020, described it, "the Harvard University study provided a credible foundation for change... it provided a major impetus for the resulting Monroe 2020 [comprehensive planning] process." According to Young, the study enabled Monroe County to enhance the "acre by acre, project by project review and made us see the interconnectedness of the natural landscape with the fiscal realities of the County.... Questions about where development is appropriate and how to balance environmental quality and growth became the issues upon which to focus. When all interests are blended, the real priority becomes quality of life and 'we all have a fundamental interest' to enhance quality of life."

Monroe 2020, begun in 1996 by the county commissioners and the planning commission, followed on the heels of *Alternative Futures for Monroe County, Pennsylvania* and the county's *Fiscal Impact Study*. The call to action afforded by the study brought together local officials, school administrators, business men and women, public interest and environmental groups, and long-time residents. The Monroe 2020 process organized into school-district-based task forces that were charged with recommending policies and practicable ways to protect the natural and built landscape in concert with sensible development and growth. A draft *Comprehensive Plan* was made public in March 1999. Recommendations from the Monroe 2020 Task Forces encouraged

- Direct growth and development along the existing infrastructure corridors and centers of population (similar to the Spine Alternative)
- Environmentally friendly economic enterprises to locate in the county
- Recycle vacant and underutilized buildings and sites
- Implementation the Open Space Policy, which specifies additional lands to be protected or added to the open-space inventory
- Improving the visual character of the county through landscaping and signage-improvement measures

The Harvard University study's effective integration of natural-resource, demographic, land-use, infrastructure, and fiscal data gave the county a unique opportunity to "see into the future," based on existing conditions and projected trends. It presented the county in a different light, showing the connection between its natural landscape features and the impacts associated with unplanned or poorly integrated policies that allowed inappropriate development (residential, industrial, and commercial) to occur. In addition, the study and its associated GIS databases that were provided to the county gave the incentive for the development of the county's own GIS capabilities. A GIS was developed and applied concurrently with the county's comprehensive planning process, Monroe 2020.

The Open Space Referendum, passed in 1998, reflects a coming together of citizens' interests. It served to recognize that Monroe County's continued wealth and well-being depend largely on the protection and conservation of its natural landscape. The Harvard University study provided a visual and analytical way for people to think differently about the natural and built environment and to see the connectedness. John Woodling, Director of the Monroe County Planning Commission, states "the Open Space Referendum's successful passage had more to do with quality of life issues rather than habitat protection. However, the resulting Open Space Policy that will direct the use of the funds will indeed go far in protecting critical and sensitive landscape elements and habitats. County leaders and residents recognize the financial and aesthetic value of clean trout streams, hiking and biking trails, healthy game and other wildlife resources, open vistas, and outdoor recreational areas." Janet Weidensaul, County Commissioner, eloquently summed up the import that the Harvard University study had on the future direction of the county, saying "we took our environment for granted and didn't recognize that it was so critical to our success... leaders are far more sensitive today to the preservation of the splendor that is Monroe County for future generations."

Acknowledgments. This study of Alternative Futures for Monroe County, Pennsylvania, is the product of a graduate-level studio at the Harvard University Graduate School of Design. It was sponsored by the U.S. Environmental Protection Agency, Region III; the Monroe County Commissioners; the Monroe County Conservation District; the Monroe County Planning Commission; and the USDA Forest Service, Pacific Northwest Research Station. The participants were Elke Bilda, John Ellis, Torgen Johnson, Ying-Yu Hung, Edith Katz, Paula Meijerink, Allan Shearer, H. Roger Smith, Avital Steinberg, and Douglas Olson. The authors especially thank Allan Shearer, Craig Todd, Janet Weidensaul, Denis White, John Woodling, and Alan Price Young for their contributions toward this report of the study. The figures in this chapter are republished from the original report, which is copyrighted by the President and Fellows of Harvard College.

References

Arendt, R.G. 1994. Rural by design: a handbook for maintaining small town character. American Planning Association, APA Planners Press, Chicago, Illinois, USA.

Arendt, R.G. 1996. Conservation design for subdivisions: a practical guide for creating open space networks. Island Press, Covelo, California, USA.

Forman, R.T.T., and M. Godron. 1986. Landscape ecology. John Wiley and Sons, New York, New York, USA.

Manning, W.H. 1923. A national plan study brief. Landscape Architecture **13**(4)(Suppl.):1–24.

Smith, C.R., and M.E. Richmond. 1994. Conservation of biodiversity at the county level: an application of gap analysis methodologies in Monroe County, Pennsylvania. New York Cooperative Fish and Wildlife Research Unit, Department of Natural Resources, Cornell University, Ithaca, New York, USA.

Steinitz, C. 1990. A framework for theory applicable to the education of landscape architects (and other environmental design professionals). Landscape Journal **9**(2):136–143.

Steinitz, C., editor, E. Bilda, J.S. Ellis, T. Johnson, Y.-Y. Hung, E. Katz, P. Meijerink, A.W. Shearer, H.R. Smith, A. Sternberg, and D. Olson. 1994. Alternative futures for Monroe County, Pennsylvania. Harvard University, Graduate School of Design, Cambridge, Massachusetts, USA. Also available on the World Wide Web at *http://www.gsd.harvard.edu/depts/larchdep/research/monroe/*.

White, D., P.G. Minotti, M.J. Barczak, J.C. Sifneos, K.E. Freemark, M.V. Santelmann, C.F. Steinitz, A.R. Kiester, and E.M. Preston. 1997. Assessing risks to biodiversity from future landscape change. Conservation Biology **11**:349–360.

9
Alternative Futures as an Integrative Framework for Riparian Restoration of Large Rivers

David W. Hulse and Stanley V. Gregory

> When creative genius neglects to ally itself to some public interest it hardly gives birth to wide or perennial influence. Imagination needs a soil in history, tradition, or human institutions, else its random growths are not significant enough and, like trivial melodies, go immediately out of fashion.
>
> George Santayana

In many ways, planning landscapes with restoration goals in mind is like solving a spatial jigsaw puzzle. The choice of which piece to restore first will influence later choices, because the characteristics of remaining pieces assist the decision maker in creating an emerging picture that is both recognizable and desired. All we ask of jigsaw puzzles is that they entertain and challenge us. People ask a great deal more from their landscapes. Fulfilling people's demand for food, water, fiber, recreation, and shelter often leads to significant changes on the land, changes that leave legacies that, in turn, affect future choices.

This chapter describes an approach for choosing among future possibilities of riparian forest restoration in large river systems. In doing so it offers an example of how ecological principles and guidelines, such as those put forward by the Ecological Society of America (ESA) (see Chapter 1), may be applied within the particular realities of local landscapes. This approach is focused on rivers and the restoration of floodplain forests. Its conception predates our awareness of ESA's Principles and Guidelines for Managing the Use of Land, but like ESA's guidelines, it asserts that the thoughtful planning of desired future landscapes requires an understanding of how past processes and choices have led to present conditions. It builds this understanding from four different disciplinary perspectives (ecology, hydrogeomorphology, demography/planning, economics) and then applies these perspectives to the task of choosing where, at three different spatial extents within the river system as a whole, to invest in restoring desired future patterns and processes. Our thesis is that the prospects for success improve as restoration projects explicitly address the connections between ecological dynamics and the local demographic, economic, and institutional dynamics in which restoration efforts are implemented.

9.1 Evolution of Riparian Zone Management

River floodplains and riparian forests throughout the world have been modified by human development, causing losses of biological diversity and ecological function (Naiman et al. 1988; Naiman and Decamps 1990; Petts 1990; NRC 1992; Gregory and Bisson 1996) and increasing the probability of loss of human life and property during floods (IFMRC 1994; Coulton et al. 1996; Galloway 1996). Planning for restoration of these floodplain and riparian landscapes is commonly based on a narrow set of disciplinary perspectives, each with a primary focus of concern: habitat and communities for ecologists, water flow for hydrologists and engineers, channel form and sediment transport for geomorphologists, human population growth and land development patterns for demographers and planners, and commodity production and trade-offs for economists. These discipline-specific perspectives are then submitted to decision-making processes, which are themselves constrained by the opportunism of landowners willing to sell or negotiate easements and the narrow mandates of agencies charged with conceiving and implementing restoration efforts. Given such a fragmented approach, it shouldn't be surprising that attempts to restore river systems have had mixed results.

Even with this fragmented approach, the United States has made great strides in environmental protection over the last 25 years. These accomplishments are based primarily on management strategies that rely on engineering and regulatory solutions. Programs that support the construction of wastewater treatment facilities and wetlands protection are examples of this approach. Yet, there is significant risk that much of what has been gained will be offset. Ecosystem degradation continues to occur and is especially evident in floodplain zones. For example, a range of water- and land-use practices continue to threaten salmon populations in Oregon despite the state having some of the most progressive land-use laws in the nation. Such a predicament has generated a growing consensus among environmental scientists, planners, and economists that increased regulatory control alone cannot produce the desired recovery of the qualities attributed to riparian corridors, threatened species, and other ecological resources. Innovative means are being explored to supplement traditional methods of environmental protection.

"Ecosystem management" is the favored term given to this new approach to environmental protection. It also is known as the "watershed protection approach," and still other groups call it "community-based environmental protection" (Noss et al. 1997; Johnson et al. 1999). Whatever the chosen phrase, the approach is a process of spatially extensive environmental assessment that is unconstrained by specific institutional mandates. It is applied to a domain of places rather than individual pollutants or other environmental stressors. Environmental results are achieved through community action designed to make progress toward publicly articulated visions of desired future landscape conditions.

9.2 Assessing Alternative Futures

During the environmental movement of the 1970s a simple, elegant innovation in ecosystem protection emerged. Embedded within the National Environmental Policy Act of 1970 (NEPA) is a concept of alternatives analysis. The idea is that the public and their officials should assess the consequences of projects that might significantly impact the environment. The twist is that, at least for projects involving significant federal action, the assessment is focused on identifying project designs and locations presumed to cause less harm to the environment. Assessment usually produces a number of project alternatives, including "no action." Decisions are made by weighing the limitations and benefits of each alternative.

Termed "alternative futures," the framework used for integration stems from land conservation and development work in the United States and rural scenario development work in Europe (Steinitz 1990; Montgomery et al. 1995; Schoonenboom 1995;). Early applications of alternative future approaches focused on rural land- and water-use planning (Murray et al. 1971; van der Ploeg 1995). Scenarios, which are written descriptions of possible future conditions of land and water use and management, have a key role in the formulation of the alternative futures in as much as they provide a condensed statement of key assumptions about the future. These scenarios are characterized as being one of two general types: either projective, meaning they project forward in time land- and water-use trends experienced over some period of the past; or prospective, meaning they anticipate some coming change that in important ways varies from the past (Harms et al. 1993; Ahern 1999). These narrative scenarios are then translated into mapped and numerical descriptions of land and water use. Together, a narrative scenario and its accompanying mapped and numerical description constitute an alternative future.

Advances in the science of landscape ecology, the art of landscape design, and the technology of geographic information systems (GIS) have reached a stage where, together with the premises of ecosystem management, the concept of alternatives analysis is ripe for enhancement. The resulting innovation is the application of alternatives analysis to entire land-use scenarios bounded at varying extents of space and time. In this manner, the environmental impacts or benefits of a specific project can be viewed in context with the effects of multiple actions. Using geographical information systems and related tools, digital and paper representations are produced depicting the past, present, and potential future conditions of a study area landscape. These are then used to identify trends over space and time in human occupancy and natural resources. Based on a set of values and desired future conditions developed through working with citizen groups, digital representations of the alternative future landscapes are evaluated for their effects on ecological quality and economic productivity using scientifically defensible evaluative models. Such approaches provide an entry point for ecological principles into land-management decision-making processes.

The success of alternative future project efforts and the efficacy of their concluding recommendations depend on broad political acceptance and on-the-ground implementation. This implementation must be accomplished through existing public and private institutions, institutions that may have been created to solve problems quite different from those now faced and those likely to come. To accomplish this acceptance and implementation, Johnson and Herring (1999) have argued for the importance of achieving a better fit between environmental recommendations and the ability of existing institutions to carry them out. An institution that is well prepared to conceive and carry out a sustained solution to an environmental problem provides an effective point of contact, what we have termed a *docking port* for reconciling the problem. Johnson and Herring make a compelling case that the absence of adequate institutional docking ports for the recommendations produced by large-scale bioregional assessments presents one of the most formidable challenges to solving the next generation of environmental problems. The ESA recommendations offer a constructive step to help address this challenge.

In our work, we have seen at least three distinct situations that limit an institution's ability to provide good docking ports to the kinds of problems that do not respond well to the command-and-control approaches typical of environmental success stories of the past few decades. The first of these is evident when an institution's territorial jurisdiction does not align with the spatial extent of the problem, that is, when the institution does not have authority to follow the guidelines and apply the actions over the full territory in which the problem is apparent. The second situation is when the source of the problem is diffuse in space and/or time, and generally caused by many small individual actions rather than by a few large ones. And the third situation is when the narrowly defined spatial jurisdiction and scope of mandate of the relevant institutions make it difficult to implement the sort of recommendations set out in chapter 1 without substantive change in these institutions and their interrelationships.

If it is true that we now have, at least at times, a poor fit between environmental problems and the capabilities of present governance systems to solve them, it may be useful to better understand the types of systems created to manage environments and human behavior in them.

9.3 Environmental Governance Systems

The comprehensive set of laws, programs, tools and institutions at work in a landscape can be conceived as a governance system. There is a continuum of such governance systems currently employed to manage land and water resources, each with varying properties, suited to different needs. Doppelt (in press) identifies five distinct environmental governance systems, ranging from total government control on one end to unfettered free markets on the other. Each is described briefly below.

1. *Total government control:* In this governance system, command flows downward, with the government unilaterally determining acceptable environmental quality, setting standards and enforcing compliance accordingly. In such a system, the government and property ownership are basically synonymous. This approach has rarely been employed in the United States, with the notable exception of classified military installations, which now serve as biodiversity strongholds in the western U.S.
2. *Government-controlled standards and enforcement with local implementation:* As with total control, in this approach command flows downward regarding acceptable environmental quality and standards, but implementation decisions are made locally, though perhaps still within a governmental structure. Property functions as a form of protection from government excess in such a system. While this system of governance deserves credit for many environmental improvements since the 1960s, questions are now being raised as to whether such command-and-control approaches have reached the point of diminishing returns, especially when employed to address non-point-source pollution and other pervasive, diffuse environmental problems.
3. *Outcome-based within a carefully monitored and regulated framework:* This approach attempts to better balance top-down direction with bottom-up creativity. The focus of government in this approach is less on standards and enforcement and more on articulating goals for environmental quality and measurable indicators of progress toward these goals. Private sector institutions play a larger role here in creating their own least-cost stewardship plans. Monitoring remains a governmental responsibility, though compliance is with self-generated as opposed to government-generated plans.
4. *Market and/or voluntary approaches with some governmental controls:* This approach casts government in the role of providing both incentives and disincentives to other sectors to achieve desired environmental results, while maintaining some regulatory controls. It is based on the assumption that as costs of polluting increase, polluters will find alternatives that decrease pollutants below government-set standards.
5. *Complete voluntary and/or market-based approaches:* This free-market option relies on self-interest and civic-minded volunteerism to achieve

desired environmental results. Much of the recent work of watershed councils in the Pacific Northwest is based on this approach. Environmental governance systems 2 and 5 are most familiar in the United States, though instances of 3 and 4 are becoming more widely discussed and tested.

We paraphrase Doppelt's typology in the hope that it may improve the discussion regarding goodness-of-fit between environmental problems and institutional capacities to solve them. Comparison of this typology with the ESA recommendations raises some interesting questions. Are all these governance systems equally adept at embracing the ESA principles, adopting the guidelines, and carrying out the actions? If not, why not? What determines or influences where better fits exist between a governance system and its ability to successfully implement the ESA recommendations? We assert that what is needed is a way to more directly tie the nature of environmental dynamics in a place to the dynamics of the environmental governance systems at work in that place.

The ESA recommendations may be regarded as ways of seeing, thinking, and acting, which, if adopted, increase awareness of the environmental dynamics in a place. As a further elaboration of our assertion that what is needed are stronger ties between environmental dynamics and environmental governance system dynamics, we argue that one of the best hopes for sustaining restoration success is to strengthen the connection between the ecological basis for restoration and the demographic and economic systems in which restoration efforts are implemented (Gregory et al. 1998). Rarely is the potential for ecological benefit explicitly linked at the scale of large river systems to the social and economic likelihood of restoration projects in planning and implementation. And when such links are made, the absence of appropriate institutional docking ports for needed actions may further frustrate attempts to accomplish the recommendations of credible assessments.

9.4 Integration of Biophysical and Socioeconomic Perspectives for River Restoration

If restoration success increases as projects explicitly address the connections between ecological and institutional dynamics, integration of disciplinary approaches for restoration next requires hypotheses of relationships between relevant processes and desired outcomes. In this approach, these hypotheses guide the development of criteria used in prioritizing areas with high potential to meet riparian restoration objectives.

Hypothesized Relationships

Hydrological: Ecological recovery will be greater in areas with hydrologic processes that couple floodplains and active channels during the appropriate seasons of the hydrologic regime and maintain the full magnitude of flow variation.

Geomorphic: Responses to restoration (physical and biotic change) will be most rapid in channels and riparian areas that are geomorphically prone to physical change.

Biological: Richness of species will be greatest where habitat is most complex or heterogeneous.

Demographic: Areas with and proximate to high rates of human population growth will experience greater pressure for land-cover conversion. It is these same areas where existing governance systems will allow and promote land conversion.

Economic: Lands with high property values, greater concentrations of long-term capital investments, high economic yields, or high subsidies and externalities present greater constraints to restoration. These constraints may be overcome, but only through greater investment per unit area restored.

The primary purposes of an integrated analysis of both biophysical and socioeconomic restoration potential are to (1) identify areas of high ecological, demographic, and economic potential for riparian restoration, (2) propose alternative future options for high-priority restoration areas, and (3) recommend changes in policies, practices, or governance systems that would increase the likelihood of restoration (Fig. 9.1).

We argue that any framework for prioritizing the expenditure of resources on restoration projects should explicitly consider both past and present patterns of ecological, demographic, and economic variables and the biophysical and institutional processes that sustain them. In our conceptual

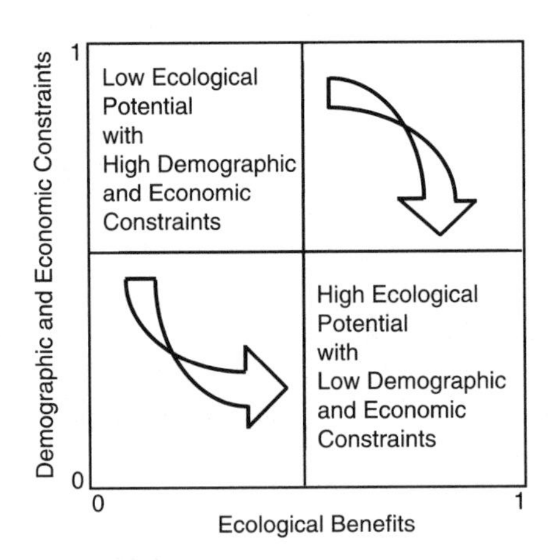

FIGURE 9.1. Matrix screening process for identifying lands with different combinations of ecological, demographic, and economic potential for restoration (from Gregory et al. 1998).

framework, patterns of critical ecosystem components of riparian areas and patterns of human land use create a spatial context within which any restorative action occurs. The processes or factors that determine the ecological and social response to restoration efforts are equally important to consider if the desired restoration effects are to persist through time. Figure 9.1 depicts a four-quadrant conceptual diagram of the interaction terms of demographic/economic constraints and potential ecological benefits. In the upper left quadrant would fall all lands where the magnitude of constraints outweighs potential ecological benefits. The lower right quadrant represents the opposite set of conditions: large restoration potential and low constraints. Recommendations for lands in these two quadrants are, conceptually at least, straightforward, and we expect existing governance systems to be better equipped to respond to recommendations for such lands. Lands falling in the other two quadrants, lower left and upper right with the curving arrows, present different challenges. Shifting lands from these two quadrants to the lower right while maintaining their economic productivity offers the chance to incrementally test techniques from less familiar governance systems, such as Doppelt's *Type 3 Outcome based within a carefully monitored and regulated framework* and *Type 4 Market and/or voluntary approaches with some governmental controls*.

9.5 Guiding Principles of Willamette River, Oregon Project

The Willamette River Basin comprises a major portion of the Columbia River system, encompassing major urban centers, residential areas, agricultural lands, and commercial and federal forests (Fig. 9.2; see color insert). In terms of total discharge, the Willamette River is the tenth largest river in the conterminous United States. As the terminus of the Oregon Trail, the Willamette Valley historically has been a center of population growth and economic development for the Pacific Northwest. It contains the richest native fish fauna in the state as well as species that are listed or considered for listing as threatened or endangered under the Endangered Species Act (e.g., Oregon chub, bull trout, spring chinook salmon, winter steelhead trout). The Willamette River Basin was the focus of a major watershed-level pollution abatement program in the 1960s that relied principally on regulatory and engineering measures.

In spite of these measures and establishment of formal conservation status as the Willamette River Greenway, riparian forests continue to decline. Human populations in the Willamette Valley are expected to double early in this century, placing tremendous demands on limited land and water resources. Effective resource management policies require explicit analysis of landscape features and integration of restoration, protection, and appropriate management practices. This requires a sound understanding of past ecosystem conditions and the extent of ecosystem change and a framework for determining the future desired condition for the landscape.

Our research focuses on the mainstem Willamette River and environs as an illustrative study area. The project seeks to balance ecological needs for restoration with social constraints on where and how to invest in restoration and efficiently use scarce resources to accomplish ecological and economic goals. The fundamental objective of the project is to develop and demonstrate an integrated system for identifying areas of greater ecological, geomorphological, demographic, and economic potential for restoration of riparian areas. We identify the following guiding principles for this project:

1. It is premised on hypothesis-based, multidisciplinary criteria for geographically prioritizing where to conduct restoration activities.
2. It applies these criteria at three spatial extents: river network, river reaches, and focal areas.
3. It produces restoration scenarios and accompanying alternative future depictions at reach and focal area extents.
4. It compares and contrasts alternative futures for (a) their riparian forest connectivity along the river network, (b) their expected increase in desired endpoint(s) following restoration, and (c) their links to upland conservation strategies at the river network extent.
5. It recommends a monitoring regime that is keyed to network (hydrological, biological, land-cover change), reach (biological, demographic), and focal area (biological, economic) processes.

9.6 Analysis of Restoration Potential

Evaluation of both biophysical and socioeconomic elements of the Willamette River and its riparian plant communities is based on a hierarchical sequence of reach delineation and spatially-explicit analyses. The river network scale is used as both the starting point for all analyses and the final integration tool. Unlike the upland terrestrial ecosystems, rivers are intricately linked along the network, and restoration efforts must account for the positive and negative consequences of this connectivity. Evaluation of restoration opportunities along a river must encompass the full extent of the riverine landscape and account for physical and biological processes that shape that landscape. After focal area selection and design, restoration plans must be reevaluated within a network context to identify interactions of riparian conditions, human impacts, and fluvial processes along the river corridor.

The public and their management agencies often restrict their notion of "the river" to the currently flowing water and ignore adjacent areas that flood. Rivers are those portions of the landscape that are shaped by fluvial

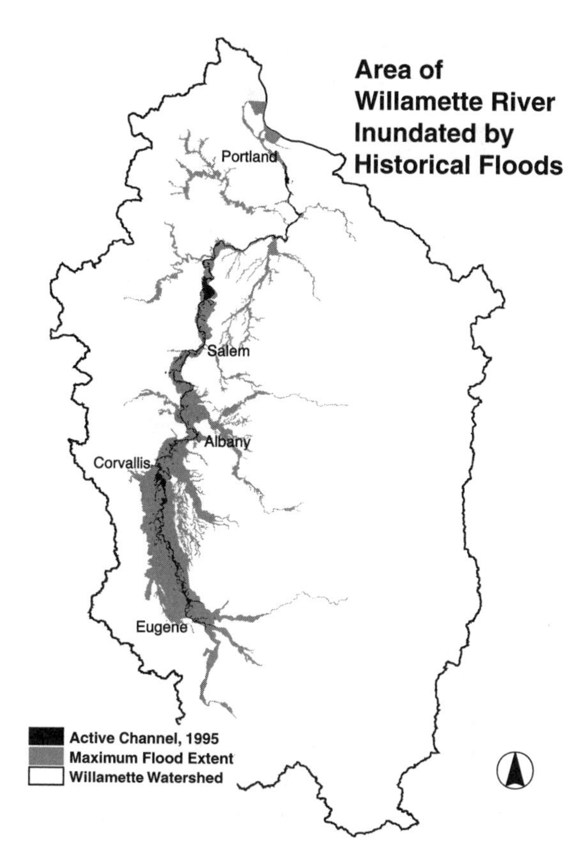

FIGURE 9.3. Map of spatial extent for Willamette River floods of record, 1850 to present.

processes during infrequent but physically powerful flood events and, therefore, include both the active channel (area inundated during average high flow) and its floodplain (alluvial surfaces deposited adjacent to the active channel) (Fig. 9.3).

Though people use and build in these floodplain portions of the river channel, these lands are part of the river. In the Willamette River, human development and flood control measures have simplified the complex river channel and eliminated floodplain storage and flood refuge for aquatic biota (Fig. 9.4). If riparian restoration is intended to be effective over the long term, the primary focus must encompass the active channel and its floodplain along the river network.

9.6.1 Reach Identification

Rivers rarely function uniformly along their entire lengths, but rather are sequences of reaches that differ in channel structure, hydrologic processes,

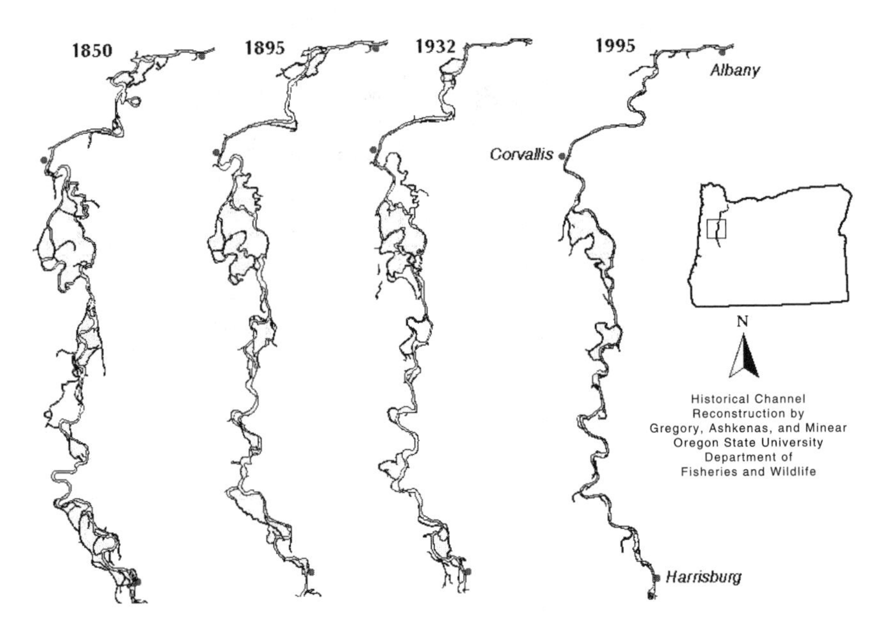

FIGURE 9.4. Time series maps of change in Willamette River channel, Harrisburg to Albany 1850–1995.

and adjacent landscape influences. Thus, the first step in restoration analysis is to divide the river network into functional units or reach types based on physical properties that are linked to hydrological and ecological performances.

Hydrologic performance (e.g., extent of historical flooding) or a management-designated floodplain (e.g., 100-year floodplain of the Federal Emergency Management Agency) provides a landscape framework for evaluation of restoration potential. The main axis of the floodplain serves as the linear context for spatial analysis rather than the river channel because the position of the channel will change and is bounded by the floodplain. The primary axis of the floodplain can be based on visual inspection or smoothing of the linear sequence of midpoints between floodplain margins. The linear floodplains can be broken into smaller segments at predetermined intervals (1 km in this study of the Willamette River) to allow for continuous linear depiction and analysis of floodplain and channel features (Figs. 9.5 and 9.6).

Major criteria for reach delineation along the floodplain include channel slope, floodplain extent, geology, constraint by adjacent landforms, discharge, and tributary influence. After major reach types are delineated, they can be crosschecked with other relevant physical or ecological information to determine whether the reaches actually function differently.

After reach types are defined, both biophysical and socioeconomic characteristics of each reach are depicted spatially. Reach types and the sequence

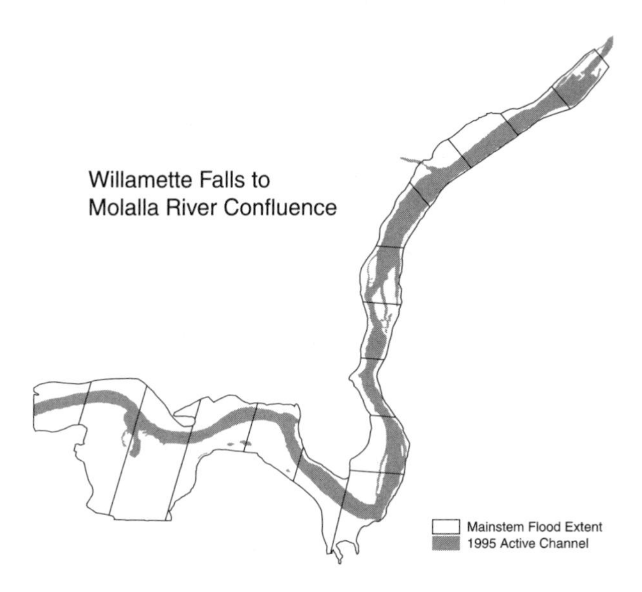

FIGURE 9.5. Detailed map showing 1995 channel and maximum historic flood extent within 1-km bands of Reach 2 of Willamette River.

of floodplain segments within them provide a common basis for linear mapping of channel conditions, riparian forests, land use, population density, or economic values and characteristics. In the Willamette River project, channel dimensions (area, length) and structure (slope, sinuosity, bifurcation, secondary channels, islands, and tributary junctions) are depicted longitudinally and analyzed both sequentially and independently. Riparian features (composition of vegetation classes, area or length of forest, patchiness) can also be analyzed in a similar fashion. Demographic features depicted for reaches include population density, urban growth boundaries, land-use zoning, structures, and roads. Simultaneously, we describe and, where feasible, depict the spatial distribution of relevant economic characteristics, including land value, economic production, costs of operations, and economic incentives such as existing restoration programs, wetland reserves, riparian conservation zones, and tax deferments.

9.6.2 Focal Area Identification

After the biophysical and socioeconomic features are spatially arrayed, criteria for restoration potential can be applied to identify areas within each reach that offer the desired set of biophysical and socioeconomic attributes for restoration. Several types of screening can occur at this step to allow for prioritization to meet specific objectives. If likelihood of long-term ecological benefit is the primary objective, the screening can highlight areas with the greatest potential to increase biological diversity or ecological

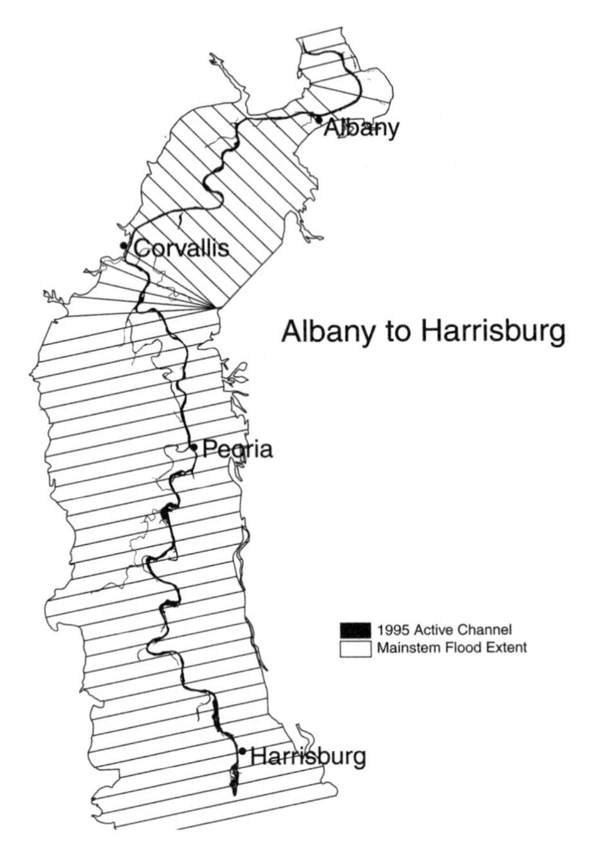

FIGURE 9.6. Detailed map showing 1995 channel and maximum historic flood extent within 1-km bands of Reach 6 of Willamette River.

function and the least socio-economic constraints (Fig. 9.1). Such efforts are more likely to provide the desired future ecological conditions with less risk of human alteration. An alternative restoration objective is to protect critical resources and habitats that are at risk because of land development and conversion. Screening for locations with combinations of high ecological potential and high socio-economic constraints can identify potential areas that could be lost or degraded in the near future. Immediate restoration or conservation efforts would be required to maintain these options. A third objective for restoration screening could be specific ownerships or land-use types, such as publicly owned lands, lands managed by specific agencies, agricultural lands for conservation incentive programs, transportation-related lands and easements, or urban lands. Specific ownerships and land-use zones can be identified through spatial analysis of tax assessment data and then prioritized by ecological potential. This spatial context for both ecological and social information allows multiple objectives for riparian restoration and conservation to be evaluated and documented.

As an example, Figure 9.7 shows a graph of riparian forest area in the floodplain of the Willamette River in 1850 (light line) compared to the area remaining in 1990 (dark line). Comparing the difference between these two lines at any river kilometer illustrates how this approach can accommodate three possible restoration prioritizations: (1) where the highest ranked site is the one exhibiting the greatest potential gain or avoided loss; (2) where it is the one with the most pristine state; and (3) where it is the one nearest some reference level of the endpoint in question (Hyman and Leibowitz 2000).

Geomorphic objectives could include reconnecting lands to the floodplain, restoring historical side channels and sloughs, and developing connections to floodplain ponds. Hydrologic objectives could involve increased storage of floodwater, returning river flows to a more natural range of variability, increased volume of floodplain lakes and ponds, restoration of wetland hydrologic conditions, or restoration of subsurface flow into off-channel habitats or critical habitats in the main channel. Objectives for riparian vegetation could include restoration of historical vegetation patterns and composition, restoration of patch structure (e.g., size, heterogeneity, frequency, richness), increased extent of floodplain forest, restoration of wetland communities, or reduction of invasive exotic plants. Aquatic ecological objectives may incorporate flood refuge for mainstem species, off-channel habitat during low flow, habitat heterogeneity, diversity (e.g., fish, macroinvertebrates, aquatic plants, amphibians), ecological processes (e.g., nutrient uptake, organic matter retention and storage, primary production, decomposition), or reduction of invasive exotic species. Demographic objectives can be related to regional or local plans for land use/ownership, design of transportation systems, minimizing the undesired effects of accommodating projected population growth and projected use of water and water resources. Economic objectives could include maximizing financial opportunities and yield to the land owner, reduction of operational costs, reduction of maintenance costs, reduction of risks of catastrophic loss (e.g., flood damage, loss of riparian land), meeting environmental requirements, or equitable distribution of economic impacts/stimuli within the community.

9.6.3 Designing Restoration Alternatives for a Focal Area

Within a given reach that encompasses both floodplain and active channel, specific locations will have different potentials to achieve either ecological or social objectives. Such areas can be ranked according to the screening criteria and then evaluated based on logistic limitations, opportunities related to public ownership or willing land owners, adjacency to the active channel, or other operational characteristics. The most highly ranked

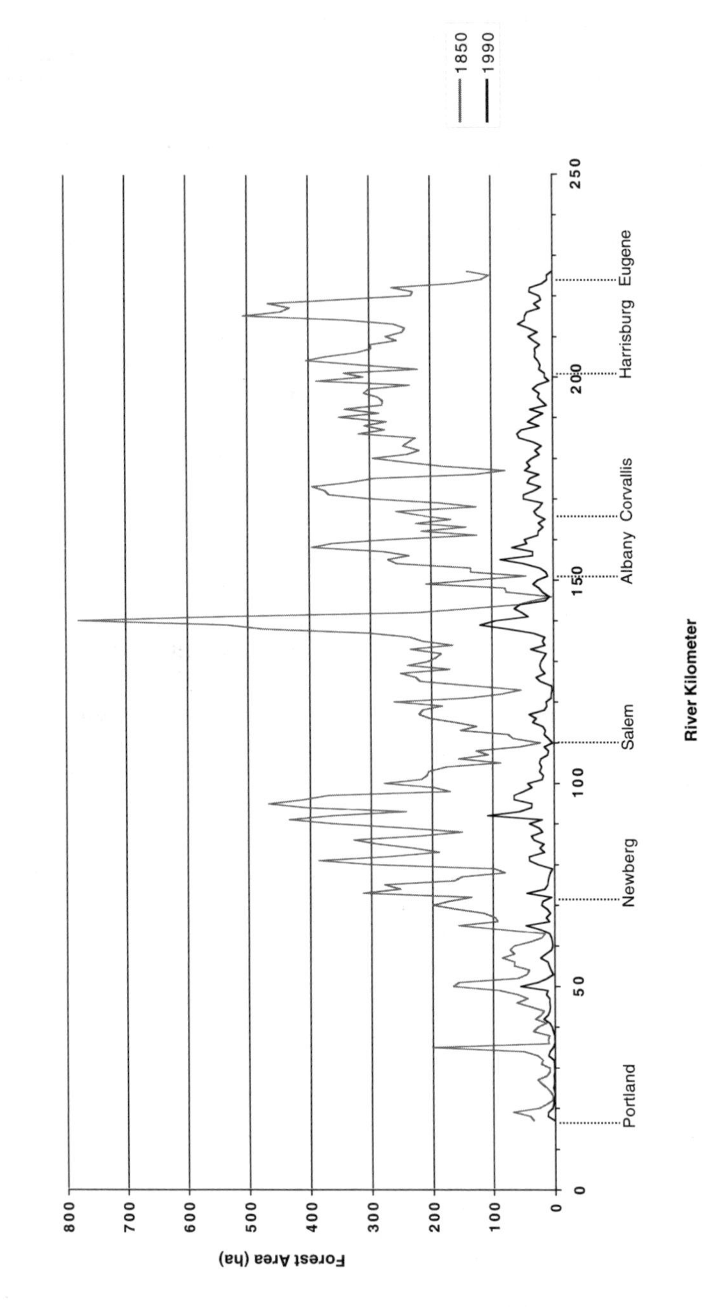

FIGURE 9.7. Graph showing change in riparian forest area by 1-km band 1850–1990.

locations after this operational screening then become focal areas for restoration planning and design.

Restoration design becomes ground based and site specific when focal areas are identified. Multidisciplinary teams review the spatial context for the focal area (e.g., topographic maps, floodplain extent, historic and contemporary maps or photos of riparian vegetation, ownership boundaries, discharge records) and work with local stakeholders to define potential biophysical and socio-economic objectives for the site. These objectives may include proposals for new or more targeted incentive programs and suggested changes to existing governance systems. Place-specific design alternatives are then created for some specified time period in the future. These design alternatives take the form of recommended changes to the amounts, patterns, and locations of specific vegetation communities and fluvial features, as well as institutional changes necessary to sustain these patterns through time. These institutional changes are often a key linkage to reach and network scale processes. The proposed focal area design alternatives respond to the restoration objectives, and are tailored to the particularities of a focal area through multidisciplinary team visits to the site to verify the analyses. The team identifies limitations and opportunities created by local conditions and modifies the focal area objectives accordingly. These on-site data are then represented in a spatially-explicit design alternative for each focal area that illustrates the ecological outcomes, socio-economic benefits and costs, and timing of expected responses. During the focal area design phase, site limitations may be revealed that prevent implementation of restoration efforts and require selection of alternative focal areas.

9.6.4 Modeling and Monitoring the Integration of Design Alternatives Into River Network Extents

A critical step in riverine riparian restoration is the integration of site-specific focal area design alternatives into a larger river network context. After design of restoration activities for focal areas, the projected outcomes are incorporated into the spatial database for the reach and entire river network. Spatial distributions are inspected to identify sections of the river that may have been overlooked or underrepresented in the analyses. Dispersal rates or estimates of longitudinal influence can be used to evaluate potential gaps or areas of interaction between the sequence of restoration activities along the river. Such analyses also offer a mechanism for communities to identify potential opportunities for private landowners to coordinate or cooperate with restoration of public riparian resources and increase the likelihood of attaining desired future conditions (see Haufler and Kernohan, Chapter 4).

While environmental monitoring has become a mantra of management, restoration efforts rarely incorporate monitoring into their planning and implementation. This river network-based approach provides a spatially-explicit framework for designing monitoring measurements linked to projected outcomes for riparian restoration activities. Where outcomes are related to landscape pattern and ecological properties that can be remotely sensed, regional monitoring of overall land cover and land use can be used to track effectiveness of site-specific restoration. Site objectives that are related to water quality or aquatic population or community responses may be more difficult to monitor, but the explicit identification of projected future conditions enhances coordination with monitoring programs at larger scales. Comparing remotely sensed representations of land-cover conditions a decade from project construction with the spatial representation of the design alternative can serve as a simple gage of success. It also points out those areas where further effort is required to achieve long-term goals. Last, explicit statement of both biophysical and socio-economic processes and objectives increases the ability to refine potential monitoring plans to a set of less complicated measures that are most closely associated to the intended outcomes for a site, reach, or entire network.

Institutions, agencies, and communities have collected extensive environmental and resource information, but our collective memory of such information is poor. Regional data, photographic records, biological and environmental samples, and background information for environmental policies are rarely archived, and centralized and accessible repositories for such information rarely exist. The integrated, spatially-explicit databases of biophysical and socio-economic information in this network approach for restoration prioritization can be electronically documented, stored, and retrieved. As such, it encourages long-term development and ongoing improvement of the databases and models used in its analysis. One of the major contributions of the approach is a framework for future information improvement and restoration design.

9.7 Conclusions

The approach presented here for the Willamette River in Oregon may be seen as a local application of the ESA recommendations, one that has the potential to make evident the current misfits between prioritized locations for restoration and the institutional capacities to implement proposed restoration designs in those locations. With these misfits identified, ecological and institutional requirements may become clearer. Once identified, this context can highlight (1) barriers that stand in the way of restoration and (2) the need to either create effective institutional docking ports where they do not now exist or strengthen them where they do.

The emergence of alternative environmental governance regimes offers the possibility of a broader set of these docking ports. A systematic review and testing of these governance regimes as they are applied to restoration opportunities at local levels may prove a useful step in more firmly grounding ecological restoration objectives in locally specific understanding of demographic and economic processes.

Acknowledgments. The authors thank project team members Ed Whitelaw, Dixon Landers, Ernie Niemi, Peter Bayley, Linda Ashkenas, Paula Minear, Allan Branscomb, Hilary Dearborn, Susan Payne, Maureen Raad, David Richey, Jorge Goicochea, and David Diethelm. The authors are also grateful for the comments of three particularly helpful anonymous reviewers. This work was funded by STAR Grant R825797 between the U.S. Environmental Protection Agency and Oregon State University. Data on historical river changes and land use/land cover were developed as part of cooperative agreement CR824682 between USEPA and OSU. This manuscript has not been subjected to U.S. EPA review. The conclusions and opinions are solely those of the authors and are not necessarily the views of EPA. Supplementary information and the digital datasets referenced herein are available via the Internet at *http://www.orst.edu/dept/pnw-erc/*.

References

Ahern, J. 1999. Spatial concepts, planning strategies, and future scenarios: a framework method for integrating landscape ecology and landscape planning. Pages 175–201 *in* J. Klopatek and R. Gardner, editors. Landscape ecological analysis: issues and applications. Springer-Verlag, Inc., New York, New York, USA.

Coulton, K.G., P. Goodwin, C. Perala, and M.G. Scott. 1996. An evaluation of flood management benefits through floodplain restoration on the Willamette River, Oregon, U.S.A. Philip Williams and Associates, San Francisco, California, USA.

Doppelt, B. In press. Emerging approaches to watershed governance: new approaches to guide sustainable ecological, economic and organizational performance within the watershed context. Island Press, Washington, D.C., USA.

Galloway, G.E. 1996. U.S. floodplain policy pursues higher ground. Pages 98–101 *in* Forum for Applied Research and Public Policy, Fall.

Gregory, S.V., and P.A. Bisson. 1996. Degradation and loss of anadromous salmonid habitat in the Pacific Northwest. Pages 277–314 *in* D. Stouder and R.J. Naiman, editors. Pacific salmon and their ecosystems: status and future options. Chapman & Hall, New York, New York, USA.

Gregory, S.V., D.W. Hulse, D.H. Landers, and E. Whitelaw. 1998. Integration of biophysical and socio-economic patterns in riparian restoration of large rivers. Pages 231–247 *in* H. Wheater and C. Kirby, editors. Hydrology in a changing environment: Proceedings of the British Hydrological Society International Conference, Exeter. John Wiley and Sons, Chicester, England.

Harms, B.H., J.P. Knaapen, and J.G. Rademakers. 1993. Landscape planning for nature restoration: comparing regional scenarios. Pages 197–218 *in* C.C. Vos and

P. Opdam, editors. Landscape ecology of a stressed environment. Chapman & Hall, London, England.

Hyman, J.B., and S.G. Leibowitz. 2000. A general framework for prioritizing land units for ecological protection and restoration. Environmental Management **25**:23–35.

Interagency Floodplain Management Review Committee (IFMRC). 1994. Sharing the challenge: floodplain management into the 21st century. Washington, D.C., USA.

Johnson, K.N., and M. Herring. 1999. Understanding bioregional assessments. Pages 352–355 *in* K.N. Johnson, F. Swanson, M. Herring, and S. Greene, editors. Bioregional assessments: science at the crossroads of management and policy. Island Press, Washington, D.C., USA.

Montgomery, D.R., G.E. Gordon, and K. Sullivan. 1995. Watershed analysis as a framework for implementing ecosystem management. Water Resources Bulletin **31**:369–386.

Murray, T., P. Rogers, D. Sinton, C. Steinitz, R. Toth, and D. Way. 1971. Honey Hill: a systems analysis for planning the multiple use of controlled water areas. Institute for Water Resources, U.S. Army Corps of Engineers. IWR Report 71-9, Vol. 1 NTIS doc. no. AD736343, Vol. 2 NTIS doc. no. AD736344.

Naiman, R.J., and H. Decamps. 1990. The ecology and management of aquatic-terrestrial ecotones. UNESCO-MAB, Paris, France.

Naiman, R.J., H. Decamps, J. Pastor, and C.A. Johnston. 1988. The potential importance of boundaries to fluvial ecosystems. Journal of the North American Benthological Society **7**:289–306.

National Research Council (NRC). 1992. Restoration of aquatic ecosystems. National Academy Press, Washington, D.C., USA.

Noss, R.F., M. O'Connell, and D. Murphy. 1997. The science of conservation planning: habitat conservation under the Endangered Species Act. Island Press, Washington, D.C., USA.

Petts, G.E. 1990. The role of ecotones in aquatic landscape management. Pages 227–261 *in* R.J. Naiman and H. Decamps, editors. The ecology and management of aquatic–terrestrial ecotones. UNESCO, Paris, France, and Parthenon, Carnforth, UK.

Schoonenboom, I.J. 1995. Overview and state of the art of scenario studies for the rural environment. Pages 15–24 *in* J.T.Th. Schoute, P.A. Finke, F.R. Veenenklaas, and H.P. Wolfert, editors. Scenario studies for the rural environment, selected and edited proceedings of the symposium Scenario Studies for the Rural Environment, Wageningen, The Netherlands, 12–15 September 1994. Kluwer Academic Publishers, Dordrecht, The Netherlands.

Steinitz, C. 1990. A framework for theory applicable to the education of landscape architects (and other environmental design professionals). Landscape Journal **9**: 136–143.

Van der Ploeg, J.D. 1995. The tragedy of spatial planning. Pages 75–90 *in* J.T.Th. Schoute, P.A. Finke, F.R. Veenenklaas, and H.P. Wolfert editors. Scenario studies for the rural environment, selected and edited proceedings of the symposium Scenario Studies for the Rural Environment, Wageningen, The Netherlands, 12–15 September 1994. Kluwer Academic Publishers, Dordrecht, The Netherlands.

10
Applying Ecological Guidelines for Land Management to Farming in the Brazilian Amazon

Virginia H. Dale

Clearing of land by small-scale farmers in the Brazilian Amazon state of Rondônia provides an example of how the ecological guidelines for land management might be applied. To understand the ecological underpinnings of this situation, a simulation model is used to project land-cover change under scenarios of extreme, typical, and sustainable uses of the land. Application of an animal-use model shows conditions under which species with different territory sizes and gap-crossing abilities might remain on the land under the three land-use scenarios. These model projections, along with an understanding of the current situation, provide a means to evaluate how well the eight guidelines suggested by the Land Use Committee of the Ecological Society of America are being applied in Rondônia. This chapter constitutes a hypothetical analysis because the guidelines were developed subsequent to the Rondônian study, and they have not yet been communicated to the landowners or government in Brazil.

Most of the guidelines are not being followed in the Brazilian Amazon by typical farmers, but the guidelines are being adhered to by farmers using sustainable practices. For example, farmer decisions are framed in the context of regional politics and economics but not in view of ecological implications. It is only from the perspective of an individual family that the long-term effects of the farming are considered. Yet, the effects of development on ecological processes are typically neither avoided nor compensated. An attempt is being made to retain large continuous areas, but modeling studies suggest that this goal will not be achieved over the long run. Where rare landscape elements are not preserved, some associated species are being eliminated in Rondônia. The guidelines that call for avoiding land uses that deplete natural resources and minimize the introduction and spread of alien species are typically not being followed in the Brazilian Amazon. This analyses points out that implementation of the guidelines depends on the farming system being implemented as well as the ability to manage land in an ecologically balanced way.

The Land Use Committee of the Ecological Society of America set forth eight guidelines for the appropriate use and management of land (Dale et al., Chapter 1):

- ☐ Examine the impacts of local decisions in a regional context.
- ☐ Plan for long-term change and unexpected events.
- ☐ Avoid or compensate for effects of development on ecological processes.
- ☐ Retain large contiguous or connected areas that contain critical habitats.
- ☐ Preserve rare landscape elements, critical habitats, and associated species.
- ☐ Avoid land uses that deplete natural resources over a broad area.
- ☐ Minimize the introduction and spread of nonnative species.
- ☐ Implement land-use and land-management practices that are compatible with the natural potential of the area.

This chapter considers how these ecological guidelines might be applied to small-scale farms in the Brazilian Amazon (even though the guidelines were developed after these decisions were made). The practices of typical farmers are compared to those of sustainable farmers (who do not burn their lots, plant a diversity of agricultural goods, and use some of the native products on the land). The examination is meant to shed the light of reality onto the guidelines and to illustrate how these guidelines may be interpreted and applied. This example is not a totally positive one, but it provides some insight into the application of the guidelines.

The land-use problem in Amazonia is severe. Local deforestation is not only changing the landscape in the Amazon but also affecting the global climate. Local changes include loss of species and habitat, shifts in hydrological regimes, and soil degradation (Hecht 1990; Shukla et al. 1990; NRC 1992; Fearnside et al. 1993; Laurance et al. 1997). Because the forested area in the Amazon is so large, its diminishment changes global climate and biodiversity (Post et al. 1990; NRC 1992).

The focus of this analysis is the Brazilian state of Rondônia (Fig. 10.1). Rondônia occupies 243,000 km^2 of largely forested land. These tropical forests have a tremendous diversity of species and very large trees. Rondônia has fairly rich soils for the Amazon. Thus, if farming is going to be successful any place in the Amazon, it would probably be in Rondônia.

10.1 Development in Rondônia

Much of the deforestation in Rondônia occurs as a result of road development. The major highway through the state was paved in the early 1980s, and a series of unpaved roads were laid out in a network around central cities with the intent to encourage farmers to migrate into this area.

FIGURE 10.1. Map of Rondônia and Brazil.

The farmer is actually the head of an extended household sometimes consisting of children and grandchildren who all live on a single lot.

The migration of farmers was promoted by the Brazilian government for at least two reasons (Forestra 1991). First, a drought had occurred in Brazil in the late 1960s, and the government wanted to ensure production of enough food for the highly populated cities in the south and coastal areas of the country in the event of a future drought. Second, the Brazilian government was concerned about a possible attempt by Peru to overtake Rondônia, which is adjacent to Peru and was once a part of that country. Having Brazilian citizens in the countryside was thought to be one way to deter such action.

The encouragement for migration was well received, and a number of people did move into the state largely from the agricultural areas of southwest Brazil. Many of these immigrants had been tenants on large farms that used heavy machinery in the production of grain and other crops that are best grown on a large scale (Dale and Pedlowski 1992). Thus, these people were poorly prepared for the life of a small, isolated farmer with no mechanized equipment.

Improvements to the road infrastructure facilitated migration into the countryside of Rondônia. People could easily travel along the improved road network and tended to settle and clear small farms along the unpaved roads. The increase in roads, thus, was accompanied by an increase in deforestation as lands adjacent to the roads were cleared for farming (Fig. 10.2) (Pedlowski et al. 1997).

Farmers move into the area with the intent of establishing agriculture, so their first action is to start clearing the forest. Usually, these forests are not used for timber or any other kind of forest product. They are merely burned to make the land available for cropping. The farmers use machetes and, more recently, chain saws to clear the vegetation and to cut down many of the trees and much of the other vegetation so they can burn it before the rainy season starts. The burning is of fairly low intensity, for it is difficult to get these forests to ignite. After a burn, much plant material is left on the ground. Some palms and other vegetation typically survive the fires and are often seen in the agricultural fields. Even while the fields are still smoking, the farmers begin planting seeds in order to get them in the ground before the rainy season starts.

Usually annual crops, such as upland rice and corn, are grown. This type of agricultural system is thought of as European in tradition rather than

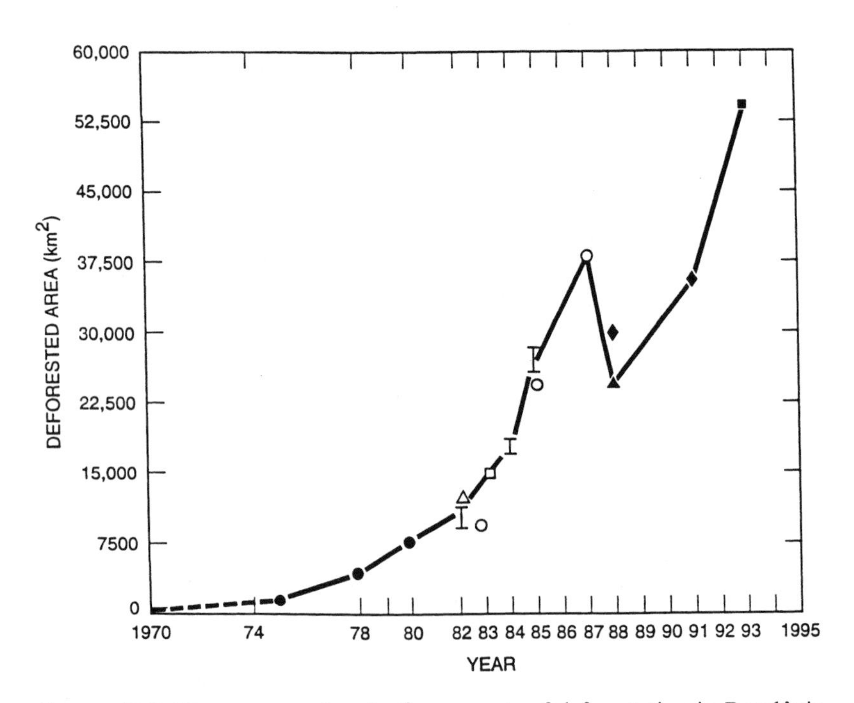

FIGURE 10.2. Changes over time in the amounts of deforestation in Rondônia.

being tropical in origin. After a few years the soil becomes depleted, causing the crop lands to decline in production, and they are often turned over to pasture. As much as 40% of the pastures do not support animals, and only about half the cows on these pastures are beef cattle. Milk is a major product in the study area because the city of Ouro Preto has milk distribution as a primary industry. A transportation system is available by which trucks come to the farms to pick up milk canisters. This system is very important, for most of the people do not have vehicles to transport their goods to market.

There was a slight decline in the rate of deforestation in Rondônia during the late 1980s that was related to precipitation events (Fig. 10.2). When the summers are wet, the people cannot establish burns and, therefore, it is very difficult to clear the forest. However, the summers usually have a fairly dry period, and the deforestation is continuing at a fairly steady rate. The impact of these activities have been so thorough that the road network and the deforestation that occurred adjacent to it can be seen from satellite images (Skole and Tucker 1993).

The deforestation that is occurring in Brazil is not an aberrant situation. Iverson (1991) compared the rate of land clearing in several locations throughout the world. The 1.47% clearing per year for Rondônia is within the range of annual clearing for Malaysia, Costa Rica, and some places in the United States. What is unique about recent deforestation in the tropics is not the rate of clearing but the scale of land-cover change. The forested area of the Amazon is currently the largest continuous forest on earth, and its diminishment has global implications.

10.2 Methods to Understand Land-Use Patterns and Changes

To understand the land-use situation and dynamics, we developed a land-use-change model (which is explained in detail in Southworth et al. [1991] and Dale et al. [1993, 1994*a*]). The model was built upon spatial data for the region. The major elements of the model are depicted in Figure 10.3. Very simply, the model contains a settlement diffusion component that deals with the human dynamics of people moving into, within, and out of the area. It has an ecological-impacts module that projects site-specific impacts. Infrastructure in the model refers to the road system and its effects. And the model includes socioeconomic aspects of land use that are concerned with such factors as how distance to market, prices, past land uses, and soil conditions affect farmer decisions about how much land is cleared and what crops to plant.

The model is initiated with spatial data layers, including several measures of soil quality, the existing road network, and the type of native vegetation.

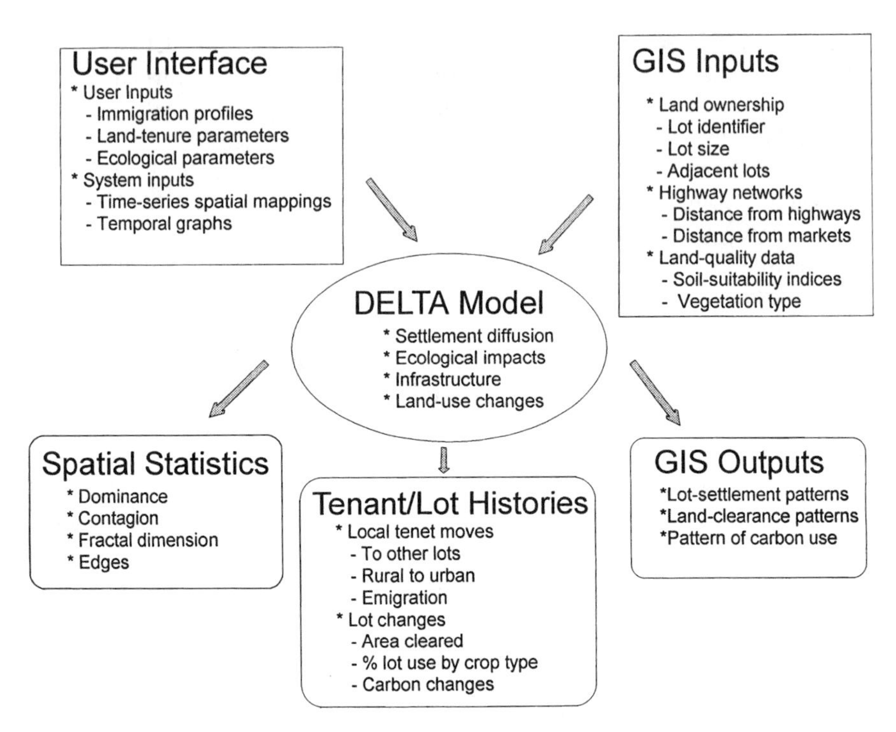

FIGURE 10.3. A simple depiction of the model used to project deforestation in Rondônia.

Farmers make decisions about where to clear, what to plant, and whether or not to remain on a lot based on a relative lot-attractiveness index, which combines 10 factors, such as distance to the market, soil conditions, and previous land uses. The data layers and the lot attraction index contribute to a decision model that estimates the probability that a farmer will select one particular lot over all other lots in the area. The probability that a farmer will select a series of lots can also be predicted.

What typically occurs in the small-scale farming in Rondônia is that a farmer establishes his family on a lot and, after an average of three growing seasons, the lot has degraded so he moves to another lot or another region of the country. The scenario under which we ran the model had a set number of families introduced over time into 294 lots near Ouro Preto. In the model, these farmers chose the appropriate lots for farming and made choices about where to plant on a particular lot, what they plant, and whether to have pasture and cattle. When farmers migrated off of the lot, we assumed that half stayed in the same 294-lot region and half moved elsewhere. For example, some of these farmers gave up farming altogether and mined gold.

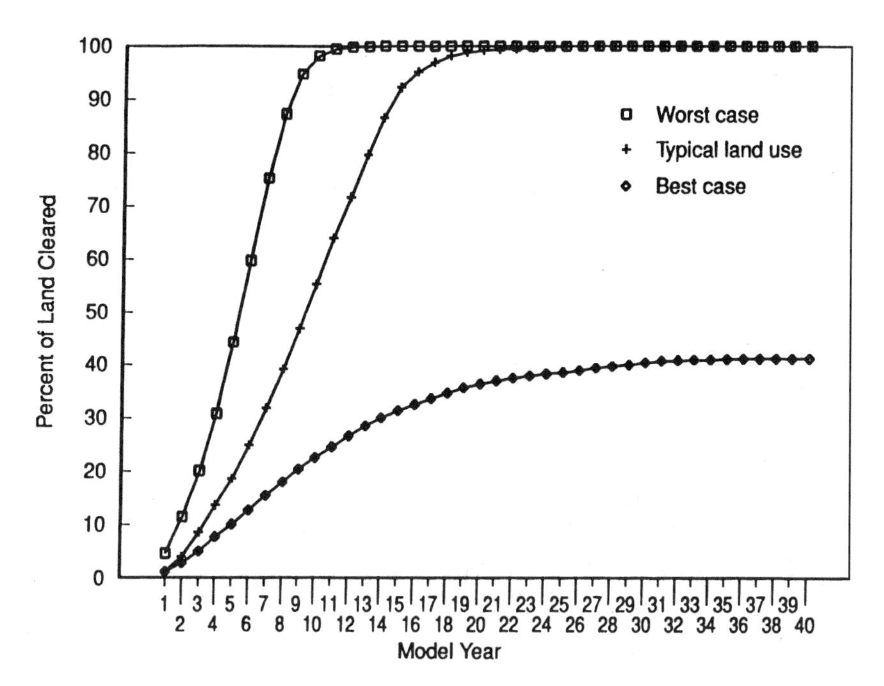

FIGURE 10.4. Projected percentage of land cleared of the 294 lots simulated in Ouro Preto, Rondônia (from Dale et al. 1994*a*).

The model projections are based on three scenarios: an extreme case, a typical case, and a sustainable situation. The extreme case simulates farming practices along the Transamazon Highway where in only a very few years farmers depleted the soils and left their lots (Moran 1981). The typical case is based on literature reports of farmer activities in Rondônia (Leite and Furley 1985; Coy 1987; Millikan 1988). The sustainable case is created as a situation in which the farmers do not burn their lots, plant a diversity of agricultural goods, and use some of the native products on the land.

The projections of the amount of land cleared in the region are shown in Figure 10.4. What is surprising is the similarity in the amount of land cleared between the extreme case and the typical case in which all of the lands are cleared fairly quickly (Dale et al. 1993, 1994*a*). In contrast, under the sustainable-case scenario, much of the land remains forested because the farmers use the natural products of the site and work with a smaller area of land.

What may be most interesting is to look at the cumulative number of farmers that leave the lots over the 40-year simulated time period under the three scenarios (Fig. 10.5). Again, the extreme and the typical cases show similar patterns in which, after only a few years, the farmers are not able to maintain themselves on the lot and quickly move out of the system (Dale et al. 1993). It is only in the sustainable-case scenario that the

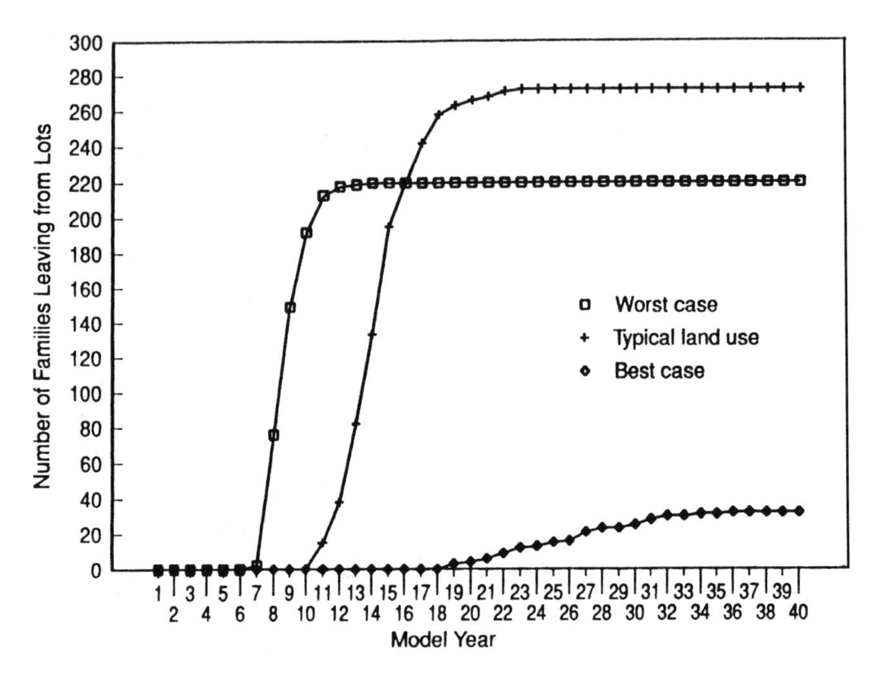

FIGURE 10.5. Projected number of families leaving the lots of the 294 lots simulated in Ouro Preto, Rondônia (from Dale et al. 1994*a*).

farmers maintain themselves on the lots for the full 40-year projection period.

Independent of these projections, we interviewed 90 farmers (Dale et al. 1993) and found only three who were using the land in a manner similar to the sustainable-case scenarios. When asked why they were employing these practices, these few farmers replied that it seemed the right thing to do. Then they would brag about the way the wild animals inhabiting neighboring areas would congregate on their land during the periods of intense burning.

Today, the state government is presenting farmers with sustainable practices as examples for other farmers in the region. The state takes farmers to visit their lands to demonstrate appropriate farming practices. This situation was quite different 25 years ago. Farmers, then, were discouraged from intercropping or using multiple crops, practices that are basic to sustainable farming in the Amazon.

10.3 Applying the Guidelines to Rondônia

The potential for land-use planning is hampered by the absence of ecological information such as is contained in the land-use report of the Ecological

Society of America (ESA) (Dale et al., Chapter 1). In fact, the ESA guidelines were developed to provide relevant information about the ecological values of the land for the landowner. But the analysis of farmers in Rondônia was done before the guidelines were available. Therefore, this chapter focuses on the how the guidelines for land use and management developed by the ESA could apply to the land practices in Rondônia even though the farmers have not yet had access to the information in the report.

The first guideline is to examine the impacts of local decisions in a regional context. Satellite images of Rondônia show a herringbone pattern of cleared land along the roads against the background of forested land, illustrating the need to consider immigration in the context of the larger region. The decision to encourage migration into Rondônia was motivated by regional and national perspectives of politics and economics (Forestra 1991) to retain the area as part of Brazil and to increase agricultural productivity. However, broad-scale ecological concerns were not considered. Yet these land-use decisions have negatively impacted the local fauna, regional patterns of precipitation, and global climate (Fearnside and Fernandes 1993). Nevertheless, sustainable farmers mimic natural processes of the region by cutting only small openings in the forests and growing a diversity of crops in any one place.

The second guideline, to plan for long-term effects on land-use change, can be illustrated by the perspective of an individual family. The typical farmer does not follow this guideline, for the land does not support a family for long. However, changes are being made in the way that farmers view the land and its resources. As mentioned previously, the few farmers who employ sustainable practices are now being promoted as examples of how long-term, profitable farming can be achieved. Also, sustainable farming practices are now being supported by such institutions as farming cooperatives, which provide economic resources that allow farmers to bridge times of low yield and also provide a market for goods (Caviglia 1999).

A third guideline is to avoid or compensate for the effects of development on ecological processes. Sustainable farmers are following this guideline by providing refuges of native vegetation where animals often hide when the typical farmers are burning their lots. However, this guideline is not generally being applied in Rondônia. Farmers typically clear large expanses of their land, thus eliminating animal habitat, and do not harvest the natural products from the land (such as cupaucu, cocoa, or brazil nuts). Local markets are not available for these products, and the many of the people have learned to prefer the more exotic imported food. Yet, there are many products from this region that are useful not only for food but also as pharmaceutical or industrial products (e.g., rubber). Another illustration of the failure to consider the effects of development on ecological processes is the road pattern. Straight roads were built in spite of the topography. The roads seem to have been laid out by an office draftsman with no first-hand knowledge of natural features in the area. The road may cross a steam several times within a kilometer. This absence of planning is costly to implement and results in roads that are very difficult to cross during the rainy season.

The model projections (Fig. 10.4) illustrate the general lack of application of the guideline of retaining large continuous areas. The attempt was made by planners to retain connectedness, for the farms were set up on large lots (averaging 101 ha). The cleared portion of each lot is typically close to the road, and the back part of each lot is kept in forest. Up to this point in time, the connected forest along the back of the lots has been retained; however, the modeling study suggests that this situation is not going to remain over the long term. The model projects these forested areas to be depleted over the 40-year time frame in both the extreme and typical cases (Dale et al. 1994*b*). Yet the sustainable case retains much of the forested land over the 40-year projection period.

Not only is loss of forest a concern, but the breaking up of the continuous forest into a mosaic of forest and nonforested land has ecological impacts. The effect of such fragmentation on animals can be analyzed by employing another model we developed that is based on two features of animals: their territory size and gap-crossing ability. Gap-crossing ability refers to the tendency of some species not to cross inhospitable habitat. For example, Mladenoff et al. (1995) found that wolves (*Canis lupus*) avoid crossing roads, and thus wolf packs in the Great Lakes region are most likely to occur in areas with road densities below 0.23 km/km^2. To illustrate the impact of fragmentation, Figure 10.6 groups Amazonian species that have similar requirements for territory size and gap-crossing ability. Figure 10.6*a* shows those species that have moderate areas and moderate gap-crossing abilities. A number of insects and some small mammals are in this category. Considering the scenarios for the extreme and sustainable cases, only the sustainable situation was able to support these organisms. But Figure 10.6*b* shows that species with large territories but small gap-crossing abilities will not be supported for long under any scenario. Jaguars (*Felis onca*) are in this category. The lesson here is that, under any of the modeled land uses, some wide-ranging species like the jaguar will not be retained. Thus, Figure 10.6 illustrates the concept that not preserving rare landscape can eliminate some species.

The guideline of avoiding land uses that deplete natural resources is typically not being considered in the Brazilian Amazon. Most of the farmers are implementing agriculture practices that are common to developed areas and not compatible with native ecosystems. These practices quickly reduce soil fertility and drastically alter water storage and quality. On the other hand, sustainable farmers are not using fire as a land-management tool. As a result the soil fertility on lands with sustainable farming is retained, and some farmers have been using the land for as long as 20 years.

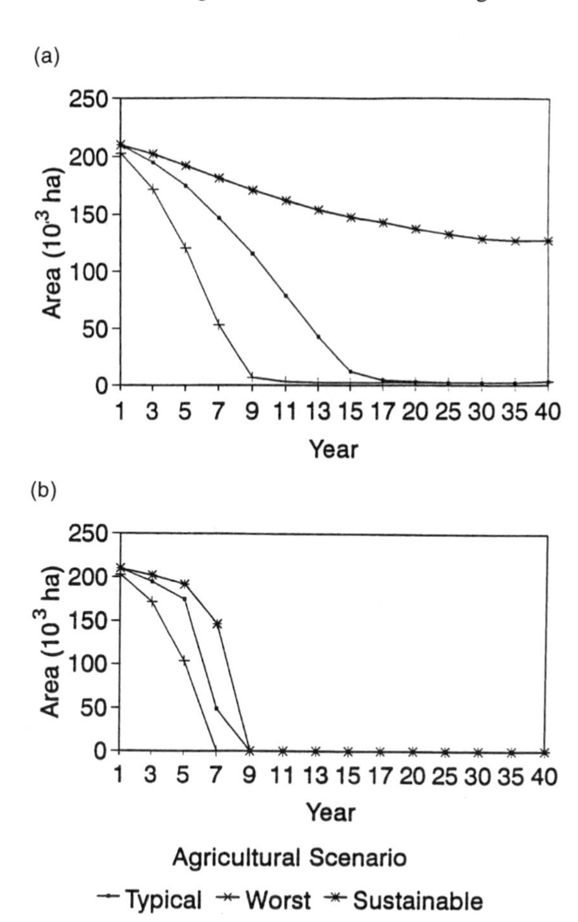

FIGURE 10.6. Projected habitat loss for species with (*a*) both moderate area requirements and moderate gap-crossing ability and (*b*) large area requirements and low gap-crossing ability (from Dale et al. 1994*b*).

A final guideline is to minimize the introduction and spread of alien species. But this guideline is not being adhered to in the Brazilian Amazon. For example, with the help of cooperatives, farmers have introduced Italian honeybees as a way of supplementing their resources (Caviglia 1999). The bees are productive, and much honey is being obtained. Furthermore, an intercropping system is being employed whereby the farmers layer trees used first for honey production and later for timber with coffee trees in the understory. This situation provides multiple crops. Nevertheless, there is concern that these introduced honeybees are competing with the native pollinators. Many of the endemic pollination systems are so specific that, if native pollinators are eliminated via competition, the plants that they pollinate may subsequently be eliminated. Sustainable farmers, on the other hand, are more likely to use the food and dyes from native plants.

10.4 Conclusions

Typical farmers tend to not practice the set of guidelines, whereas sustainable farmers do. Implementation of the guidelines depends on the farming system. Yet, it is important to consider the individual landowner and the importance of that person's personal goals for the land. Based on our interviews with a few farmers, it seems the farmers have to accept ecological values before they can consider implementing these guidelines. However, farmers do not always know how to implement ecological guidelines even if ecological goals are a part of their value system. Scientists, therefore, have an obligation to impart the ways and tools with which people can manage land in an ecologically balanced manner. This book is a start along that communication task, but it is going to take a long-term and committed effort to maintain the regular transfer of the most up-to-date scientific information to land managers.

Acknowledgments. Fred O'Hara edited the manuscript. Reviews of an earlier draft by Cynthia Botteron, Richard Haeuber, Dale Huff, and Bob O'Neill have been quite helpful. This is publication number 4987 of the Environmental Sciences Division of Oak Ridge National Laboratory (ORNL). ORNL is managed by Lockheed Martin Energy Research Corp. for the U.S. Department of Energy under contract DE-AC05-96OR22464. The submitted manuscript has been authored by a contractor of the U.S. Government under contract No. DE-AC05-96OR22464. Accordingly, the U.S. Government retains a nonexclusive, royalty-free license to publish or reproduce the published form of this contribution, or to allow others to do so, for U.S. Government purposes.

References

Caviglia, J.L. 1999. Sustainable agriculture in Brazil: Economic development and deforestation. New Horizons in Environmental Economics Series. Edward Elgar, Cheltenham, England.

Coy, M. 1987. Rondônia: Frente pioneira e programa POLONOROESTE. O processo de diferenciação sócio-econômica na periferia e os limites do planejamento público. Pages 53–270 *in* G. Kohlepp and A. Schraeder, editors. Homen e natureza na Amazonia simposio international e interdisciplinar, Blaubeuren, Arbeitsgemeinschaft Deutsche Lateinamerika Forschung (AD-LAF). Tübingen Geographische Studien. Tübingen, Germany.

Dale, V.H., and M.A. Pedlowski. 1992. Farming the forests. Forum for Applied Research and Public Policy **7**:20–21.

Dale, V.H., R.V. O'Neill, M.A. Pedlowski, and F. Southworth. 1993. Causes and effects of land-use change in central Rondônia, Brazil. Photogrammetric Engineering & Remote Sensing **59**:997–1005.

Dale, V.H., R.V. O'Neill, F. Southworth, and M.A. Pedlowski. 1994*a*. Modeling effects of land management in the Brazilian settlement of Rondônia. Conservation Biology **8**:196–206.

Dale, V.H., S.M. Pearson, H.L. Offerman, and R.V. O'Neill. 1994*b*. Relating patterns of land-use change to faunal biodiversity in the Central Amazon. Conservation Biology **8**:1027–1036.

Fearnside, P.M., N. Leal, and F.M. Fernandes. 1993. Rainforest burning and the global carbon budget: biomass, combustion efficiency, and charcoal formation in the Brazilian Amazon. Journal of Geophysical Research **98**:16733–16743.

Foresta, R.F. 1991. The limits of providence: Amazon conservation in the age of development. University of Florida Press, Gainesville, Florida, USA.

Hecht, S.B. 1990. Indigenous soil management in the Latin American tropics: neglected knowledge of native peoples. Pages 151–158 *in* M.A. Altieri and S.B. Hecht, editors. Agroecology and small farm development. CRC Press, Boca Raton, Florida, USA.

Iverson, L.R. 1991. Forest resources of Illinois: What do we have, and what are they doing for us? Illinois Natural History Survey Bulletin **34**:361–374.

Laurance, W.F., S.G. Laurance, L.V. Ferreira, J.M. Rankin-de Merona, C. Gascon, and T.E. Lovejoy. 1997. Biomass collapse in Amazonian forest fragments. Science **278**:1117–1118.

Leite, L.L., and P.A. Furley. 1985. Land development in the Brazilian Amazon with particular reference to Rondônia and the Ouro Preto colonization project. Pages 119–140 *in* R. Hemming, editor. Change in the Amazon Basin, Vol. II: the frontier after a decade of colonization. Manchester University Press, Manchester, England.

Millikan, B.H. 1988. The dialectics of devastation: tropical deforestation, land degradation, and society in Rondônia, Brazil. Thesis, University of California, Berkeley, California, USA.

Mladenoff, D.J., T.A. Sickley, R.G. Haight, and A.P. Wydeven. 1995. A regional landscape analysis of favorable gray wolf habitat in the northern Great Lakes region. Conservation Biology **9**:279–294.

Moran, E.F. 1981. Developing the Amazon. Indiana University Press, Bloomington, Indiana, USA.

National Research Council. 1992. Conserving biodiversity: a research agenda for development agencies. National Academy Press, Washington, D.C., USA.

Pedlowski, M.A., V.H. Dale, M.A.T. Matricardi, and E.P. da Silva Filho. 1997. Patterns and impacts of deforestation in Rondônia, Brazil. Landscape and Urban Planning **38**:149–157.

Post, W.M., T.H. Peng, W. Emanuel, A.W. King, V.H. Dale, and D.L. DeAngelis. 1990. The global carbon cycle. American Scientist **78**:310–326.

Shukla, J., C. Nobre, and P. Sellers. 1990. Amazon deforestation and climate change. Science **247**:1322–1325.

Skole, D., and C. Tucker. 1993. Tropical deforestation and habitat fragmentation in the Amazon: satellite data from 1978 to 1988. Science **260**:1905–1910.

Southworth, F., V.H. Dale, and R.V. O'Neill. 1991. Contrasting patterns of land use in Rondônia, Brazil: simulating the effects on carbon release. International Social Sciences Journal **130**:681–698.

11
Applying Ecological Principles to Land-Use Decision Making in Agricultural Watersheds

Mary Santelmann, Kathryn Freemark, Denis White, Joan Nassauer, Mark Clark, Brent Danielson, Joseph Eilers, Richard M. Cruse, Susan Galatowitsch, Stephen Polasky, Kellie Vache, and Junjie Wu

The use of ecological principles and guidelines in land-use planning, as advocated by the Ecological Society of America Committee on Land Use (Dale et al., Chapter 1) will be critically important to achieving sustainable ecosystems in the next few decades as the world's human population continues to grow and land area under human management increases. Definition of these principles and articulation of guidelines for use by planners and decision makers is an important first step, but there are many obstacles to the application of ecological guidelines in the land-use planning process. The use of alternative future scenarios can help overcome some of the difficulties associated with application of ecologically healthy land-use practices in agricultural watersheds. With the future scenario approach, abstract goals such as enhancing water quality and restoring biological diversity are translated into specific land-use practices (wetland restoration, riparian buffers, alternative cropping practices, preserves) expected to help achieve these goals. Maps and Geographic Information Systems (GIS) databases of the future alternatives become the spatial data used to evaluate the responses of the various modeled endpoints as well as the response of human perceptions of changes in land use. Alternative futures can be used to frame landscape ecological hypotheses (cf. Ahern 1999); models can then be employed to test those hypotheses and focus additional research on components that are poorly understood. In addition to the important principles put forth by Dale et al. (Chapter 1) we suggest three additional principles and guidelines: (1) consider land-use impacts on ecosystem processes such as biogeochemical cycles, (2) incorporate an understanding of human values and cultural practices to determine and implement land-use and land-management practices compatible with these cultural practices and values, and (3) explore and express levels of uncertainty associated with potential effects of land-use alternatives in communications with decision makers, and use knowledge of uncertainties and their relative importance to focus additional research.

When we see land as a community to which we belong, we may begin to use it with love and respect. There is no other way for land to survive the impact of mechanized man, nor for us to reap from it the esthetic harvest it is capable, under science, of contributing to culture. That land is a community is the basic concept of ecology, but that land is to be loved and respected is an extension of ethics. That land yields a cultural harvest is a fact long known, but latterly often forgotten....our bigger-and-better society is now like a hypochondriac, so obsessed with its own economic health as to have lost the capacity to remain healthy.

Aldo Leopold, 1949
Foreword to *A Sand County Almanac*

The Ecological Society of America Committee on Sustainable Land Use has put together a set of ecological principles and guidelines to help in land-use decision making (Dale et al., Chapter 1). The practical application of these principles and the associated guidelines to planning efforts in real landscapes will require the development and use of strategies and tools to translate them into specific sets of land-use practices (riparian buffers, floodplain and wetland restorations, etc.) and ways to place these practices effectively on the landscape. In this chapter, we discuss the use of future scenarios coupled with Geographic Information Systems (GIS) based evaluative models as a methodology for effective land-use planning (cf. Steinitz et al. 1994; White et al. 1997; Hulse et al. 2000). We demonstrate the application of this method to explore alternative futures for two agricultural watersheds in the U.S. Corn Belt.

11.1 Background and Relevant Concepts

Even before Leopold, naturalists and biologists of the nineteenth century realized that the pace of agricultural change was so rapid as to consume or fundamentally alter the natural vegetation communities of the Midwest. In the late 1800s, Thomas MacBride, Bohumil Shimek, Louis Pammel, and others called for preservation of prairie and forest tracts to conserve the natural heritage of Iowa. However, conservation plans written in the early twentieth century did not translate into land set aside for preservation until the mid-1940s when, under the leadership of Ada Hayden, action was taken to preserve some remnants of the native vegetation of Iowa (Roosa 1981).

In spite of more than a century of recognition of the importance of conservation, and decades of efforts to develop sustainable agricultural practices, the pace of agricultural change and industrialization has outstripped the pace of the adoption of conservation practices (Farrar 1981; Roosa 1981). Iowa has been ranked 50th among the 50 United States in the amount of remaining intact natural habitat (Klopatek et al. 1979). Tilling and cropping, removal of riparian forest, draining of wetlands, introductions of nonnative species, use of agrichemicals, and resulting pollution and soil erosion have contributed to declines in water quality and loss of biodiversity in farmland (Crosson and Ostrov 1990; Pimentel et al. 1991; Freemark 1995; Schwartz 1995). Agricultural land use is the primary cause of surface water quality impairment in the United States today (Puckett 1994; Alexander et al. 1996; Runge 1996). In 1995, the U.S. Office of Technology Assessment (OTA 1995) designated the Corn Belt region as the top priority region for action to improve surface water quality. Estimates of the sources of nutrient pollution to the Louisiana Gulf Coast estuaries (Alexander et al. 1996) indicate that 70% of the total nitrogen delivered to the Gulf originated above the confluence of the Ohio and Mississippi Rivers, with 39% of the total coming from watersheds in the Upper and Central Mississippi basins. Regardless of regional source, the United States Geological Survey has estimated that about 90% of the nutrients entering the Gulf originate from non-point sources, primarily agricultural runoff and atmospheric deposition.

11.1.1 Obstacles to the Use of Ecological Principles in Land-Use Decision Making

The concept of using ecological principles to guide land use is not new, although it is still revolutionary. On private lands, private economic returns to landowners rather than ecological principles tend to drive land-use and land-management decisions. Private landowners often ignore the harm that their actions cause to ecosystem processes, especially if those processes play out on a larger spatial or temporal scale. The costs of ecosystem damages may fall principally on others besides the landowner, both in current and in future generations, thereby giving the landowner little direct incentive to prevent the damage from occurring.

The underlying motivation for using ecological principles to guide land use is rooted in a land-use ethic that sees man as dependent on nature for sustainable existence rather than its master (Leopold 1949). By their nature, principles broad enough to be generally applicable are abstract. Land-use and land-management decisions, however, are very specific. To be most effective, ecological principles must be applied over spatial scales on the order of thousands of hectares or more, and time scales on the order of decades and centuries. Decisions that determine whether an agricultural operation is profitable are made on the spatial scale of a farm (hundreds of hectares) and often on a temporal scale of seasons or years.

Even when concrete goals and time frames have been established for improving ecosystem health, and specific practices to achieve these goals are accepted as ecologically sound, economic barriers can be difficult to overcome. Some ecologically sound practices may not pay for themselves in an economic sense, at least in the short to medium term (Walpole and Sinden 1997). Even if ecologically sound practices are economically superior to current practices, sufficient benefits must accrue to landowners to convince them to adopt the practices, or government regulation must make such practices mandatory. Often many benefits of ecologically sound land-use decisions accrue outside the ownership of the land. Some means, such as tax incentives or subsidy programs, has to be found to give landowners sufficient incentive to choose ecologically sound practices. There is also a skepticism of adopting new methods when the tried and true methods seem to work, at least from the landowners's point of view. For example, research results have demonstrated that certain conservation tillage practices are both economically and environmentally sound, but it has taken years before these practices have been accepted and adopted by a significant fraction of farmers. Finally, the benefit of adopting ecologically sound practices often depends on cooperation among multiple landowners. Ecologically meaningful units of land-use planning (such as watersheds) often include a number of landowners, who by their participation or nonparticipation in the process can enhance or negate the efforts of other landowners in their watershed. Getting multiple landowners to participate and gaining agreement on how to share the responsibilities and costs or benefits often involves a difficult negotiation process.

In summary, for success in practical application of ecological principles in land-use decision making:

1. Decision makers must understand the need and share the goal.
2. Abstract principles must be translated into specific land-use decisions.
3. Responsibility for associated costs (which tend to occur up front and are specific to place) must be assigned and acceptable to landowners and decision makers.
4. Benefits (which tend to be realized in the longer term and diffuse in space) must be understood and shown to have enough value to outweigh immediate and specific costs.
5. Practices must be culturally acceptable (this includes respect for the rights of property owners).

Until these conditions are met, it is unlikely that ecological principles will have substantial influence on land-use decision making.

11.1.2 Future Scenario Approach: An Effective Strategy for Land-Use Planning

Alternative future scenarios of landscape change can help decision makers visualize and evaluate alternative choices in a way specific to time and place (Harms et al. 1993; Steinitz et al. 1994; Schoonenboom 1995; Freemark et al. 1996; Ahern 1999). Models that explore effects of different land-use practices on species and key ecosystem processes can be effective tools for evaluation of alternative scenarios (Donigian and Huber 1991; Pulliam and Danielson 1991; Dunning et al. 1995; Holt et al. 1995; White et al. 1997). Equally important are assessments of economic impacts (e.g., Williams et al. 1988; Walpole and Sinden 1997) and human perceptions of the alternative choices (Nassauer 1988). Finally, methods must be developed for summarizing results of various evaluative tools in a coherent way (e.g., Anselin et al. 1989; Heathcote 1998).

We present here a methodology for the use of future scenarios as a tool for guiding land-use decisions for agricultural watersheds in the U.S. Corn Belt. The goals of this project have been

1. Generation of designed alternative futures that explore a range of human land-use and management choices for watersheds in the U.S. Corn Belt
2. Development and calibration of models to evaluate the alternative scenarios and compare potential impacts of future change on water quality, biodiversity, and human perceptions of the landscape
3. Evaluation of the scenarios using the models developed and/or calibrated for these watersheds
4. Summarizing and publishing these results in an integrative assessment

Toward these goals, three alternative future scenarios have been designed for two different agricultural watersheds to represent potential landscape composition 25–30 years into the future from each of three different sets of human land-use management priorities: (1) a continuation of present trends, with food production and economic profit given the highest priority, using the existing regulatory framework (i.e., deregulation); (2) an effort to preserve biodiversity and improve water quality using conventional methods and within the existing regulatory framework; (3) incorporating a greater range of innovative agricultural practices coupled with efforts to preserve and restore native biodiversity and improve water quality. This research is intended to inform decision makers, from farmers who make the day-to-day decisions on the use and management of their land to those who must develop policy to encourage the incorporation of ecological principles into land-use decision making.

11.2 Applying the Future Scenario Approach to Agricultural Watersheds

11.2.1 Study Watersheds

The watersheds chosen for this study are two Iowa watersheds used in the Midwest Agrichemical Surface–subsurface Transport and Effects Research (MASTER) program. Both are within 120 km of Ames, Iowa (Fig. 11.1). Watershed area and some land-use characteristics are summarized in Table 11.1. Land-use classes in Table 11.1 were summarized from land-cover data generated under MASTER (Freemark and Smith 1995). These land-cover data were digitized from 1:20,000 aerial photographs taken in 1990, and ground-truthed in 1993–1994.

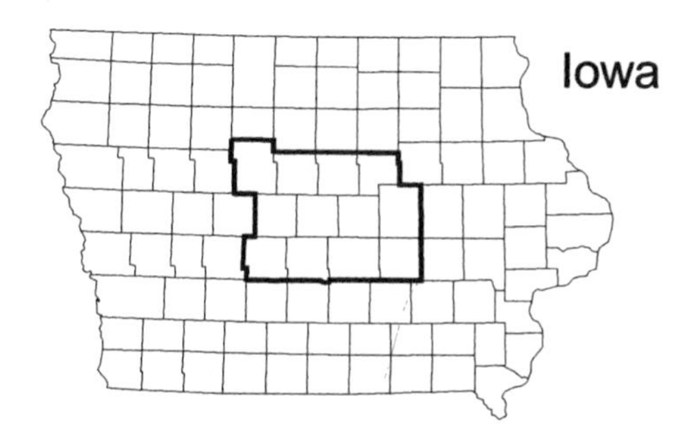

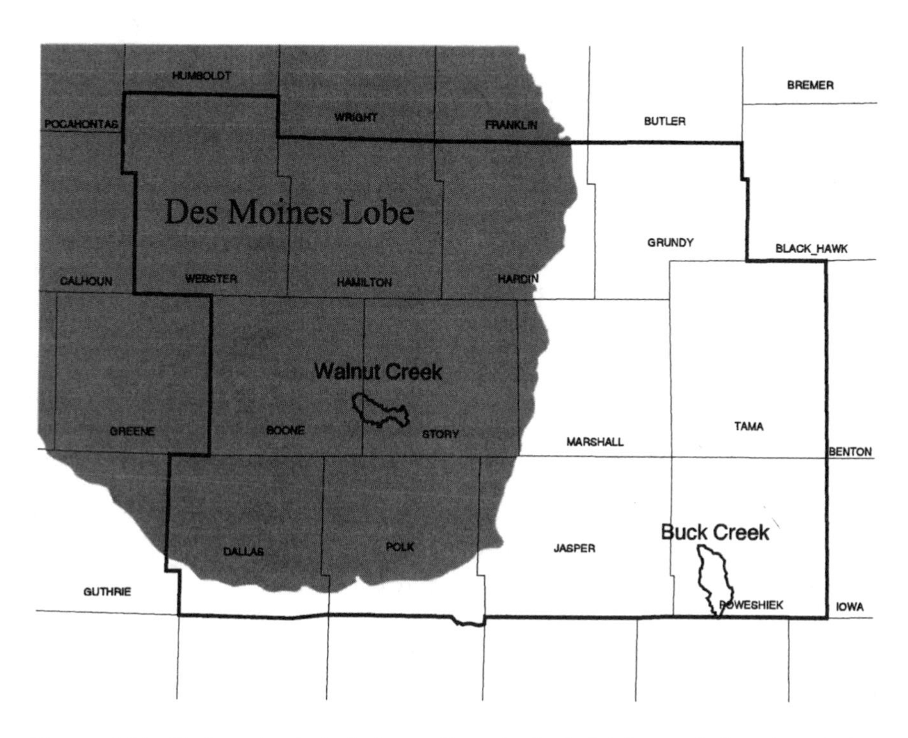

FIGURE 11.1. Study watersheds in Iowa. Counties included in our delineation of the Central Iowa region are outlined in bold. Shaded area shows the extent of the Des Moines Lobe Landform Region; unshaded areas are part of the Southern Iowa Drift Plain (Prior 1991).

Most of the land area in central Iowa lies within two landform regions, the Des Moines Lobe and the Southern Iowa Drift Plain (Prior 1991). To represent the way the same land-use priorities might be realized in different landform regions, alternative future scenarios were designed for two watersheds, one (Walnut Creek, Story Co.) on the Des Moines Lobe, the other (Buck Creek, Poweshiek Co.) in the Southern Iowa Drift Plain. These

TABLE 11.1. Characteristics of study watersheds.

	Walnut Creek Story/ Boone Cos.	Buck Creek Poweshiek Co.
Physiographic region	Des Moines Lobe	Southern Iowa Drift Plain
Total land area (ha)	5130	8790
Percentage of land area in following land uses:		
Row crops	83	45
Pasture/grassland	4	20
CRP	0	16
Woodland/savanna	5	9
Alfalfa/hay	2	4
Other	6	6

Source: Land-use data are summarized from Freemark and Smith (1995) for 1994 land cover of the study watersheds. Physiographic regions are described in Prior (1991).

landform regions vary in their topographic relief and current land use. Walnut Creek, like most of the Des Moines Lobe, is relatively flat with rich, productive soils. Its land cover is dominated by corn and soybean row crops (Fig. 11.2*a* and 11.3). Buck Creek has a more rolling topography, a highly branched stream network, and more varied land cover (Table 11.1, Fig. 11.2*b* and 11.4).

Historically, the Walnut Creek watershed, like most of Story County, was dominated by prairie, dotted with prairie pothole wetlands (Hewes 1951), most of which have now been drained for row crops. Traces of former wetlands can still be seen in spring (Fig. 11.3). Buck Creek, located on an older glaciated surface, had soils that were better drained. Its hills and valleys, particularly in the lower, southern end of the watershed, provided firebreaks that allowed the growth of more extensive riparian forest. Today, Buck Creek has soils more prone to erosion, hence more of the land cover in this watershed is in pasture and Conservation Reserve Program (CRP) set asides. In addition, there is more forest along riparian channels and in upland woods (Table 11.1, Fig. 11.4).

FIGURE 11.2. Current land cover in study watersheds: (*a*) Walnut Creek, Story and Boone Counties, and (*b*) Buck Creek, Poweshiek County. Land-cover data interpreted from aerial photographs 1:20,000 and ground-truthed under the MASTER research program (Freemark and Smith 1995). (*Note:* The strip intercropping land-cover class is not found in Figure 11.2, which shows the current land cover in these watersheds. It occurs only in Scenario 3 (Figs. 11.5*c* and 11.6*c*, which also refer to the legend in Figure 11.2.) Also, the water/wetland classes can be distinguished from farmsteads (though shading in legend is similar) by their elliptical shape and location (away from the road network). Farmsteads and towns are rectangular or angular in shape, and occur along roads.

11.2.2 Scenario Design

Alternative future scenarios were designed by the landscape architecture team at the University of Michigan, (Nassauer et al., *http://www.snre.umich.edu/faculty-research/nassauer/*; (sidebar: Rural watersheds and

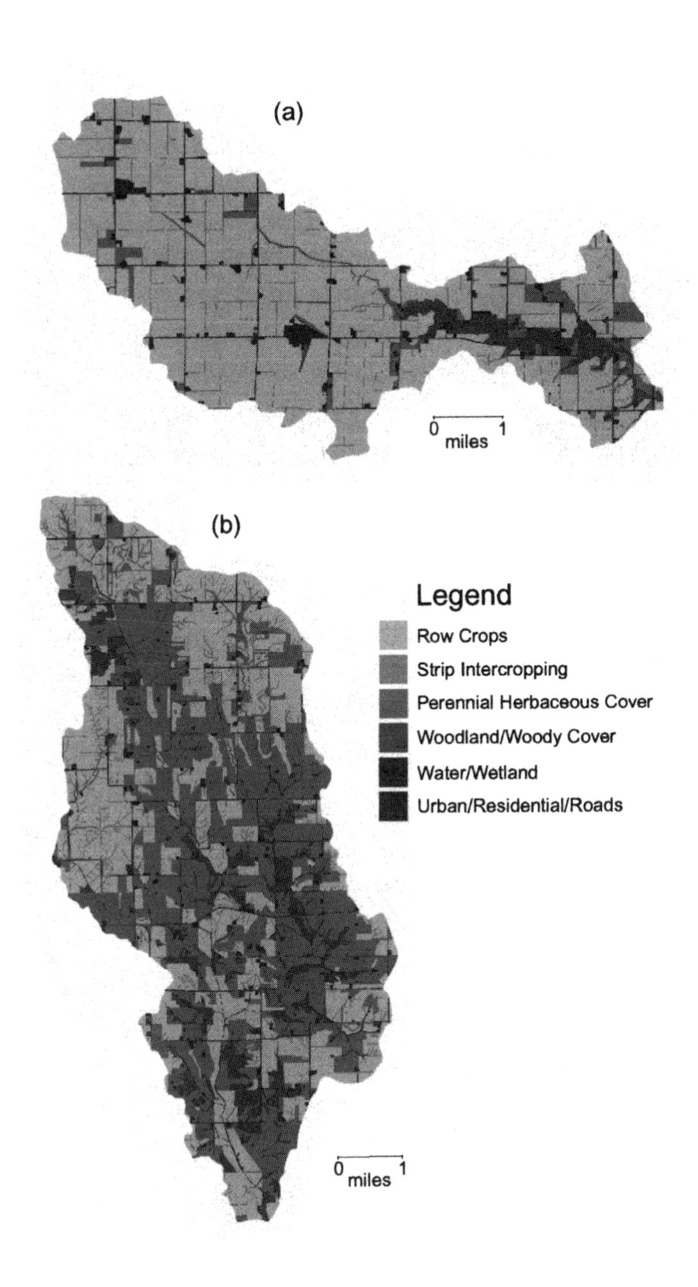

FIGURE 11.3. Photo of Walnut Creek Watershed in spring 1998. (Photo courtesy of Rob Corry.)

11.2.2 Scenario Design

Alternative future scenarios were designed by the landscape architecture team at the University of Michigan, (Nassauer et al., *http://www.snre.umich.edu/faculty-research/nassauer/*; (sidebar: Rural watersheds and policy)) in consultation with disciplinary experts in the fields of agronomy, vertebrate ecology, plant ecology, wetlands ecology, water quality, hydrology, agricultural policy, agricultural extension, and Geographic Information Systems, including scientists from the region as well as collaborators on the project.

Scenario 1 (Figs. 11.5*a* and 11.6*a*) assumes that profitable agricultural production is the dominant objective of landscape management and that profit is perceived as short-term economic return. This scenario assumes high demand for Corn Belt grain crops by world markets, high use of fossil fuel, high use of chemical and technological inputs, and public support for large-scale, industrial agriculture. Scenario 1 also assumes that public trust in the quality of food produced by industrial agriculture is high, that the public perceives the landscapes resulting from industrial agriculture to be environmentally acceptable, that the fossil fuel necessary to industrial

FIGURE 11.4. Photo of Buck Creek Watershed in fall 1998. (Photo courtesy of Rob Corry.)

agriculture remains economical or that alternative fuels emerge, and that the public remains willing to support industrial agriculture (through research, direct payments to farmers, crop insurance, etc.) at levels similar to those of the 1990s. It assumes public incentives for conservation at a level that encourages widespread adoption of the types of best management practices existing in 1994. Woodlands disappear as more land is converted to cultivation. The Corn Belt landscape has been depopulated by 50% compared with 1994. Most farmers do not live on their farms through the winter months. Many farmsteads have been demolished and groves cut down. Farm size has doubled, and field size has increased up to 320 acres. Crops are corn and soybeans. Livestock are raised almost exclusively in confinement feeding operations in a few counties of the state. Few people visit the rural landscape for recreation.

Scenario 2 (Figs. 11.5*b* and 11.6*b*) assumes that agricultural enterprises and practices have changed in response to federal policy, which has enforced clear, measurable water quality performance standards for pollution of surfacewater and groundwater, on a farm by farm basis. Under this scenario, public environmental concerns are assumed to focus on clean water. Public support for agriculture is targeted to practices that efficiently reduce soil erosion, reduce sediment delivery to streams, prevent the impacts on riparian systems) have been widely adopted as profitable enterprises that help to meet water-quality performance standards on rolling or erodible land. Woodlands have been widely maintained for grazing. Both

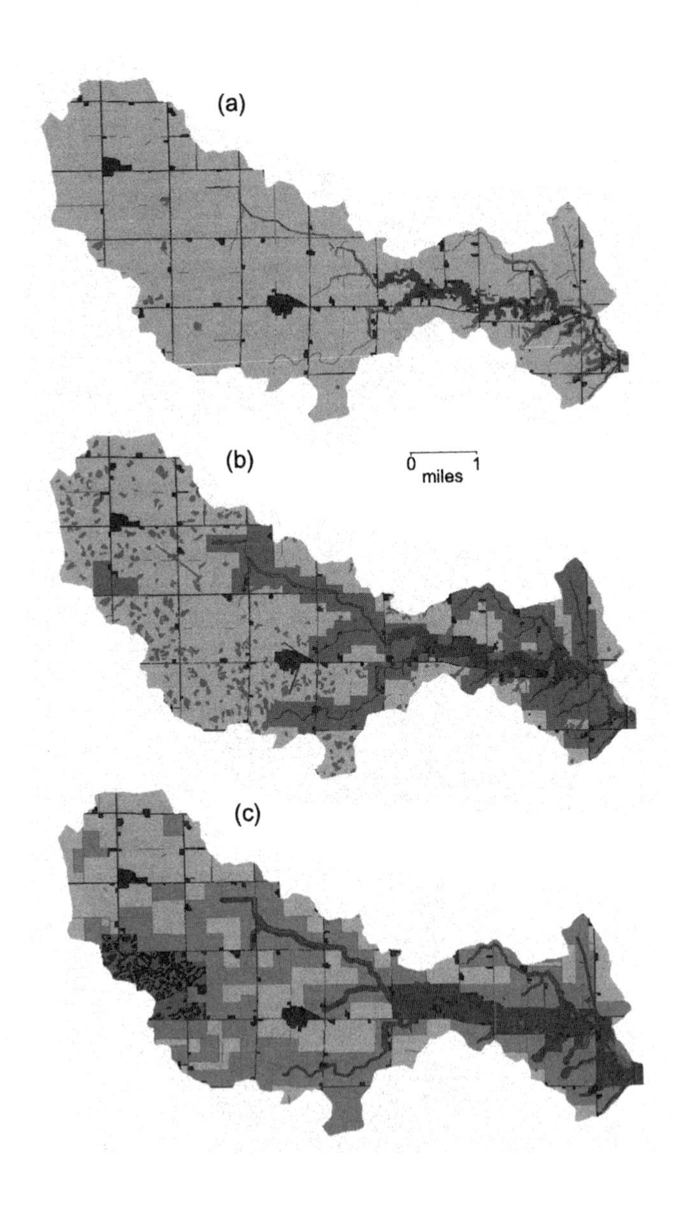

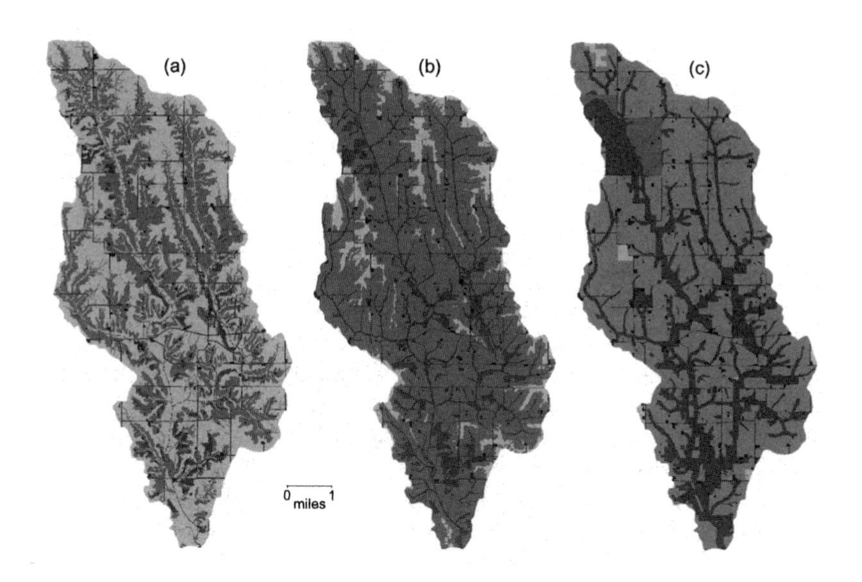

FIGURE 11.6. Designed alternative scenarios for Buck Creek watershed: (*a*) Scenario 1, (*b*) Scenario 2, and (*c*) Scenario 3 (legend as in Fig. 11.2.). Note scenario features similar to those for Walnut Creek (see caption for Fig. 11.5) but applied in a different landscape context.

urban and rural citizens appreciate the pastoral appearance of agricultural landscapes, where animals can be seen grazing on green hills. Farm vacations and countryside second homes regularly bring urban people into rural areas. To manage livestock operations and to respond to rural

FIGURE 11.5. Designed alternative scenarios for Walnut Creek watershed: (*a*) Scenario 1, (*b*) Scenario 2, and (*c*) Scenario 3 (legend as in Fig. 11.2). Note the increase in land area in row crops at the expense of perennial cover for Scenario 1; the increased amount of land in perennial cover (pasture and forage crops) as well as wider riparian buffers in Scenario 2; and the biodiversity reserves, wide riparian buffers and extensive prairie, forest and wetland restorations in Scenario 3. Features of Scenarios 2 and 3 not visible in these maps are the use of crop rotations that include alfalfa and oats, as well as small detention pond wetlands at the outlets of tile drains in Scenario 2, and the incorporation of organic crops and a type of strip intercropping that includes a strip of native perennials as an agricultural innovation in many fields for Scenario 3 (see descriptions in Nassauer et al., *http://www.snre.umich.edu/faculty-research/nassauer/*) for more detail on the future scenario designs).

recreation demand, 50% more farmers live in Corn Belt agricultural landscapes in 2025 than under Scenario 1.

Scenario 3 (Figs. 11.5*c* and 11.6*c*) assumes that technology and agricultural practices have dramatically responded to federal incentives to increase indigenous biodiversity across the nation. Public investment maintains and restores native flora and fauna through a comprehensive system of reserves. It also supports profitable agricultural production with new technologies that enhance biodiversity within agricultural production. Public ecological perceptions and concerns drive federal investment in agriculture, which is targeted to ecological results and long-term economic return. Under this scenario the world grain market is robust but continues to produce a comfortable surplus. Public health perceptions and environmental concerns have affected global dietary choices, and the global market for beef and pork has not dramatically increased as predicted in 1998 (Worldwatch Institute 1998). Livestock enterprises have continued to trend toward confinement feeding operations, which are constructed according to rigorous standards for sewage treatment in a few counties. Federal land purchase programs have established at least one indigenous ecosystem core reserve of at least 640 acres in many Iowa watersheds. Federal support for innovative, biodiversity best management practices (e.g., perennial strip intercropping and agroforestry) has been targeted to landscapes that connect and buffer the new reserves and riparian corridors. The substantial public investment in core reserves and corridors invites public enjoyment of the rural landscape. Trail systems connect the corridor system and the reserves. While farm size increases as in Scenario 1, and the number of farms decreases to about 50% of the number present in 1994, nearly all of the farmsteads present in 1994 remain inhabited in 2025. Many nonfarmers who enjoy the beauty of biodiversity in the rural landscape live on farmsteads no longer occupied by farmers.

Once designed, the alternative futures were digitized into representations of future land cover in a GIS. The GIS provides spatially explicit input data to a variety of models that can be used to evaluate the impacts of the various land-use practices in the scenarios. In addition, historic vegetation coverages were generated from the Iowa Soil Properties and Interpretations Database (ISPAID; *http://icss.agron.iastate.edu/*), based on characteristics of soils formed under prairie and forest vegetation and wetland types associated with hydric soils (Galatowitsch and van der Valk 1994). Historic and current land cover for each watershed thus provide landscapes for relative comparison with the alternative futures.

11.2.3 Scenario Evaluation

A suite of modeling approaches were then used to evaluate and contrast the response of water quality, economic profitability, and plant and animal biodiversity in these two agricultural watersheds to changes in land use and management. Landscape components that were explored included riparian buffers of varying width, establishment of large patches of restored native habitat in reserves, and changes in the agricultural matrix itself (field size, cropping practices, interpolation within fields of perennial cover such as grass waterways, filter strips, field skips, etc.).

The use of several different water-quality models to evaluate the scenarios for water quality response was explored. The SWAT (Soil and Water Assessment Tool); (Arnold et al. 1997) was chosen as best suited to these watersheds and the available data. Calibration of the model to the study watersheds has been an important outcome of this effort in addition to the ranking of the scenarios.

The EPIC (Erosion Productivity Index Calculator) (Williams et al. 1988) is being used to simulate the impact of alternative conservation practices on crop yields. Economic returns from agricultural production under alternative scenarios are then evaluated by combining the yield impacts with data on prices and production costs. Ecological benefits can then be related to their economic cost to the producer, helping policymakers understand, for example, the relative magnitude of incentives or subsidies that would be required to entice landowners to adopt a given set of practices. In addition, spatial representations of the alternative futures along with computer-generated images of the way the scenarios might look in simulations of the landscape have been used in farmer interviews intended to discern cultural acceptability of specific agricultural practices (Corry and Nassauer, in press; also *http://www.snre.umich.edu/faculty-research/nassauer/*).

For modeling risk to biodiversity, our approach was to combine the use of a heuristic model (White et al. 1997) that responds primarily to change in habitat area for species as a "coarse-filter," with the use of Spatially-Explicit Population Models (SEPMs) as "fine-filter" approaches to assess the impact of changes in land use and management on species of interest (Rustigian 1999).

For some species, spatial patterns of habitat fragmentation may be very important in determining persistence of populations; however, some results suggest that degree of fragmentation can be a much smaller effect than the amount of habitat itself (Fahrig 1997; Trzcinski et al. 1999). Multiple modeling approaches can help to elucidate these issues and provide several perspectives on landscape effects on biodiversity. The combination of simple habitat loss models for a wide range of species (White et al. 1997) with spatially-explicit demographic modeling of species selected for representative of a larger group in life history traits or other factors (Dunning et al. 1995) provides a broad first-order analysis of effects, with details on how these effects may be modified for certain classes of species. In addition, where habitat loss itself is the primary determinant of species declines, simple, empirically based models can be used for rapid evaluation of multiple alternatives and in exploration of the sensitivity of various taxa or groups of taxa to elements of landscape change.

Multispecies models and detailed landscape structure analysis offer promising methods for examining ecological community processes. Simulation provides a systematic, quantitative approach to assessing factors affecting communities and populations within those communities. With spatially-explicit models, comparisons can be made between effects of landscape and species interactions, both of which may influence community diversity and dynamics (Palmer 1992; Korpimäki and Norrdahl 1998; Losey and Denno 1998). When comparing landscapes with different amounts of habitat or differing physiognomy (e.g., grassland vs. forested habitat) it is important to determine how efficiently habitat is exploited, in addition to differences in abundances and densities. One metric providing a normalized measure of the efficiency of habitat exploitation is relative density (population density in a patch in proportion to patch carrying capacity). Relative densities can be compared for different species as they respond to a given landscape, for a single species in response to different landscapes, or for species in a community or alone. Clark et al. (*http://www.public.iastate.edu/~meclark/*) compared relative densities predicted from spatially explicit single and multispecies models of mammals applied to the Buck Creek and Walnut Creek watersheds to assess the relative importance of habitat loss and species interactions. Their results suggest that for species with relative densities unaffected in different landscapes, populations change linearly with the abundance of suitable habitat, similar to their finding that diversity increases nearly linearly with noncrop, grassland habitat. However, their results also indicate that species interactions can play as important a role as landscape in affecting biodiversity.

The comparison of spatially-explicit alternative scenarios with models is a first step in quantifying the economic and ecological costs of continuing current practices as well as the benefits of some potential changes. In many cases, models can only approximate the watershed response to changes in land use, ranking the futures with respect to their effectiveness in achieving the desired objectives. However, ranking of alternatives is only one function of models (Starfield 1997). An equally important role of modeling is to explore the state of our current understanding of existing processes, locate critical areas where data are needed to improve assumptions and effectively parameterize models, and focus additional research efforts to help fill important gaps in our understanding. In addition, as with land-use plans that are implemented and then evaluated (Ahern 1999), the application of evaluative models to alternative futures can facilitate further discussion, collaboration, and knowledge exchange among landscape planners, policy makers, and ecologists. This approach provides a means to evaluate a broad range of innovative alternatives (which might be costly to implement in experimental studies), to winnow out those that are clearly unfeasible or undesirable, and to focus important experimental research on options with the greatest promise.

11.2.4 Contrasting Approaches: Exploring Effects of Specific Practices or Targeting Specific Environmental Goals?

One major decision in the generation of alternative scenarios is between the choice to explore the implications of implementation of a particular set of practices—e.g., How much improvement in water quality could we achieve by establishing riparian buffers along the entire stream network in this watershed?—compared to an approach that attempts to explore various ways to meet an environmental standard—e.g., We need to reduce nitrate export from this watershed to (some target level); What land-use or management changes could help achieve this? How far toward the target goal might each take us?

In the first case, all that is needed is an understanding of the sets of practices that are likely to be favored by policy and a means to locate these effectively on the landscape (knowledge of soils, topography, geomorphology, existing land cover). Whether the estimated improvements in water quality are minimal or substantial, one will be able to address the research question with respect to the effects of implementation of a given policy.

In the second case, at the very outset of scenario design, thought must be given to "mass balance" considerations, to the magnitude of measures needed to achieve desired results. If, for example, the goal of prospective land-use and management practices is 75% reduction in nutrient export from the watershed, and filtering of runoff through restored wetlands is suggested as a land-use mechanism to achieve this goal, the areal extent of such wetlands must be substantial to achieve such a substantial reduction. In addition, the type of wetland restored must be appropriate to the desired function. Small depression wetlands in headwater areas may have little opportunity to assimilate nutrients because they tend to have a small catchment areas and low flows. Riverine wetlands, which are linked more extensively with uplands and riparian systems, can have greater capacity for altering water quality (Brinson 1988). Detention ponds at outlets to tile drains in the watershed may be effective for nutrient removal under low flow conditions, but may be overwhelmed by storm events or spring snowmelt. Single storms in a given year can be responsible for most of the sediment and nutrient export from fields adjacent to riparian areas (Daniels and Gilliam 1996), thus understanding dynamics of nutrient export and the hydrology of the system is important to effective restoration design.

Much more watershed-specific information is needed at the outset when the research objective of the scenario design process is achievement of a targeted goal, and the process will likely require several iterations to identify cost-effective alternatives that achieve the desired goals. Choice of modeling approaches used will also be critical; tradeoffs may have to be made between precision and flexibility of the model(s) used if there is a need to explore multiple landscape-level solutions. If the research objective is to explore alternative ways to achieve a targeted goal, at least some of the alternatives must reach the target. This may not happen in the initial designs.

11.3 Application of ESA Principles and Guidelines in the Context of This Project

11.3.1 Principles

In many ways, the principles outlined by the ESA Committee on Land Use are applicable to this project, although both efforts evolved independently. The use of alternative future scenarios makes land-use choices both spatially and temporally specific (principles 1 and 3). We use a "twenty-five years out" perspective, and contrast the landscape level effects of alternative future choices designed to restore native biodiversity and improve water quality with those resulting from perpetuation of current trends. Effects of landscape change from the historical past to the present can also be effectively illustrated, using historic landscapes as the basis for comparison. Thus, designed alternative future landscapes can be contrasted with three "standard" landscapes for the same watersheds: a historic landscape, the current landscape, and the landscape that may result from perpetuation of current trends.

Because they are based on real watersheds, future scenarios and the tools used to evaluate them consider the influences of and impacts on the species assemblages characteristic of the region (principle 2). In our study, for example, the goal given highest priority in designed future Scenario 3 was restoration of native biodiversity. We are employing several modeling approaches based on plant and animal species of the region and their response to habitat loss or restoration as well as to their population dynamics in combination with other species in the watershed (Rustigian 1999).

In the future scenarios, we specify the intense disturbance regimes associated with agricultural practices, as well as practices such as prescribed burning (intended to emulate historic natural disturbance regimes) to be used in management of nature reserves. These disturbances are incorporated into some of the models used to evaluate the futures (principle 4).

Landscape characteristics (principle 5) such as the size, shape, and relationships among habitat patches were important considerations in the choice of study watersheds, the design of the future scenarios, and in the models used to evaluate them. For example, increases in field size and the resulting coarse-grained agricultural matrix associated with current trends for industrialization of agriculture (represented by Scenario 1, the perpetuation of current trends) are contrasted with smaller field sizes and more diverse types of land cover in the designed futures (Scenarios 2 and 3).

In addition, the scenarios were designed to allow comparison of the effects of riparian corridors of varying widths, and the creation of large reserves for the restoration of native biodiversity.

11.3.2 Guidelines

Several examples of how the guidelines suggested by Dale et al. (Chapter 1) might be applied to real landscapes are illustrated by the designed scenarios. The coupling of future scenarios with GIS-based evaluative models allows both the examination of the impacts of land-use decisions in a spatial context and exploration of the effects of long-term change (guidelines 1 and 2). The scope of the study determines the regional context and time frame in which these impacts can be evaluated. In Scenario 3, the use of native plants along roadsides, in farmsteads, and in strip intercropping, as well as the establishment of biodiversity reserves, represent efforts to restore regionally rare landscape elements and associated species (guideline 3). The adoption of innovative agricultural practices such as strip intercropping, the use of filter strips, and specifications that highly erodible land and land adjacent to streams be in perennial crops or land cover (Scenarios 2 and 3) are elements designed to reduce soil erosion and loss of organic carbon (i.e., guideline 4, avoid land uses that deplete natural resources). The establishment of biodiversity reserves connected to a wide riparian network in Scenario 3 embodies the guideline for large contiguous or connected areas of critical habitat. Finally, in each of the designs, GIS data bases on soils and current land cover were used to guide allocation of future land-use and land-management practices compatible with the natural potential of the area (guideline 8). For example, in Scenario 3, wetland restorations were located in areas with hydric soils suitable for prairie pothole wetland restorations, and additional areas of forest for reserves were added adjacent to existing forested land cover.

11.4 ESA Guidelines: Implications and Additions

11.4.1 Implications of Existing Guidelines

The principles and guidelines developed by the ESA Committee on Land Use are an essential starting point for the incorporation of ecological thought into land-use planning. Not only has the committee articulated a set of principles and defined guidelines that derive from those principles, it has summarized the voluminous, controversial, and sometimes contradictory ecological literature in the context of these principles (Dale et al., Chapter 1). Their efforts provide a foundation for ecologists to work from and build on in collaboration with others working in the areas of land-use planning. To that end, we suggest three additional principles and accompanying guidelines: (1) consideration of land-use impacts on ecosystem processes, (2) consideration of the human dimension of effective planning, and (3) understanding and communication of uncertainty in impacts of land-use practices and associated risks.

11.4.2 Ecosystem Processes

Principle: Substantial land-use impacts on some systems, particularly lakes, streams, and riparian and coastal marine systems, often originate outside the boundaries of that system.

Guideline: Consider effects of land use on ecosystem processes both within and across ecosystem boundaries.

A critical ecological principle incorporated into our research is the importance of understanding effects of land use and management on ecosystem processes, such as exports of nutrients and sediment from the watershed, or alterations of the hydrologic regime. In agricultural regions, processes such as soil erosion, drainage of wetlands, channeling of streams and rivers and accompanying hydrologic changes, and nutrient and sediment exports from the system have long been recognized as problems. Human alteration of the nitrogen cycle and non-point-source pollution of surface waters with phosphorus and nitrogen are having severe, global environmental impacts (Vitousek et al. 1997). Land-use and land-management practices that reduce nutrient exports must be incorporated into land-use planning for agricultural regions in particular if we are to address some of the most serious environmental problems that stem from agricultural land use.

The principle differes from the first guideline of Dale et al. (examine the impacts of local decisions in a regional context) primarily in that it specifies the need to examine the impacts of land-use and land-management decisions *across boundaries* of study area and region; to consider impacts on ecosystems that are linked to but may be far removed from, the systems under study. Land-use and land-management decisions on nutrient management, for example, have far-reaching implications not only within our watersheds (local context) or even within the U.S. Corn Belt (regional context), but also on systems that are geographically far-removed from the local and regional unit of study (e.g., the "dead zone" in the Gulf of Mexico).

To incorporate considerations of ecosystem processes into land-use planning, we must first determine which processes and cycles (hydrologic, carbon, nitrogen, phosphorus, etc.) in the planning entity are most affected by land-use and land-management practices. The most critical impacts of a given practice (for example, cropping and tillage) may differ from one watershed or region to another. Our two study watersheds differ in the degree to which sediment and nutrients are the primary water-quality problem. Buck Creek, with its rolling topography, highly erodible soils, and extensive livestock production, has substantial problems with sediment, which during major storm events can be measured in grams per liter rather than the usual milligrams per liter for streams (Eilers and Vache, unpublished data). Although nitrate levels in Buck Creek are higher than might be desired, they tend to be lower than the levels in Walnut Creek, where nitrate–nitrogen concentrations average near the drinking water standard of 10 mg L^{-1}, with peak concentrations higher than the standard (USDA ARS, unpublished data).

Components of the designed scenarios such as riparian buffers, wetland restorations filter strips, and some of the alternative cropping practices were intended to explore alternatives that are currently part of federal programs (Conservation Reserve program [CRP], Conservation Reserve Enhancement Program [CREP], Wetland Reserve Program [WRP], and Environmental Quality Incentive Program [EQIP]) to control soil erosion and improve water quality, and compare their effectiveness to landscapes in which innovative agricultural practices and permanent reserves are used to achieve similar purposes.

Proposed land-use or land-management practices must target specific environmental goals. In land-use planning, there are multiple ways of addressing the same broad goal in alternative future scenarios. For example, in the Netherlands, where conservation and restoration of habitat for native species was a primary goal of landscape planning efforts, a number of configurations of habitat have been designed, each benefiting different sets of species with different life histories (cf. Harms et al. 1993). Similarly, if improvement in surface water quality is a desired goal, there are multiple ways to achieve improvements on agricultural land by changing land-use or land-management practices. Different practices will be most effective for different water-quality problems. If soil and stream bank erosion that delivers sediment to streams is the primary water quality problem in a watershed, vegetated filter strips and riparian buffers may be extremely effective solutions (Daniels and Gilliam 1996). If dissolved nutrients that reach the stream primarily in flow from tile drains are the primary problem, riparian buffers may be a smaller part of the solution.

11.4.3 Human Dimension

Principle: To be sustained, land uses must be recognized as valuable by landowners and the larger community (Nassauer 1997).

Guideline: Implement land-use and land-management practices compatible with human economic and cultural practices and values.

Designing new land-use patterns that recognizably fulfill human aspirations is as important to sustainability as is designing patterns that are compatible with the natural potential of the area. Gaining acceptance for ecologically sound land-use and land-management practices requires that local landowners and decision makers find these practices attractive, both financially and otherwise. Innovative practices and landscape patterns that people do not like or that do not fulfill economic needs are not sustainable over time (Nassauer 1997). Landscape design, planning, economics, and social science can help policy makers anticipate human perception and behavior in response to new land-use patterns. Ecology can suggest the ecological purposes for innovative landscape patterns, but it alone cannot tell us what the new patterns should be.

We acted on this principle in designing three alternative scenarios for Corn Belt agricultural watersheds in 2025. The landscape architects drew on the expert knowledge of colleagues in several disciplines, and drew on environmental design research to shape innovative landscape patterns that local people would be likely to find acceptable. Each scenario was actively informed by consultation with colleagues in ecology and hydrology as well as agronomy and economics. The scenarios were designed to be recognizable and plausible as agricultural landscapes that would be economically viable under the assumptions of each scenario and would be desirable as a place to live and to farm. Each scenario, then, was a hypothesis about cultural acceptability as well as a response to expert knowledge about how the landscape could function—agronomically, hydrologically, and ecologically.

These hypotheses were formally tested in on-farm interviews with Iowa farmers (Corry and Nassauer, in press and *http://www.snre.umich.edu/faculty-research/nassauer/* [select sidebar on Rural Watersheds and Policy, then select Cultural Acceptability]). The real test of cultural acceptability will be in individual farmers' responses to any new policy, market, or technology that would intersect with existing cultural values and practices to meet the assumptions of any of the scenarios.

11.4.4 Understanding Uncertainty

Principle: Future predictions always carry some degree of uncertainty.

Guideline: The magnitude of uncertainty associated with potential effects of land-use alternatives should be explored and expressed in communication with decision makers, and used to focus additional research.

Another important issue ecologists and modelers must deal with is the uncertainty inherent in modeling. However, ecologists and modelers often do a poor job of communicating levels of uncertainty to decision makers and planners. Sensitivity analysis of models used to evaluate future scenarios can focus additional research by identifying input parameters to which the model is sensitive. Those parameters for which there is the greatest amount of uncertainty at present, for which reductions in uncertainty would most improve the accuracy of model predictions should become the targets of additional research.

One of the most intractable problems in the use of future scenarios is that there is an inverse relationship between the degree of innovation in a land-use practice or scenario and our ability to model it accurately. The innovative land-use practices that may be of the greatest interest in the long run are also the ones whose effects on ecological endpoints are most difficult to estimate, particularly at the spatial scale of watersheds 5000 or 10,000 hectares in area instead of fields on farms or agricultural experiment stations a few hundred hectares in area. Models and sensitivity analyses can be used to explore the potential impacts of innovations in small watersheds such as Buck Creek or Walnut Creek, but the results will always have a high level of uncertainty associated with them until they have been tried at the spatial scale of such watersheds. Extrapolating results of experiments carried out on 100-ha plots to watersheds 5000 to 10,000 ha in area or more ignores a hierarchy of processes and interactions that can qualitatively alter the expected outcome. The sustainable production of agricultural commodities, preservation of biodiversity, clean water, and healthy aquatic systems in the future are all critically important components of agricultural ecosystem health. Great risks may be posed to both the agricultural economy and agricultural ecosystems if model predictions are not accurate. Thus, long-term ecological research on the application of innovative agricultural practices at multiple spatial scales (i.e., field, farm, small [5000–10,000 ha] watershed, as well as larger hydrologic units) should be a funding priority.

The research described here is a first step on the road to incorporating ecological principles and ecologically sound practices into land use and management in the U.S. Corn Belt. The designation of long-term agroecological research sites at least the size of these watersheds would be an important step in the advancement of ecological research and the adoption of ecological principles to guide land use in agricultural regions. Long-term ecological research in agricultural watersheds will be important not only for model validation but as demonstrations that real watersheds can have working farms and substantial environmental results (improved water quality and stream health, enhanced biodiversity, abundance of wildlife, and aesthetic appeal). Such watersheds will be the strongest advocates we could have to induce change.

Our next steps will be to rank the future scenarios with respect to their effectiveness in achieving the goals of the designs, explore ecosystem response to specific land-use practices using spatial models, describe the outcomes of the modeling efforts, and integrate the multiple modeled endpoints for assessment of the alternatives. We hope the results will inspire and guide further watershed-level research in agricultural ecosystems.

11.5 Conclusions

More than 52% (398 million ha) of the land area in the continental United States is in farmland (USDA 1992). Ecological principles and guidelines must be incorporated into land-use and management decisions on agricultural land or agricultural ecosystems and the aquatic systems to which they are linked will be neither healthy nor sustainable. Central Iowa, in the heart of the U.S. Corn Belt, is a highly productive agricultural region dominated by private land ownership and agricultural land use. Changes in land use must be acceptable to private landowners and economically feasible. Understanding the human dimension of changes in agricultural land use and adoption of conservation practices will be a critical step in the development of ecologically healthy agricultural landscapes. Partnerships between those who have developed and studied innovative agricultural practices and those who must use them on their farms will be important. Economic analyses can help policy makers understand the magnitude and extent of the incentives or agricultural subsidies that might be required to make a given practice economically viable in the economic context of agriculture today. If the costs of implementing ecologically sound practices cannot be recouped by landowners within a time frame they find acceptable, then agricultural policy makers will need to develop alternatives for financing these practices.

The ecological impacts of innovative agricultural practices need to be studied not only at the field scale (hundreds of hectares) but also at the small watershed scale (thousands of hectares) and above in order to improve our understanding of their effectiveness at the scale on which they might someday be implemented. Research on biogeochemical processes that influence biogeochemical cycles in agricultural watersheds must be a component of such whole-watershed research. In addition, responses of plant and animal species will be among the most sensitive indicators of ecological response to changing land use, and are affected by processes at multiple spatial scales (Pratt and Cairns 1992). Responses of species to interspecific interactions may be as important as their responses to changes in land use.

Success in the application of ecologically sound land-use decisions will require strong local leadership and broad, community-based planning. The US-EPA Office of Wetlands, Oceans and Watersheds web site, "Top Ten Watershed Lessons Learned" (*http://www.epa.gov/owow/lessons/*), presents case studies that illustrate applications of watershed-level planning with both successes and failures. Research projects such as the one described here can provide inspiration for local communities to envision more innovative practices than might otherwise be the case, as well as providing information and evaluative tools for community-based planning efforts.

Partnerships among ecologists, planners, and local communities in application of ecological principles and guidelines to land use and management will require substantial effort, and challenge all of us to listen and learn from other perspectives. Only through such efforts will we be able to realize the vision for the land articulated by Leopold and others who have spoken out for ecologically healthy and sustainable land use.

Acknowledgments. We acknowledge support from the U.S. EPA/NSF Partnership for Environmental Research STAR grants program, Grant R825335-01. We thank J. Hatfield of the USDA ARS Tilth Laboratory for generously sharing with us water quality data collected on the Walnut Creek, Story County watershed, and Rob Corry for photographs of the watersheds. We are grateful to the many individuals at Iowa State University, University of Iowa, University of Minnesota and their associated agricultural extension services, as well as the USDA NRCS and Iowa Geological Survey, who contributed their time and knowledge to this project, and whose work provides the foundation for the development of ecologically healthy agricultural ecosystems. Thanks to P.J. Wigington, George King, and three anonymous reviewers whose review and comments on the manuscript improved it. Thanks also to John Bolte, Court Smith, and Jennifer Gilden for informative discussions on application of ecological goals in land-use planning. This manuscript has been subjected to review by the US EPA and approved for publication. The conclusions and opinions are solely those of the authors and are not necessarily the views of the Agency.

References

Ahern, J. 1999. Spatial concepts, planning strategies and future scenarios: a framework method for integrating landscape ecology and landscape planning. *In* J.M. Klopatek and R.H. Gardner, editors. Landscape ecological analysis: issues and applications. Springer-Verlag, New York, New York, USA.

Alexander, R.B., R. Smith, and G. Schwartz. 1996. The regional transport of point and nonpoint-source nitrogen to the Gulf of Mexico. *In* Proceedings of the Gulf of Mexico Hypoxia Management Conference, 5–6 December, 1995, Kenner, Louisiana. U.S. Geological Survey, Reston, Virginia, USA.

Anselin, A., P.M. Meire, and L. Anselin. 1989. Multicriteria techniques in ecological evaluation: an example using the analytical hierarchy process. Biological Conservation **49**:215–229.

Arnold, J.G., J. Williams, and D.R. Maidment. 1997. Continuous time water and sediment routing model for large basins. Journal of Hydrologic Engineering **121**: 171–183.

Brinson, M.M. 1988. Strategies for assessing the cumulative effects of wetland alteration on water quality. Environmental Management **12**:655–662.

Corry, R.C., and J.I. Nassauer. In press. Cultural possibilities for small patch patterns: Corn Belt watershed examples. *In* J. Liu, editor. Integrating landscape ecology into resource management. Cambridge University Press, Cambridge, England.

Crosson, P., and J.E. Ostrov. 1990. Sorting out the environmental benefits of alternative agriculture. Journal of Soil and Water Conservation **48**:34–41.

Daniels, R.B., and J.W. Gilliam. 1996. Sediment and chemical load reduction by grass and riparian filters. Soil Science Society of America Journal **60**: 246–251.

Donigian, A.S., Jr., and W.C. Huber. 1991. Modeling of nonpoint source water quality in urban and non-urban areas. EPA/600/3-91/039. Environmental Research Laboratory, Athens, Georgia, USA.

Dunning, J.B., D.J. Stewart, B.J. Danielson, B.R. Noon, T.L. Root, R.H. Lamberson, and E.F. Stevens. 1995. Spatially-explicit population models: current forms and future uses. Ecological Applications **5**:3–11.

Fahrig L. 1997. Relative effects of habitat loss and fragmentation on population extinction. Journal of Wildlife Management **61**:603–610.

Farrar, D. 1981. Perspectives on Iowa's declining flora and fauna- A symposium. Proceedings of the Iowa Academy of Science **88**:1–6.

Freemark, K. 1995. Assessing effects of agriculture on terrestrial wildlife: developing a hierarchical approach for the U.S. EPA. Landscape and Urban Planning **31**:99–115.

Freemark, K., and J. Smith. 1995. A landscape retrospective for Walnut Creek, Story County. Technical report to the U.S. EPA.

Freemark, K.E., C. Hummon, D. White, and D. Hulse. 1996. Modeling risks to biodiversity in past, present and future landscapes. Technical Report No. 268. Canadian Wildlife Service Headquarters, Environment Canada, Ottowa K1A 0H3.

Galatowitsch, S.M., and A. van der Valk. 1994. Restoring prairie wetlands: an ecological approach. Iowa State University Press, Ames, Iowa, USA.

Harms, B., J.P. Knaapen, and J.G. Rademakers. 1993. Landscape planning for nature restoration: comparing regional scenarios. *In* C. Vos and P. Opdam, editors. Landscape ecology of a stressed environment, Chapman & Hall, London, England.

Heathcote, I. 1998. Integrated watershed management: principles and practice. John Wiley and Sons, New York, New York, USA.

Hewes, L. 1951. The northern wet prairie of the United States: nature, sources of information, and extent. Annals of the Association of American Geographers **41**:307–323.

Holt, R.D., S.W. Pacala, T.W. Smith, and J. Liu. 1995. Linking contemporary vegetation models with spatially-explicit animal population models. Ecological Applications **5**:20–27.

Hulse, D., J. Eilers, K. Freemark, C. Hummon, and D. White. 2000. Planning alternative future landscapes in Oregon: evaluating effects on water quality and biodiversity. Landscape Journal **19**:1–19.

Klopatek, J.M., R.J. Olson, C.J. Emerson, and J.L. Joness. 1979. Land-use conflicts with natural vegetation in the United States. Environmental Conservation **6**:191–199.

Korpimäki, E., and K. Nordahl. 1998. Experimental reduction of predators reverses the crash phase of small-rodent cycles. Ecology **79**:2488–2455.

Leopold, A. 1949. A Sand County almanac. Oxford University Press, New York, New York, USA.

Losey, J.E., and R.F. Denno. 1998. Positive predator–predator interactions: enhanced predation rates and synergistic suppression of aphid populations. Ecology **79**:2143–2152.

Nassauer, J.I. 1988. Landscape care: perceptions of local people in landscape ecology and sustainable development. Landscape and Land-use Planning 8. American Society of Landscape Architects, Washington, D.C., USA.

Nassauer, J.I. 1997. Cultural sustainability. *In* J.I. Nassauer, editor. Placing nature: culture in landscape ecology. Island Press, Washington, D.C., USA.

Office of Technology Assessment (OTA). 1995. Targeting environmental priorities in agriculture: reforming program strategies. Congress of the United States, OTA-ENV-640. U.S. Government Printing Office, Washington, D.C., USA.

Palmer, M.W. 1992. The coexistence of species in fractal landscapes. American Naturalist **139**:375–397.

Pimentel, D., L. McLaughlin, A. Zepp, B. Latikan, T. Draus, P. Kleinman, F. Vancini, W.J. Roach, E. Grapp, W.S. Keeton, and G. Selig. 1991. Environmental and economic effects of reducing pesticide use. BioScience **41**:402–409.

Pratt, J.R., and J. Cairns, Jr. 1992. Ecological risks associated with the extinction of species. *In* Predicting ecosystem risk, J. Cairns, Jr., B.R. Neiderlehner, and D.R. Orvos, editors. Advances in modern ecotoxicology. Princeton University Scientific Publishing Co., Inc., Princeton, New Jersey, USA.

Prior, J. 1991. Landforms of Iowa. University of Iowa Press, Iowa City, Iowa, USA.

Puckett, L.J. 1994. Nonpoint and point sources of nitrogen in major watersheds of the United States. U.S. Geological Survey, Water Resources Investigations Report 94-4001, USGS Washington, D.C., USA.

Pulliam, R., and B.J. Danielson. 1991. Sources, sinks, and habitat selection: a landscape perspective on population dynamics. American Naturalist **137**:S50–S66.

Roosa, D.M. 1981. Iowa natural heritage preservation: history, present status, and future challenges. Proceedings of the Iowa Academy of Science **88**:43–47.

Runge, C.F. 1996. Agriculture and environmental policy: new business or business as usual? Working Paper No. 1. *In* Environmental reform: the next generation project. Yale Center for Law and the Environment, New Haven, Connecticut, USA.

Rustigian, H. 1999. Assessing the potential impacts of alternative landscape designs on amphibian population dynamics. Master of Science Research Paper, Department of Geosciences, Geography Program, Oregon State University, Corvallis, Oregon, USA.

Schoonenboom, I.J. 1995. Overview and state of the art of scenario studies for the rural environment. Pages 15–24 *in* J.T.Th. Schoute et al., editors. Scenario studies for the rural environment. Kluwer Academic Publishers, Dordrecht, The Netherlands.

Schwartz, M., editor. 1995. Conservation in highly fragmented landscapes. Chapman & Hall, New York, New York, USA.

Starfield, A. 1997. A pragmatic approach to modeling for wildlife management. Journal of Wildlife Management **61**:261–270.

Steinitz, C., E. Bilda, J. Ellis, T. Johnson, Y. Hung, E. Katz, P. Meijerink, D. Olson, A. Shearer, H. Smith, and A. Sternberg. 1994. Alternative Futures for Monroe County, PA. Harvard University Graduate School of Design, Cambridge, Massachusetts, USA.

Trzcinski, M.K., L. Fahrig, and G. Merriam. 1999. Independent effects of forest cover and fragmentation on the distribution of forest breeding birds. Ecological Applications **9**:586–593.

USDA. 1992. Agricultural Statistics 1991. USDA, Washington, D.C., USA.

Vitousek, P.M., J.D. Aber, R.W. Howarth, G.E. Likens, P.A. Matson, D.W. Schindler, W.H. Schlesinger, and D.G. Tilman. 1997. Human alteration

of the global nitrogen cycle: sources and consequences. Ecological Applications **7**:737–750.
Walpole, S.C., and J.A. Sinden. 1997. BCA and GIS: integration of economic and environmental indicators to aid land management decisions. Ecological Economics **23**:45–57.
White, D., P.G. Minotti, M.J. Barczak, J.C. Sifneos, K.E. Freemark, M. Santelmann, C.F. Steinitz, A.R. Keister, and E.M. Preston. 1997. Assessing risks to biodiversity from future landscape change. Conservation Biology **11**:349–360.
Williams, J.R., C.A. Jones, and P.T. Dyke. 1988. EPIC, the Erosion Productivity Index Calculator, Model Documentation, Vol. 1, Temple, Texas. USDA ARS.
Worldwatch Institute. 1998. State of the world: a Worldwatch Institute report on progress toward a sustainable society. Norton, New York, New York, USA.

World Wide Web Citations

Iowa Cooperative Soil Survey; Iowa Soil Properties and Interpretation Database. 3 March 2000. *http://icss.agron.iastate.edu/*
Rural Watersheds and Policy. Nassauer, J.I. 3 March 2000. *http://www.snre.umich.edu/faculty-research/nassauer/*(sidebar: Rural Watersheds and Policy).
Simulations of future alternatives for agriculture. Clark, M.E. 3 March 2000. *http://www.public.iastate.edu/~meclark/*(sidebar: Baseline Simulations, Conservation Simulations).
US-EPA Office of Wetlands, Oceans, and Watersheds; Top Ten Watershed Lessons Learned. 3 March 2000. *http://www.epa.gov/owow/lessons/*

Part IV
Making Decisions About the Land

12 Many Small Decisions: Incorporating Ecological Knowledge in Land-Use Decisions in the United States

Richard A. Haeuber and N. Thompson Hobbs

Few alterations of land exert effects that are as pervasive as those caused by human settlement (Douglas 1994). The conversion of natural landscape to areas dominated by human development can cause dramatic changes, such as paving over land to construct a shopping mall and parking lots, or changes that are more subtle, such as fragmenting the landscape through low-density housing and roads. Decisions about how people occupy the land can have far-reaching effects on ecological systems, including simplification of the landscape, modification of natural disturbance patterns, changes in soil and water quantity and quality, and altered movement of nutrients, organisms, and other elements of ecological systems.

Landowners and land managers are challenged to provide a land-use policy that sustains natural systems in the face of rapid development of land. In this chapter, we develop the idea that meeting this challenge requires a broader conservation focus, extending beyond the emphasis on conservation of publicly owned lands that has dominated many past conservation efforts in the United States. In the future, the key will be to bolster and support conservation efforts on private lands as a complement to past and future conservation efforts on public lands.

In making the case for a broadened environmental conservation focus, we explore how available tools and mechanisms can be used to stimulate increased private land conservation efforts by implementing the ecological principles and guidelines discussed in Chapter 1 of this volume (Dale et al.). We approach this task by first providing an overview of trends, challenges, and opportunities that characterize the current state of land-use policy and management in the United States. We then examine the framework under which land-use policy and management decisions are implemented in the United States. More specifically, we describe a range of land-use planning and management tools employed in the United States, with an eye to their usefulness in implementing specific ecological principles. We offer examples that explore the potential for further developing the interface between ecology and planning.

12.1 Land-Use Trends, Challenges, and Opportunities

12.1.1 Land-Use Change Is Diffuse in Space

Urbanization is a significant trend in the United States, with the proportion of the population considered "metropolitan" increasing from 65.9% to nearly 80% between 1960 and 1990. During the same period, however, the population of central cities grew by only 15.2%, while the suburbs grew by 71.1% (Platt 1996). Development at the suburban–rural interface consumes land as residents flee inner cities, fueling extensive development in areas ever further from the urban core. In the United States, for example, the total amount of land under urban use grew from 21 million hectares in 1982 to 26 million hectares in 1992. In a single decade, 2 million hectares of forestland, 1.5 million hectares of cultivated cropland, 0.9 million hectares of pastureland, and 0.7 million hectares of rangeland came under urban uses (WRI 1997).

Although urbanization is a strong agent of landscape change, low-density development may exert even greater effects, particularly in the western United States. For example, National Resource Inventory (NRI) data for Colorado reveal that 30 square miles of forest and 62 square miles of rangeland and pasture were urbanized in Colorado during 1982 to 1992. Although these statistics reveal a substantial conversion of habitat, the NRI data do not account for diffuse development that cannot be characterized as "urban." This development occurs at low density on the periphery of urban areas as well as in areas of the state remote from any urbanization. To examine the effects of such development, we overlaid U.S. Bureau of Census maps of housing density on U.S. Geological Survey maps of land cover. Combining these data provides an estimate of the area of broad habitat types that were developed across a range of housing densities. Nineteen hundred square miles of forest, range, and wetland habitat were developed at exurban or higher densities during 1960–1990 (Table 12.1). We predict that rapid development of forest, range, and wetlands observed during 1960–1990 will continue. The area developed in these habitat types will almost double during 1990 to 2020 (Table 12.1).

In Colorado, much of this development is occurring in areas that have traditionally been remote from cities and rural in character. Moreover, some of the most rapid growth is occurring adjacent to national parks, national forests, and other conservation areas (Howe et al. 1997). Development adjacent to protected areas creates a high contrast between natural systems within conservation areas and the human-dominated systems surrounding them. The spread of low-density development in rural areas is a significant departure from traditional development patterns, characterized by urban areas surrounded by less developed rural landscapes.

Departure from traditional land-use patterns is likely to accelerate as a result of the rise of service- and information-based economies, as well as

TABLE 12.1. Data and projections of area (sq mi) of habitat types in Colorado developed at densities ranging from rural to urban during 1960 to 1990.

Development category (acres/unit)	Forest			Range			Wetlands		
	1960	1990	2020	1960	1990	2020	1960	1990	2020
Urban (≤2)	0	5	17	2	41	73	0	1	2
Suburban (>2–10)	6	34	77	11	107	182	1	3	7
Exurban (>10–40)	138	931	1638	191	1112	2242	17	36	44
Ranch (>40–80)	269	728	925	198	1053	1556	15	19	27
Rural (>80)	6577	5372	4450	26397	25291	23894	188	165	146

Note: Each cell in the table gives the area in each habitat and density class during a given year. See Table for definitions of development densities. Data sources are U.S. Bureau of Census Block Group Statistics and U.S. Geological Survey Land-Use Land Cover data.

regional mobility fostered by expanding unearned income and retirements. Historically, urban concentrations of population were encouraged by the need to be near centers of production and commerce. Enterprises based on information and technology are not constrained by requirements based on the geographic location of production and commerce centers. Workers in these industries are free to live and work in areas remote from cities (Riebsame et al. 1996). Moreover, the rise in unearned income resulting from retirements of baby boomers will dramatically increase regional mobility and will likely accelerate migration to areas offering environmental amenities (Bennett 1993; Riebsame et al. 1996; Carlson et al. 1998).

12.1.2 Land-Use Change Is Diffuse in Time

A significant challenge to stimulating change in land-use policy and management results from the way that changes to ecological systems accumulate through land-use and land-management decisions. Cumulative impacts generally refer to effects arising from multiple activities in a given region that overlap spatially, or multiple perturbations from a single, repeated activity over time (Beanlands 1995; Lane 1998). The decisions giving rise to cumulative impacts are many in number and diffuse in time. Their effects tend to accumulate slowly, leading to a phenomenon sometimes referred to as the "tyranny of small decisions."

Whether manifested through atmospheric nutrient deposition from multiple sources, "leap-frog development" patterns in a forested landscape, or repetitive cutting of a forest plantation over time, cumulative impacts are one of the most pervasive avenues of human-induced stress on ecological systems. Unlike a clear-cut, where the spatial extent of the change is obvious and other impacts are observable directly, such as soil erosion into nearby streams, cumulative impacts are indirect and often not immediately recognizable. When many decisions are made one at time and each choice affects a small area of land, it is easy to overlook incremental impacts. For planners, the challenge lies in understanding the complexity of interconnected causal pathways, the feedback among and between human activities and environmental responses, and the additive effects of human actions over large spatial scales and long temporal scales.

12.1.3 Evolving Landscape of Land-Use Policy

Throughout the United States, politicians, planners, and citizens increasingly recognize the negative aspects of extensive, sprawling development patterns. These aspects include not only environmental impacts, but overcrowded schools, increasing traffic congestion, and strained city and county budgets (see Bank of America et al. 1995; Beatley and Manning 1997). The once arcane issue of land use clearly is gaining salience around the nation as the public recognizes the direct connections between development and land-use patterns, environmental concerns, and quality of life. Changes in the state and local political climate around the country create fertile ground for changing land-use patterns in keeping with greater attention to ecological principles.

One clear indicator of change in the national political climate involves the agendas of the nation's governors, who wield significant power as the elected chief executives of the states. In their 1998 "State of the State" or inaugural addresses, 15 incumbent or newly elected governors from both parties mentioned growth management issues, such as preservation of open space and maintaining the character of rural communities, as central components of their agendas for the coming year. By 1999, the policy agendas of 20 governors—Democrat, Republican, and Independent, from Maine to Utah—included land-use planning issues of various sorts, primarily clustered under the heading of sprawl or growth management initiatives.

The appearance of land use on governors' policy agendas is easily understood, given the recently exhibited preferences of voters. In the 1998 elections, voters approved nearly 200 state and local ballot initiatives dealing with land-use issues, such as preservation of open space and loss of agricultural lands (Egan 1998). Taken together, the initiatives approved in November 1998 dedicated more than $7 billion to purchasing land or development rights (Goodman and Eggen 1998). In the state of New Jersey, for example, voters in 43 cities and six counties agreed to raise taxes to buy land and preserve open space. Statewide, New Jersey voters approved Republican Governor Christie Whitman's plan to spend nearly $1 billion over 10 years to buy half of the state's remaining open space.

The groundswell of popular interest in land-use issues is manifest at the national political level as well. Throughout 1999, Vice President Gore adopted land use as one of his top environmental policy agenda issues, perhaps perceiving that it is more accessible and relevant to the general public than longer-term issues like global warming. In addressing audiences such as the May 1999 National Town Meeting for a Sustainable America, Mr. Gore elevated land use to the level of a campaign issue, ensuring that it figured prominently in the 2000 presidential election.

As a central component of efforts to address land-use issues, and capture those issues for electoral purposes, the Clinton–Gore Administration has proposed a "Livability Agenda." Intended to "build more livable communities," the initiative proposes $700 million in tax credits for state and local bonds to preserve open space, redevelop brownfields, and improve parks; a $6.1 billion investment in public transit; a regional planning effort funded at $50 million; and, an additional $100 million to assist communities in education, fighting crime, and community planning. This initiative comes on top of the January 1999 announcement of the "Lands Legacy Initiative," a $1 billion program dedicated to land acquisition and protection.

12.1.4 Ecological Science and the Future of Land-Use Policy and Management

Until recently, ecologists tended to avoid human-dominated systems as locations for study (McDonnell and Pickett 1993). Although this tendency is changing, the investment of ecologists in understanding human impacts has largely focused on effects of extraction of materials from the environment (agriculture, logging, energy development) and the anthropogenic addition of materials to it (air and water pollution, carbon additions to the atmosphere) (Fig. 12.1). The effects of human settlement, effects distinct

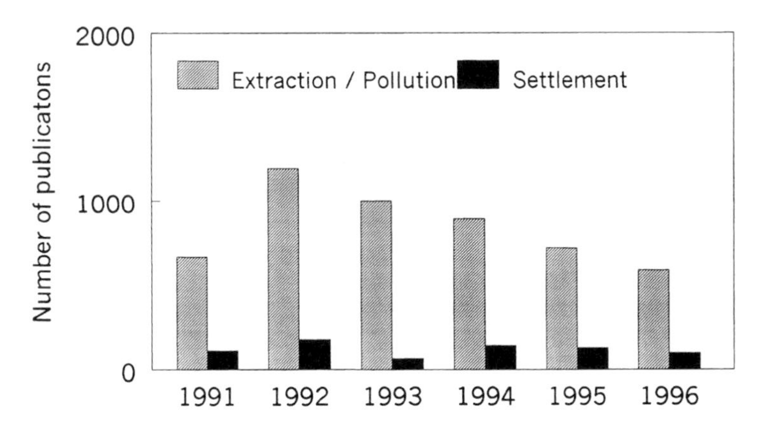

FIGURE 12.1. Number of citations in biological abstracts dealing with ecology and effects of extraction or pollution compared with those dealing with ecology and effects of human settlement.

from extraction and pollution, have not been studied extensively (Fig. 12.1). Yet, as described above, the effects resulting from residential development and expansion of infrastructure throughout the United States are accumulating rapidly and may be a greater source of environmental change than either extraction or pollution.

As a result of the potential magnitude of environmental change involved, local governments urgently need sound information on ecological effects of development. Such information is needed to develop master plans for entire landscapes and to review development proposals at site scales (Albrecht et al. 1995; Duerksen et al. 1997). The recent developments at local, state, and national levels described above illustrate that fertile ground exists for incorporating an explicitly ecological approach into land-use planning and management. In fact, the attention and funding dedicated to land acquisition and protection may constitute the most important opportunity in many years to employ ecological principles toward the end of "inventing landscapes."

The recent white paper commissioned by the Ecological Society of America (Chapter 1 and Dale et al. 2000) demonstrates one way that ecological science can contribute to the continuing evolution of land-use policy and management in the United States. Other recent reviews also offer data and understanding that can inform ecologically sound land-use decisions at the local level (Beatley 1994; Duerksen et al. 1997; Noss et al. 1997; Peck 1998; Bourgeron et al., Chapter 13; Haufler and Kernohan, Chapter 4). These papers are complemented by interactive sites on the internet that enable planners and decision makers to obtain the best current data on biological resources that may be impacted by development at a specific site (e.g., see *http://ndis.nrel.colostate.edu*). Such internet-based tools are likely to multiply rapidly in the future, and will increase the availability and utility of ecological data in supporting land-use decisions.

It is clear that choices on future uses of land in the United States will have profound impacts on the integrity of its ecosystems and the quality of life of its people. Recent political and policy developments provide a significant and unusual opportunity to demonstrate how ecological principles and guidelines can improve land-use decisions in a way that matters to people making those decisions on a daily basis. The next section explores the compatibility between ecological principles and guidelines and specific land-use decision-making tools and approaches.

12.2 Implementation of Land-Use Decisions in the United States

Land is most commonly understood in terms of physical and biological characteristics, such as geology, hydrology, topography, and vegetation. These characteristics constrain the use of land. Additional constraints result from economic incentives and disincentives for certain uses of land, as well as from legal and political authority (Platt 1996). Because this legal authority can encourage or prohibit particular land uses, it also can structure the physical and biological terrain. In a fundamental sense, then, the current appearance of any landscape results from a complex interaction between natural features and a history of human influences expressed in economic, political, legal, and socio-cultural decisions.

Two important aspects characterize governmental roles in land-use decision making. First, governments possess both direct and indirect power over land use by virtue of their ability to enact legislation, promulgate regulations, and own land. Second, different levels of government possess differing types and degrees of influence over land use.

12.2.1 Government Powers

American planning traditions allow rather weak direct government control over decisions by private landowners. Still, governmental constraints and incentives at various levels do modify economic forces and individual landowner prerogatives. Land-use planning is the process through which government exerts this modifying influence, guiding and regulating land development to meet societal goals (Smith 1993). The term "planning" often refers to general guides for private land use that channel patterns of residential and commercial development, as well as detailed plans for public facilities like roads and schools. Plans establish the need for facilities and timetables for their creation, and set forth principles for zoning and building code ordinances. In practice, land-use planning typically entails creating maps of preferred land uses in an area and developing the detailed regulations that would achieve the desired landscape.

Governmental authority for planning and zoning rests on the government's "police power," which refers to "the right of the community to regulate the activities of private parties to protect the interests of the public" (Levy 2000). The concept of police power includes three basic functions that constitute the primary sources of direct government control over land use: (1) the role of government in reducing harm and nuisances; (2) the role of government in protecting the public good; and (3) the need for government to provide for orderly timing of development and associated services (Platt 1996). As we discuss below, planning and its corollary activity, zoning, most frequently are exercised by local level governments.

The roster of potential indirect government influences on land-use patterns is quite long (Jackson 1981). It is difficult to characterize with certainty the degree of influence over land use exerted by indirect means (GAO 1999). However, an array of federal and state transportation, tax, agriculture, and other policies clearly affect land use indirectly to some degree. For example, policies of the Federal Housing Administration, in concert with development of federal and state highways, fueled patterns of urban and suburban development in the post-WWII decades (Smith 1993; Levy 2000). According to some observers, these tax and transportation policies aided the emerging pattern of exurban development in previously rural areas (Kunstler 1993). Current estate and capital gains tax laws further tend to encourage large-lot/large-house residential development.

On the other hand, an argument can be made that various federal policies act in the opposite direction as well, constraining rather than facilitating sprawling development patterns. The combined influences of federal air and transportation policy provisions provide an interesting example of such indirect influence on land use. Federal and state highway development has combined with extensive suburban development patterns to increase overall levels of emissions from mobile sources. The increase in emissions has placed many urban areas (e.g., Dallas–Fort Worth in Texas) at or near ozone nonattainment status under federal air quality standards. Urban areas experiencing increases in mobile source emissions are threatened with loss of federal transportation funds (Mintz 1999). Because development is inextricably intertwined with transportation infrastructure, loss of transportation funds indirectly influences future land-use patterns.

12.2.2 Levels of Government and Land-Use Powers

In the United States, the federal government exerts only limited direct authority over private land use. Most land-use authority resides with the states, except in the case of federally owned or controlled lands (Platt 1996). Most states, in turn, used Standard State Zoning Enabling Acts to delegate direct land-use control powers to county and municipal governments, though a few pursue statewide land-use planning and require local governments to comply with state standards (Garrett 1987; Burby and May 1997). While it has had a roller-coaster history in the United States, effective regional land-use planning is practically nonexistent (Platt 1996). Thus, most land-use and zoning authority rests with local government (e.g., towns and counties), although some states (e.g., Oregon, Vermont, Florida, and Hawaii) have strong statewide planning legislation.

12.2.2.1 The Federal Role

The direct federal role in American land-use policy is weak outside of the federal lands. To the extent that laws and regulations confer direct federal influence over private land use, this control generally involves a specific type of activity or a narrowly prescribed spatial domain. For example, federal law prohibits dredging and filling wetlands without an environmental review and permit. Similarly, the Endangered Species Act prohibits the "taking" of federally listed species on private lands. The Wild and Scenic Rivers Act of

1968 is an example of a spatially-delimited policy with direct impact on private land use. The Act protects limited areas adjacent to river segments designated under the Act through a combination of land purchases, prohibitions against damaging activities, and guidelines for local zoning on riverfront lands. Regardless of such statutes and policies, however, the federal government generally takes only a limited direct role in influencing private land-use patterns.

In contrast with its weak role in private land use, federal agencies have primary authority over federally owned lands through statutes such as the National Forest Management Act and Federal Land Policy and Management Act. Despite control granted through ownership and statutory authority, however, federal land-use planning operates in a pluralistic political environment. The frequently held notion that federal bureaucrats in Washington, D.C. wield absolute land-use authority on the federal lands ignores reality. All federal land-use decisions are hotly contested in multiple public arenas, and so take place in a glass house rather than an isolated bunker (Meidinger 1997).

Federal agencies face two fundamental problems with large-scale land-use planning. First, there is little coordination among agencies and other landowners. In an area like the Greater Yellowstone Ecosystem, for example, lands are owned and managed by multiple entities, including the U.S. Forest Service, Bureau of Land Management, Fish and Wildlife Service, state and local governments, and private landowners (Clark and Harvey 1990; Clark et al. 1991; Goldstein 1992; Glick and Clark 1998). The second difficulty with federal land-use planning is that it is largely separate from adjacent and nearby private land-use planning. While federal agencies may have the statutory authority to undertake cross-boundary stewardship efforts, they generally hesitate to get involved in land-use planning beyond the edges of federal lands (Meidinger 1998). Yet, their actions affect nearby private lands. Thus, for example, private land-use patterns may well be influenced by federal land activities such as forest planning, ski area permitting, and recreation planning. Similarly, adjacent development on private lands has significant impact on federal lands. Despite this interplay between decisions and actions taken on federal and private lands, federal agencies largely are unable to directly address the impact on federal lands of decisions taken on private lands, except through utilizing statutes such as the Endangered Species Act.

12.2.2.2 The State Role

Historically, most states simply delegated power over land-use regulation to local governments through "planning and zoning enabling acts," giving local government authority to zone and set subdivision regulations (Platt 1996). The reemergence of a stronger state role in land-use planning became evident with the strengthening of Hawaii's Land-Use Law in the 1960s.

From 1965 to 1975, professional planners pushed for a stronger state role to achieve better coordination among local governments and to attain the holy grail of planning: regional master planning that transcends fragmented local jurisdictions. Many states expanded their land-use planning and management roles in reaction to widespread lack of planning at the local level and disregard of statewide and regional consequences of local land-use decisions (DeGrove 1992; Callies 1996). At the same time, many county and municipal governments adopted policies meant to channel land-use and development into prescribed patterns (Platt 1996).

Despite advances achieved by state governments in the 1960s and 1970s, as well as the progress of local growth management initiatives, many state governments found it necessary to escalate efforts to manage land-use and land-development patterns. Continuing problems included the unevenness and fragmentation of local growth management efforts; the spatial distribution and cumulative nature of development impacts; and, a mismatch between the level of regulation and level of consequences. As a result, the last 25 years have seen a spate of state-legislated growth-management programs intended to create statewide consistency in land-use and land-development patterns. Starting with Oregon in 1973, eight states have enacted growth-management legislation, and several others have enacted legislation aimed at structuring regional growth patterns (Bollens 1992; DeGrove 1992; Gale 1992).

12.2.2.3 The Role of Local Governments

Local governments are truly on the front lines in terms of shaping uses of private lands in the United States. In 1997, there were 87,453 local governmental units in the United States, including 3043 counties, 19,372 municipal governments, and 16,629 townships and towns (U.S. Bureau of Census 1997). Local governments are legally creatures of the states in which they are located and are allowed to exercise only those powers granted to them by state governments. Most states have passed legislation specifically enabling and, in some cases, requiring local governments to directly control private land use through planning and other functions. State enabling acts generally are procedural rather than substantive, prescribing a process that local governments must follow in adopting land-use regulations but not the content of the regulations themselves (Platt 1996).

The most common tools for exerting local government influence over private lands are comprehensive plans and zoning ordinances. Comprehensive plans (also known as general or master plans) are the most basic plan prepared by staff of local government planning agencies to guide the future development of a community. The comprehensive plan spatially encompasses the entire community and is prepared to define development trajectories over a period up to 20 years. The nature of the comprehensive plan has evolved significantly over the last 50 years, and the contemporary comprehensive plan includes maps that depict land uses, community facilities, and transportation systems, as well as a set of general policies meant to guide and manage development (Kaiser and Godschalk 1995).

In general, comprehensive plans are offered as advice to planning commissions in making land-use decisions. Actual decisions about land use are undertaken through regulatory actions such as zoning. Conventional zoning is based on use of the "police power" to regulate how land in a community may be developed and used, including the use of private land, density of structural development per unit of land, and the dimensions of buildings (Platt 1996; Levy 2000). Zoning ordinances consist of two parts. The zoning map divides the community into a number of zones for specified land uses, and includes enough detail to determine where actual parcels of land lie. The zoning text provides a detailed description of what may be constructed in each zone and the sanctioned uses for those structures. Zoning ordinances are prepared by planning staff and acquire their legal authority only after being formally enacted by the community's legislative body (Levy 2000).

Zoning comes in various forms and is supplemented by a variety of other tools. Relevant tools for land-use planning in regions experiencing population growth include subdivision and "planned unit development" ordinances, specifying exactly how development will play out on the landscape. Such regulatory tools are developed at the local level but are quite similar among local jurisdictions because they are based on model ordinances created by professional organizations. In the next section, we describe a variety of these tools and discuss their utility in implementing an ecological approach to land-use planning.

12.3 Land-Use Planning Tools and Implementing Ecological Principles

Under the circumstances described above, local governments regulate some aspects of private land use but can impose restrictions only through limited constraints on development. Few land-use plans for American communities created before the late 1970s to early 1980s address issues such as maintenance of open space, agricultural lands, or wildlife habitat (Kaiser and Godschalk 1995). Neglect of these issues was evident particularly in relation to detailed zoning and subdivision ordinances. In this section, we explore how planners and land managers can address open space and similar issues by practicing land-use planning in keeping with the ecological principles and guidelines discussed in Chapter 1 (see Table 12.2).

Attention to issues such as open space, agricultural land protection, and wildlife habitat planning was increasingly apparent as plans were revised in the 1990s. One of the most significant opportunities for incorporating

TABLE 12.2. Ecological principles and guidelines for managing the use of land.

Ecological principles	Ecological guidelines
Time: Ecological processes function at many time scales, some long, some short; and ecosystems change through time.	Examine the impacts of local decisions in a regional context.
Species: Particular species and networks of interacting species have key, broad-scale ecosystem-level effects.	Plan for long-term change and unexpected events.
Place: Local climatic, hydrologic, edaphic, and geomorphologic factors as well as biotic interactions strongly affect ecological processes and the abundance and distribution of species at any one place.	Preserve rare landscape elements, critical habitats, and associated species.
Disturbance: The type, intensity, and duration of disturbance shape the characteristics of populations, communities, and ecosystems.	Avoid land uses that deplete natural resources over a broad area.
Landscape: The size, shape, and spatial relationships of land-cover types influence the dynamics of populations, communities, and ecosystems.	Retain large contiguous connected areas that contain critical habitats.
	Minimize the introduction and spread of nonnative species.
	Avoid or compensate for effects of development on ecological processes.
	Implement land-use and land-management practices that are compatible with the natural potential of the area.

ecological knowledge in land-use decisions occurs in the goal-setting phase of local planning processes. Ecological goals must be made explicit in community comprehensive plans. Ecological principles can assist communities in identifying and expressing their goals and vision for maintaining intact habitat, preserving biological diversity, and assuring ecosystem function by illuminating the ways in which such goals are affected by specific land use decisions.

The comprehensive planning process allows for input and public review and informs property owners of likely restrictions on development. Increasingly, public input and participation is structured using sophisticated modeling and visualization techniques (Levy 2000). Such approaches enable the community to understand the potential impacts of alternative land-use scenarios and allow explicit evaluation of the outcomes produced by applying ecological principles and guidelines in planning (Hulse and Gregory, Chapter 9; Steinitz and McDowell, Chapter 8). Judicial review of land-use regulation stipulates that governments can legitimately regulate development to achieve various community goals (e.g., open space protection), but only if those goals have been incorporated into comprehensive plans. Backed by informed citizen input, the plan legally and technically supports regulatory decisions.

Utilizing a set of ecologically-based principles and guidelines can help communities identify and map components of landscapes that need to be protected to sustain critical ecosystem components and processes. Wetlands are an excellent example in which applying ecological principles and guidelines assists in understanding an important landscape feature and placing it in the context of future development. Because wetlands often constitute large intact patches of native habitat and dispersal corridors for populations, the Species, Place, and Landscape principles are particularly helpful in understanding special considerations that apply to development near wetlands. To account for how wetlands contribute to habitat and species dispersal, planning in wetlands areas should be based on ecological guidelines that suggest preservation of rare landscape elements, retaining large contiguous connected areas, and minimizing the introduction of nonnative species. Moreover, the ecological processes at work in wetlands now are recognized as an important source of ecological services to society (Daily 1997; Ewel 1997). Thus, in the case of wetlands, planners should pay special attention to the ecological guideline that suggests avoiding or compensating for the effects of development on ecological processes.

Community ecological resources should be inventoried, including mapping of land for ecological values, prioritization of land uses to protect those values, and creation of detailed plans explicitly linking development regulations with resources and resource protection goals. Such documentation assists property owners in understanding the potential restrictions on their land as they apply for development permits. Ecological resources information also helps to ensure that ecological values will be considered more thoroughly in decisions regarding government land uses and infrastructure development (e.g., road design, siting of schools, and water treatment plants). Plans that develop such layers of information, over short to long term, and link the information to development regulation and capital improvement decisions are increasingly used in contemporary community planning (Kaiser and Godschalk 1995).

Land-use plans must be coordinated across jurisdictions, in keeping with the Landscape and Disturbance ecological principles. Given the fragmented nature of land-use planning, this is essential for managing larger landscape units. For example, plans for maintaining open space should be coordinated across county, municipal, and other jurisdictional lines. In creating landscape features, such as greenways, planners should pay special attention to the ecological guidelines that suggest examining the impact of local decisions in a regional context and retaining large contiguous connected areas that contain critical habitat. Despite the difficulties inherent in working across public–private boundaries, managers of public lands also can work to implement these ecological principles and guidelines by coordinating their land-use actions with local land-use planners and by accounting for actions on adjacent or even distant private lands that affect the quality of public lands.

Planning also must account for the cumulative effects of development on ecosystems. As discussed earlier, residual environmental changes accumulate over the long term as a consequence of previous development actions

(Beanlands 1995). In addressing cumulative impacts, planners and land managers should pay particular attention to the ecological principles and guidelines that emphasize managing and planning at broader spatial and temporal scales (Forman 1995). Given the permit-by-permit pattern of development, however, anticipating and planning for cumulative effects is a difficult task. Because efforts to reduce cumulative impacts may result in differential treatment of development proposals over time, local governments must support their decisions with well-developed planning and analysis to avoid legal problems.

Good planning involves using a range of tools to achieve landscape goals, allowing planners to modify development commensurate with habitat importance, as well as the preferences and values of local citizens (e.g., see Soule 1991; Grant et al. 1996; Duerksen et al. 1997). Professional planners should give elected officials a full range of options for responding to the inevitable demand for relief from particular land-use restrictions. Single approaches limit planning options, creating landowner frustration and resulting in demands for exemptions and variances. States with strong land-use legislation (e.g., Oregon, Florida, Vermont) provide local government with tools for open space and agricultural protection, and the most effective local land-use planning and management action is often evident in those states.

Even without state support, however, local governments can utilize many diverse strategies and tools for achieving specific environmental goals, such as maintaining wildlife habitat and open spaces (Duerksen et al. 1997). Table 12.3 demonstrates the broad compatibility between ecological principles and a wide range of planning tools and strategies. Such tools can be regulatory in nature, or they may offer incentives for landowners to use and manage land in a desired way. Alternatively, local governments may choose to influence land-use patterns through a variety of land acquisition approaches. Table 12.4 provides several examples to illustrate how specific regulatory tools can be applied in implementing an ecological approach to land-use planning and management.

The Maryland Smart Growth Initiative (SGI) provides a recent example of a growth-management approach that relies on a combination of regulatory and incentive-based approaches to effect land-use change (Haeuber 1999). In early 1997, Governor Parris Glendenning introduced a legislative package designed to influence land-use patterns through growth management. Of several separate pieces, the Smart Growth Areas Act (also known as the Priority Funding Areas Act) and the Rural Legacy Act are the most significant legislative components.

The Smart Growth Areas Act reflects a desire to couple urban revitalization with conservation, using an incentive-based approach. The heart of the act involves using state expenditures for capital improvements to influence economic growth and development. Beginning 1 October, 1998, the state must direct funding for "growth-related" projects to Priority Funding Areas (PFAs). As defined in the legislation, growth-related projects involve state

TABLE 12.3. Planning tools and strategies for implementing ecological principles.

	Time	Species	Place	Disturbance	Landscape
Regulatory tools					
Zoning texts and maps	X	X	X	X	X
Phasing of development	X	X			
Special overlay districts	X	X	X	X	X
Agriculture and open space zoning	X	X	X		X
Performance zoning		X	X		
Subdivision review standards		X	X		
Sanctuary regulations	X	X	X	X	X
Urban growth boundaries	X	X	X		X
Targeted growth	X	X	X		X
Capital improvements programming	X		X		X
Incentive-based tools					
Density bonuses		X	X		
Clustering	X	X	X	X	X
Transferable development rights	X	X	X	X	X
Preferential tax treatment		X	X		
Acquisition approaches					
Fee simple purchase	X	X	X	X	X
Sellbacks and leasebacks	X	X	X	X	X
Intergovernmental agreements	X	X	X	X	X
Easements, purchase of devlopment rights	X	X	X	X	X
Land dedication, impact fees		X	X		
Development agreements		X	X		
Land trusts	X	X	X	X	X
Limited conservation development		X	X		

programs that encourage or support growth and development, such as construction of highways, sewer, and water systems; economic development assistance; and state leases or construction of new office facilities. The Act specifically identifies as PFAs those locations that constitute the state's urban-development core, and allows counties to designate PFAs that meet certain minimum criteria regarding land use, water and sewer service, and residential density. The Smart Growth Areas Act is an excellent example of an incentive-based program that uses capital improvement expenditures to promote development in keeping with several of the ecological guidelines in Table 12.2. These guidelines include examining the impact of local decisions in a regional context, avoiding land uses that deplete natural resources over a broad area, and avoiding development that impacts ecological processes.

Enacted on 22 May, 1997, the Rural Legacy Program is the second significant component of the Smart Growth Initiative. The legislation is designed to establish greenbelts by protecting rural areas from sprawl through a grant program funded by a portion of the real estate transfer tax and state bonds. Grants are awarded on a competitive basis to protect "Rural Legacy Areas," or regions rich in agriculture, forestry, and natural

TABLE 12.4. Implementing ecological principles in land-use planning.

Planning tool	Application
Zoning texts and maps: Basic tools used to implement local zoning regulations	• Tree protection and vegetation management can be used to maintain the feel of a "place" or a particular assemblage of species. • River corridor protection standards assist in species protection and sustaining landscape integrity by creating corridors. • Buffer zones can support disturbance regimes (e.g., flooding), as well as landscape level ecological processes, such as nutrient and material cycling.
Phasing of development: Requires construction of new development in specific phases	• Helps to slow the rate of change, and so increase the timeframe over which landscape change occurs. • Lengthening the development timeframe supports the adaptive capability of species populations in the area.
Special overlay districts: Supplement basic zoning regulations applicable to a property by adding specific restrictions as an overlay to existing zoning	• Allows preservation of processes such as disturbance regimes over large areas encompassing several districts (e.g., tied to floodplain management). • Enables focus on specific features that are important for the feel of a place (e.g., wetlands). • Supports landscape conservation approaches (e.g., creation of wildlife corridors by linking landscape elements).
Agricultural and open space zoning: Maintains low-density and agricultural areas as zoned land uses	• Establishes large minimum lot sizes or areas designated exclusively for agriculture and helps maintain connectivity of the landscape and a sense of place. • Can be used to provide habitat protection for existing species and communities.
Performance zoning: Establishes required goals in terms of habitat or wildlife protection, but allows developers to choose how to achieve those outcomes	• Can be used to maintain habitat for individual species of importance and assemblages of species. • Supports landscape elements unique to a place.
Subdivision review standards: Establishes specific characteristics of lots that are available for development	• Lot size, shape, and buffer requirements can assist in supporting habitat requirements for particular species of importance, as well as maintaining elements important for the integrity of a place.

and cultural resources. Funded at $71.3 million for FY 1998 through 2002, the program encourages local governments and private land trusts to apply for funds to complement existing conservation areas or establish new areas.

Competing for $29 million in state funds, the first solicitation drew 23 applicants seeking $123 million to preserve 53,000 acres of land (Shields 1998). Grants will enable the purchase of conservation easements and development rights through voluntary agreements entered into with private landowners. The Rural Legacy Program focuses on protecting large contiguous tracts of land characterized by mosaics of forests, farms, and other open lands (e.g., historic sites, such as Civil War battlefields). Thus, the program uses several of the tools listed in Table 12.3 to influence land-use patterns in ways that reflect many of the principles and guidelines in Table 12.2, particularly principles and guidelines that emphasize protecting landscape elements and habitat connectivity.

12.4 Conclusion

Efforts to conserve biological diversity in the United States traditionally focused on lands owned or controlled by the federal government. In large measure, these efforts concentrated on establishing protected areas, such as parks and preserves, and on regulating extractive activities like logging and grazing. Although the focus on federal lands produced significant conservation gains in the past, it promises to be less effective in the future. The decreasing effectiveness of conservation approaches focused primarily on federal lands is due, in part, to the simple fact that the preponderance of land in the United States is controlled by private landowners. In the United States, private and tribal lands account for 72.7% of the land base, while federal lands comprise only 21.6% and state and local lands account for 5.7% of the total (USDA 1996). As a result of this pattern of ownership, more than 90% of federally listed threatened and endangered species in the United States in 1993 had some or all of their habitat on nonfederal lands (GAO 1994).

More importantly, private lands and the activities undertaken on them create a landscape context that impacts ecological systems on public lands. For example, because people are attracted to areas with large amounts of public land, the presence of public land is one of the best predictors of population growth rates at the county level in the Rocky Mountains and Great Plains (Case and Alward 1997). Moreover, the trend toward service- and information-based economies will likely reduce the importance of public land extractive activities as a concern for conservationists. Instead, issues such as the impact of low-density residential development on agricultural lands and other open space areas, as well as areas adjacent to protected areas, are likely to be the focus of conservation concern in the future (Knight et al. 1995; Gersh 1996; Howe et al. 1997).

A new understanding of the connectedness of various elements of a landscape at different spatial scales is an important emerging lesson of conservation science that calls into question the adequacy of conservation approaches that focus primarily on publicly owned lands (Pickett and Thompson 1978; Wiens 1997). By this statement, we do not mean to deny the importance of large protected areas such as Yellowstone National Park. As we have discussed throughout this Chapter, however, ecological processes function at large spatial scales, and this reality ultimately constrains the effectiveness of a "cordon and conserve" strategy focusing primarily on establishing conservation areas such as parks and preserves (Pickett and Thompson 1978). Even protected areas the size of Yellowstone National Park exist in the context of a Greater Yellowstone Ecosystem whose boundaries are not contiguous with those of the park itself (Glick and Clark 1998). In essence, vast amounts of energy and attention are devoted to promoting conservation on a relatively small portion of U.S. lands. Because federal and other agencies have little or no influence on activities beyond the boundaries of their lands, it is difficult to manage publicly owned lands with sufficient regard for the regional context and the influence of surrounding privately held areas.

Unless we place federal conservation efforts in a larger regional context, learning to understand and compensate for the changing context, we risk turning our protected areas into islands (Pickett and Thompson 1978). Moving in this direction necessitates addressing conservation issues on private lands (Bean and Wilcove 1997; Knight 1999; Haufler and Kernohan, Chapter 4). In the future, effective efforts to conserve ecological states and processes will require attending to environmental effects of land-use change on lands other than publicly owned or controlled federal lands. However, this does not mean developing an unwieldy new regulatory structure that is economically burdensome and politically unrealistic. Instead, as we described throughout this chapter, significant conservation gains can be achieved on private land and other areas under the jurisdiction of local government simply by utilizing existing planning tools and incentives to implement ecological principles and guidelines.

References

Albrecht, V., J. Barone, R.C. Einsweiler, H. Grossman, W. Hare, R. Lilieholm, W.K. Olson, T. Quimby, A. Randall, and C. Whiteside. 1995. Managing land as ecosystem and economy. Lincoln Institute of Land Policy, Cambridge, Massachusetts, USA.

Bank of America, Greenbelt Alliance, California Resources Agency, and Low Income Housing Fund. 1995. Beyond sprawl: new patterns of growth to fit the new California. Greenbelt Alliance, San Francisco, California, USA.

Bean, M.J., and D.S. Wilcove. 1997. The private-land problem. Conservation Biology **11**:1–2.

Beanlands, G. 1995. Cumulative effects and sustainable development. Pages 77–87 *in* M. Munasinghe and W. Shearer, editors. Defining and measuring sustain-

ability: the biogeophysical foundations. The World Bank, Washington, D.C., USA.
Beatley, T.J., 1994. Habitat conservation planning: endangered species and urban growth. University of Texas Press, Austin, Texas, USA.
Beatley, T.J., and K. Manning. 1997. The ecology of place: planning for environment, economy, and community. Island Press, Washington, D.C., USA.
Bennett, D. 1993. Retirement migration and economic development in high-amenity, nonmetropolitan areas. Journal of Applied Gerontology **12**:446–452.
Bollens, S.A. 1992. State growth management: intergovernmental frameworks and policy objectives. Journal of the American Planning Association **58**(Autumn): 454–466.
Burby, R.J., and P.J. May. 1997. Making governments plan. Johns Hopkins University Press, Baltimore, Maryland, USA.
Callies, D.L. 1996. The quiet revolution revisited: a quarter century of progress. Pages 19–26 *in* American Planning Association, editors. Modernizing state planning statutes: the growing smart working papers, Vol. 1. American Planning Association, Washington, D.C., USA.
Carlson, J.E., V.W. Junk, and S.E. Cann. 1998. Factors affecting retirement migration to Idaho: an adaptation of the amenity retirement migration model. The Gerontologist **38**:18–24.
Case, P., and G. Alward. 1997. Patterns of demographic, economic, and value change in the western United States. Report to the Western Water Policy Review Advisory Commission. National Technical Information Service, Springfield, Virginia, USA.
Clark, T.W., and A.H. Harvey. 1990. The Greater Yellowstone Ecosystem policy arena. Society and Natural Resources **3**:281–284.
Clark, T.W., E.D. Amato, D.G. Whittemore, and A.H. Harvey. 1991. Policy and programs for ecosystem management in the Greater Yellowstone Ecosystem: an analysis. Conservation Biology **5**:412–422.
Daily, G.C., editor. 1997. Nature's services: societal dependence on natural ecosystems. Island Press, Washington, D.C., USA.
Dale, V.H., S. Brown, R.A. Haeuber, N.T. Hobbs, N. Huntly, R.J. Naiman, W.E. Riebsame, M.G. Turner, and T.J. Valone. 2000. Ecological principles and guidelines for managing the use of land. Ecological Applications **10**:639–670.
DeGrove, J.M. 1992. Planning and growth management in the States. Lincoln Institute of Land Policy, Cambridge, Massachusetts, USA.
Douglas, I. 1994. Human settlements. Pages 149–169 *in* W.B. Meyer and B.L. Turner, editors. Changes in land use and land cover: a global perspective. Cambridge University Press, New York, New York, USA.
Duerksen, C.J., D.L. Elliot, N.T. Hobbs, E. Johnson, and J.R. Miller. 1997. Habitat protection planning: where the wild things are. American Planning Association Planning Advisory Service Report Number 470/471.
Egan, T. 1998. Dreams of fields: the new politics of urban sprawl. New York Times, 14 November, E1.
Ewel, K.C. 1997. Water quality improvement by wetlands. Pages 329–344 *in* G.C. Daily, editor. Nature's services: societal dependence on natural ecosystems. Island Press, Washington, D.C., USA.
Forman, R.T.T. 1995. Land mosaics. Cambridge University Press, New York, New York, USA.

Gale, D.E. 1992. Eight state-sponsored growth management programs: a comparative analysis. Journal of the American Planning Association **58** (Autumn):425–439.

GAO. 1994. Endangered Species Act: information on species protection on nonfederal lands. GAO/RCED-95-16. General Accounting Office, Washington, D.C., USA.

GAO. 1999. Community development: extent of federal influence on "urban sprawl" is unclear. GAO/RCED-99-87. General Accounting Office, Washington, D.C., USA.

Garrett, M.A., Jr. 1987. Land use regulation: the impacts of alternative land use rights. Praeger, New York, New York, USA.

Gersh, J. 1996. Subdivide and conquer: concrete, condos, and the second conquest of the American West. The Amicus Journal **18**(3):14–20.

Glick, D.A., and T.W. Clark. 1998. Overcoming boundaries: the Greater Yellowstone Ecosystem. Pages 237–256 *in* R.L. Knight and P.B. Landres, editors. Stewardship across boundaries. Island Press, Washington, D.C., USA.

Goldstein, B. 1992. The struggle over ecosystem management at Yellowstone. Bioscience **42**:183–187.

Goodman, P.S., and D. Eggen. 1998. A vote to keep sprawl at bay. Washington Post, 5 November, B1.

Grant, J., P. Manuel, and D. Joudrey. 1996. A framework for planning sustainable residential landscapes. APA Journal **62**:331–344.

Haeuber, R.A. 1999. Sprawl tales: Maryland's Smart Growth Initiative and the evolution of growth management. Urban Ecosystems **3**:303–319.

Howe, J., E. McMahon, and L. Propst. 1997. Balancing nature and commerce in gateway communities. Island Press, Washington D.C., USA.

Jackson, R.H. 1981. Land use in America. John Wiley and Sons, New York, New York, USA.

Kaiser, E.J., and D.R. Godschalk. 1995. Twentieth century land use planning. APA Journal **61**:365–385.

Knight, R.L. 1999. Private lands: the neglected geography. Conservation Biology **13**:223–224.

Knight, R.L., and T.W. Clark. 1998. Boundaries between public and private lands: defining obstacles, finding solutions. Pages 175–191 *in* R.L. Knight and P.B. Landres, editors. Stewardship across boundaries. Island Press, Washington, D.C., USA.

Knight, R.L., G.N. Wallace, and W.E. Riebsame. 1995. Ranching the view: subdivisions versus agriculture. Conservation Biology **9**:459–461.

Kunstler, J.H. 1993. The geography of nowhere: the rise and decline of America's manmade landscape. Simon Schuster, New York, New York, USA.

Lane, P. 1998. Assessing cumulative health effects in ecosystems. Pages 129–153 *in* D. Rapport et al., editors. Ecosystem health. Blackwell Science, Oxford, England.

Levy, J.M. 2000. Contemporary urban planning, 5th edition. Prentice Hall, Upper Saddle River, New Jersey, USA.

McDonnell, M.J., and S.T.A. Pickett, editors. 1993. Humans as components of ecosystems: the ecology of subtle human effects and populated areas. Springer-Verlag, New York, New York, USA.

Meidinger, E.E. 1997. Organizational and legal challenges for ecosystem management. *In* K.A. Kohm and J.F. Franklin, editors. Creating a forestry for the 21st century: the science of ecosystem management. Island Press, Washington, D.C., USA.

Meidinger, E.E. 1998. Laws and institutions in cross-boundary stewardship. *In* R.L. Knight and P.B. Landres, editors. Stewardship across boundaries. Island Press, Washington, D.C., USA.

Mintz, J. 1999. Texas environment could work against Bush. Washington Post, 15 October, A1.

Noss, R.F., M.A. O'Connell, and D.D. Murphy. 1997. The science of conservation planning: habitat conservation under the Endangered Species Act. Island Press, Washington, D.C., USA.

Peck, S. 1998. Planning for biodiversity: issues and examples. Island Press, Washington, D.C., USA.

Pickett, S.T.A., and J.N. Thompson. 1978. Patch dynamics and the design of nature reserves. Biological Conservation **13**:27–37.

Platt, R.H. 1996. Land use and society: geography, law, and public policy. Island Press, Washington, D.C., USA.

Riebsame, W.E., H. Gosnell, and D.M. Theobald. 1996. Altlas of the new west: portrait of a changing region. Norton, New York, New York, USA.

Shields, T. 1998. Maryland embraces its rural legacy. Washington Post, 15 March, B4.

Smith, H.H. 1993. The citizen's guide to planning. American Planning Association, Chicago, Illinois, USA.

Soule, M.E. 1991. Land use planning and wildlife maintenance: guidelines for conserving wildlife in an urban landscape. APA Journal **57**:313–323.

U.S. Bureau of the Census. 1997. 1997 census of governments, government organization. U.S. Department of Commerce, Washington, D.C., USA.

U.S. Department of Agriculture. 1996. A Geography of Hope. U.S. Department of Agriculture, Washington, D.C., USA.

Wiens, J.A. 1997. The emerging role of patchiness in conservation biology. *In* S.T.A. Pickett et al., editors. The ecological basis of conservation. Chapman & Hall, New York, New York, USA.

World Resources Institute (WRI) 1997. World resources: 1996–1997. WRI, Washington, D.C., USA.

13
Integrated Ecological Assessments and Land-Use Planning

Patrick S. Bourgeron, Hope C. Humphries, Mark E. Jensen, and Bennett A. Brown

A requirement for ecological information to identify, quantify, and evaluate the potential impact of land-use decisions on ecosystems has been recognized for some time (Everett et al. 1994; O'Callaghan 1996; Boyce and Haney 1997; Lyle 1999; Treweek 1999; Jensen et al. 2001). The process of integrated ecological assessment (IEA) has been developed to provide a comprehensive description of the ecosystem patterns, processes, and functions, including relevant socio-political factors, needed to synthesize our knowledge of ecological and human systems. IEA techniques are rooted in ecological, social, and economic sciences. IEAs incorporate evaluation of the implications of human activities, including production of land-management scenarios. They should be part of integrated systems of environmental regulation that include strategic planning, objective setting, performance standards, monitoring, and review of the entire process (Treweek 1999). Recent applications of IEAs (see case studies in Johnson et al. [1998] and Jensen and Bourgeron [2001]) represent the integration of a number of approaches developed in response to specific problems, socio-political contexts, national legislation, and international accords (see review in Treweek [1999]; see also English et al. [1999]; and Lyndon [1999]). Various terms have been used to describe all or part of the process of assessing the state of the environment and its relationship to economic development.

Ecological assessments initially were developed during the early twentieth century to evaluate the capacity of the land to sustain various human activities. These methodologies were directly aimed at land-use planning (e.g., Klingebiel and Montgomery 1961; Beek and Bannema 1972; Olson 1974; FAO 1976, 1993; Zonneveld 1988). In 1969, the U.S. National Environmental Policy Act (NEPA) established the first legislative requirement for the use of some type of environmental impact assessment, also called environmental assessment, as a prerequisite for effective environmental planning and management (Robinson 1992; Treweek 1999). An environmental impact assessment is defined broadly as the process of assessing the environmental consequences of current or proposed management/development scenarios (Vanclay and Bronstein 1995; Treweek 1999). The specific methodology used in an environmental impact assessment varies with the category of impact considered, such as human health or economic development (see summary of various kinds of environmental impact assessments in Treweek [1999, her Table 2.1, p. 16]).

In this context, ecological risk assessment (US EPA 1992, 1998; NRC 1994) is a specific category of environmental impact assessment designed to meet specific regulatory mandates. Ecological risk assessment focuses on the relationship between one or more stressors and ecological effects in a way that is useful for environmental decision-making. Such assessments usually are geographically restricted and their scope narrowly defined to satisfy the regulation that triggered them (but see Hunsaker et al. [1990] and Graham et al. [1991], for a regional approach). Environmental impact assessment legislation currently exists in over 40 countries (Robinson 1992), and legislative tools have been developed for international assessments of problems that have multiple causes (e.g., Convention on Biological Diversity, Burhenne-Guilmin and Glowka 1994). IEA is the mechanism "for considering the ecological implications of development in any context and at any scale, ensuring that ecological factors are included in multidisciplinary appraisals of economic, social, and technical issues" (Treweek 1999, p. 44). IEAs can be implemented through legislation as a special category of environmental impact assessment or through any other mechanism.

IEAs can be project-based or strategic (Treweek 1999). Project IEAs, the most common IEA type, are used to evaluate individual proposals or actions to ensure that their ecological implications are understood before a decision is made. Because they are confined to the project level, these IEAs have serious problems and limitations (Treweek 1999, p. 21, her Table 2.2) that prevent the implementation of some of the guidelines proposed by Dale et al. (Chapter 1). For example, failure to go beyond the project site boundary precludes examining local decisions in a regional context; localized, short-term results prevent planning for long-term and unexpected events. A family of IEAs called strategic IEAs has been conducted for strategic planning and policy making (Ortolano and Shepherd 1995) at large scales. Strategic IEAs provide a framework for proactive assessment of the impact of human activities and environmental changes at broad scales, and placement of site-specific impacts within a regional/national context (Thérivel et al. 1992; Thérivel and Rosário Partidário 1996; Thérivel and Thompson 1996; Kessler 1997; Kessler and Van Dorp 1998; Jensen et al., in press). In the United States, implementation of ecosystem management by many land-management agencies (Grumbine 1993; Slocombe 1993*a*; Christensen et al. 1996) has triggered a series of strategic regional IEAs that integrate biophysical, biological, and human factors over large areas (Southern Appalachian Man and Biosphere [SAMAB 1996], Sierra Nevada Ecosystem Project [SNEP 1996], and Interior Columbia Basin Ecosystem Management Project [Quigley and Arbelbide 1997*a*]). Johnson et al. (1998) summarize seven IEA case studies.

The purpose of this chapter is to present a framework for conducting IEAs at multiple spatial scales. The chapter discusses the following aspects of IEAs: (1) their purpose and the role of issues and objectives in framing the process, (2) basic components and methods, and (3) integration into the planning process. Several different examples of IEAs are presented.

13.1 General Properties of IEAs

13.1.1 What Is an IEA?

An IEA is a process for answering the questions "what do we have, how did we get here, and where are we going" about any ecosystem of interest at all appropriate scales; ideally, IEA is implemented early in the decision-making process (Zonneveld 1988; Lessard 1995; Lessard et al. 1999; Treweek 1999; Jensen et al. 2001). IEA includes characterization and integration of the biological, physical, and human dimensions of ecosystems, their status, functions, interactions, and limits at spatial and temporal scales appropriate to the objectives of the assessment (Slocombe 1993*a*, 2001; Lessard et al. 1999; Jensen et al. 2001). Past, present, and future states of components as well as all assumptions and estimates of uncertainty must be documented.

13.1.2 What Is the Purpose of an IEA?

IEAs are intended to provide the data and interpretive tools needed to address planning objectives and place land-use planning decisions for specific areas in the appropriate context (Haynes et al. 1996). IEAs are indicative rather than prescriptive. They are not intended to resolve issues themselves or to give direct answers to policy questions. The information contributed by an IEA provides the context for decisions made by managers and their implementation. An IEA should be the template for future planning and for the iterative process of adaptive management (e.g., Slocombe 1993*b*). Information from an IEA can be used to link national and local planning levels.

13.1.3 When Is an IEA Conducted?

IEAs are the platform upon which planning activities are built. A variety of management and conservation issues can be addressed during an IEA. There are several specific circumstances that may trigger an IEA:

- *Making new decisions and assessing previous decisions.* An assessment is usually conducted when major land-management decisions are required or when the results of previous decisions are reanalyzed. For example, regional assessments of existing conservation areas may be triggered to assess the status of these areas, their sustainability, and how well they represent targets of conservation to decide upon future conservation actions (Bourgeron et al. 2001*b*).

- *Changes in organizational priorities.* Concern about such issues as preservation of biodiversity, ecosystem health, and effects of climate change has led to a reconsideration of land allocation, regulation, and management strategies in many U.S. and international organizations (Lessard 1995; Johnson et al. 1998). The resulting commitment to ecosystem management by 18 U.S. Federal agencies and many other institutions (Christensen et al. 1996) required a new information base at scales appropriate to the new issues being addressed. Therefore, a series of IEAs has been implemented (e.g., SAMAB 1996; SNEP 1996; Quigley and Arbelbide 1997*a*) to provide the required knowledge.
- *Occurrence of significant change in an ecosystem.* Significant changes in an ecosystem may occur as a result of unexpected events (Holling 1986). Such occurrences trigger an IEA when they have not been anticipated in a management or conservation plan. Unexpected events such as large-scale fires, windstorms, or floods could result in a reconsideration of the appropriate scales for addressing pattern–process relationships in an ecosystem. For example, the 1988 Yellowstone National Park fire was the impetus for conducting an assessment to answer questions about ecosystem patterns and dynamics, as well as future directions for park management (Christensen et al. 1989; Knight and Wallace 1989; Romme and Despain 1989). In another example, an assessment conducted after the Valdez oil spill in 1989 led to designing a restoration, protection, and monitoring program in Prince William Sound, Alaska (Cooney 1995).
- *Changes in understanding of ecosystems due to new information.* Our understanding of how any particular ecosystem functions is frequently incomplete, even when all currently available information has been analyzed and incorporated into a management or conservation plan. New knowledge may arise as new sources of information become available, or as a result of advances in scientific approaches to conservation problems. As our understanding improves, it may become necessary to revisit local land-use decisions.

13.1.4 What Level of Assessment Is Appropriate?

The objectives determine the type and level of an IEA, i.e., the nature of the components of an IEA and amount of detail in their representation, by defining the spatial scales, patterns and processes of interest, information needed, and the kinds of analyses to be conducted. The scales addressed in an assessment are objective-driven and as such are not bounded a priori within specific areas. Three attributes of the required information impact the level of an assessment and the application of the assessment results to land-use planning at landscape scales: the grain of the data, i.e., the finest level of spatial resolution (Turner and Gardner 1991); the extent, which has both a spatial dimension, i.e., the size of the area encompassed by a data set or the population described by the data (Wiens 1989), and a temporal dimension, i.e., the time period covered; and the interval between samples (Legendre and Legendre 1998). The grain and extent of a data set determine the lower and upper limits, respectively, to our ability to detect pattern in an analysis (Wiens 1989). For example, during the ICBEMP (Quigley and Arbelbide 1997*a*), a map of land-cover types with a 1-km resolution was used to assess broad-scale current ecological conditions across the entire assessment area. The grain of the map constrained the ability to characterize patterns occurring at scales finer than 1 km. The spatial extent may affect the grain (e.g., the broader the extent, the coarser the grain). The sampling interval may influence the ability to extrapolate from sampled to unsampled areas and to construct ecological models.

In practice, it may not be technologically possible or cost-effective to collect or acquire the most appropriate data for an assessment. Therefore, available data are used as surrogates for those that cannot be acquired. Surrogacy has been widely utilized in large-scale biodiversity assessments because of the scarcity of complete data. For example, biophysical data have been used as surrogates for biological data in the establishment of conservation areas (e.g., Mackey et al. 1989; Belbin 1993). Also, the USGS GAP program (Scott and Jennings 1998) uses richness of vertebrate species as a surrogate for total species richness. Use of surrogate data may often be the only way a project can be undertaken, but also potentially may lead to loss of accuracy in characterizing patterns and processes. Consequently, surrogates should be used with full recognition of potential problems, including quantification of such problems whenever possible (Faith and Walker 1996).

The linkage between the objectives of an assessment and the data used to address the objectives is critical to the success of an assessment, because the data constrain applications and analysis techniques (Dale and O'Neill 1999; Bourgeron et al. 2001*c*). The objectives drive database content, i.e., the attributes of the information, and conversely, database content constrains the types of objectives that can be effectively addressed. For example, a database with information about rare species would allow meeting the guideline of rarity (Chapter 1), but would not provide suitable information for evaluating species diversity or keystone species, or for determining how well the species composition of an area represents a regional complement of species.

13.1.5 Scientific Principles of IEA

Three areas of science provide the principles underpinning IEAs (Haynes et al. 1996), which IEAs share with ecosystem management (Christensen et al. 1996; Jensen et al. 1997), and land-use planning (Chapter 1; Dale et al., 2000).

- *Hierarchy theory:* Hierarchy theory provides a framework for characterizing ecosystem components and the linkages between different scales of ecological organization (e.g., the approach taken by Peterson et al. [1998] to ecological resilience, biodiversity, and scale; but also see O'Neill and King [1998]). In hierarchical systems, higher levels provide the context or environment within which lower levels evolve (Allen and Starr 1982). For example, knowledge of hierarchical ecosystem linkages allows implementation of the first land-use guideline, examining impact(s) of local decisions in a regional context (Chapter 1).
- *System dynamics:* Assessments of the spatial relations of ecosystems cannot ignore temporal dynamics. Although it may be convenient for some scientific exercises (e.g., modeling) and planning to assume equilibrium at a given scale, ecosystem dynamics may include succession along multiple pathways, discontinuities, unexpected changes (called surprise events by Holling [1978]), and the tracking of constantly changing environmental conditions (e.g., Holling 1986; Kay 1991; Costanza et al. 1993). Ecosystems have limits (e.g., productivity, accumulation of biomass), which should be recognized by decision-makers. This principle informs four land-use guidelines (Chapter 1): planning for long-term change and unexpected events, avoiding depletion of natural resources, avoiding or compensating for effects of development on ecological processes, and implementing practices compatible with the natural potential of an area.
- *Limits to predictability:* The existence of complex dynamics, hierarchical structuring, and multiple optimum operating points in ecosystems, most far from equilibrium, raises important questions about the predictability of ecosystem behavior (Costanza et al. 1993). Predictability varies over spatial and temporal scales. Characteristics of some events are predictable, but others may not be. In some cases, a small fluctuation in an ecosystem component can lead to a large change in the functioning of the system as a whole (May 1976). Land-use guidelines concerning planning for long-term change/unexpected events, preserving rare landscape elements, avoiding depletion of natural resources, and retaining large connected or contiguous areas containing critical habitats (Chapter 1) address the principle of limits to predictability.

13.1.6 Is IEA Relevant to Land-Use Planning?

Because IEAs may be conducted at extremely large scales (e.g., the 58 million ha area of the ICBEMP), the grain of the data, and therefore the results of the analyses, may seem irrelevant to those making effective decisions at the landscape level (e.g., a landowner or big game wildlife manager). IEA does not lead to decisions about landscape planning in the traditional sense. It indicates areas of concern (e.g., loss of natural processes such as historical fire regimes) or interest (e.g., suitability for a given use) on a broad scale. The landscapes contained in these areas should be analyzed with the appropriate data for making local land-use decisions. However, information from the IEA should be used to guide these analyses and decisions.

The relationships between the eight proposed guidelines for land-use planning (Chapter 1) and IEA components (described in Section 13.2) are summarized in Table 13.1. The first guideline, concerning examination of impacts of local decisions in a regional context, explicitly requires the regional information provided by all IEA components (Table 13.1). Successful implementation of other guidelines is also enhanced by information from IEA components, such as regional biological distribution data, which enable identification of rare landscape elements (guideline 3) and large areas containing critical habitat (guideline 5).

For example, if one of the goals of an IEA is to choose a network of conservation areas that is representative of all species found in the region, the land systems selected during the assessment will not be the final conservation areas. The selected land areas and their surroundings are

TABLE 13.1. Relationships among eight guidelines for land-use planning (Chapter 1; Dale et al. 2000) and the components of an IEA.

	IEA Components					
Guideline	Biophysical environments	Biological and socio-economic data	Integrated response units	Spatial variability	Temporal variability	Ecological conditions and pattern/process persistence
1	X	X	X	X	X	X
2			X	X	X	X
3	X	X	X	X	X	
4			X			X
5	X	X	X	X	X	X
6	X	X	X			X
7	X	X	X			X
8			X	X	X	

Note: The components with an X contribute useful information for implementing a guideline. For example, guideline 4 requires the assessment of ecological conditions and persistence of patterns and processes over a large area to make a decision. *Guidelines:* (1) examine impacts of local decisions in a regional context; (2) plan for long-term change and unexpected events; (3) preserve rare landscape elements and associated species; (4) avoid land uses that deplete natural resources; (5) retain large contiguous or connected areas that contain critical habitats; (6) minimize the introduction and spread of nonnative species; (7) avoid or compensate for the effects of development on ecological processes; (8) implement land-use and land-management practices that are compatible with the natural potential of the area.

analyzed in a site conservation planning process, such as the framework developed by Poiani et al. (1998), which is concerned with the actual delineation of the conservation area boundaries, identification and involvement of stakeholders, and development of management and conservation strategies at the landscape level. The IEA provides the information to design the conservation planning and management activities of individual conservation areas in the context of the entire network, to ensure the sustainability of the conservation targets within it. An example of an application of this information in the context of a regional network is allowing landscapes in each conservation area to fluctuate within the bounds of natural disturbance regimes and processes. Conservation targets may be locally lost within some areas but maintained or acquired within others.

13.2 Implementation of IEA

13.2.1 Conceptual Tools for the Implementation of IEAs

The implementation of an IEA can be very complicated, with a variety of issues, factors, scales, and analyses that can be considered. Therefore, three major steps should be clearly articulated at the onset of an assessment: integration of biophysical, biological, land-use, and socio-economic data; the degree of abstraction and relationships between different levels of activities for various planning and analytical purposes; and delineation of spatial boundaries for different aspects of an IEA. While these three steps are treated separately here, in reality, they are usually considered together during an assessment.

First, the integration of biophysical, biological, land-use, and socio-economic data (Herring 1998; Haeuber 2001; Slocombe 2001) requires formulating clear relationships among ecosystem components (Berry et al. 1996). Figure 13.1 represents the relationships among regional biophysical, biological, land-use, and socioeconomic systems (after Messerli and Messerli [1979] and Haber [1994]). Complexity can be added to this simplified representation by including realistic cause–effect relationships among all components of a specific application.

Second, the scaled relationships between the different activities of an IEA must be recognized. The IEA process involves increasing levels of abstraction and assumptions as it proceeds from direct measurements of ecosystem and landscape attributes to scenario planning. Figure 13.2 (adapted from Haber [1994]) represents the scaled relationships of the major activities of an IEA: data acquisition to describe landscapes; characterization of the spatial variability of the biophysical, biological, land-use, and socio-economic elements; analyses and modeling; interpretations; and strategic scenario planning. As activities progress from data acquisition to the strategic level, the degree of abstraction increases and validation of results is achieved by

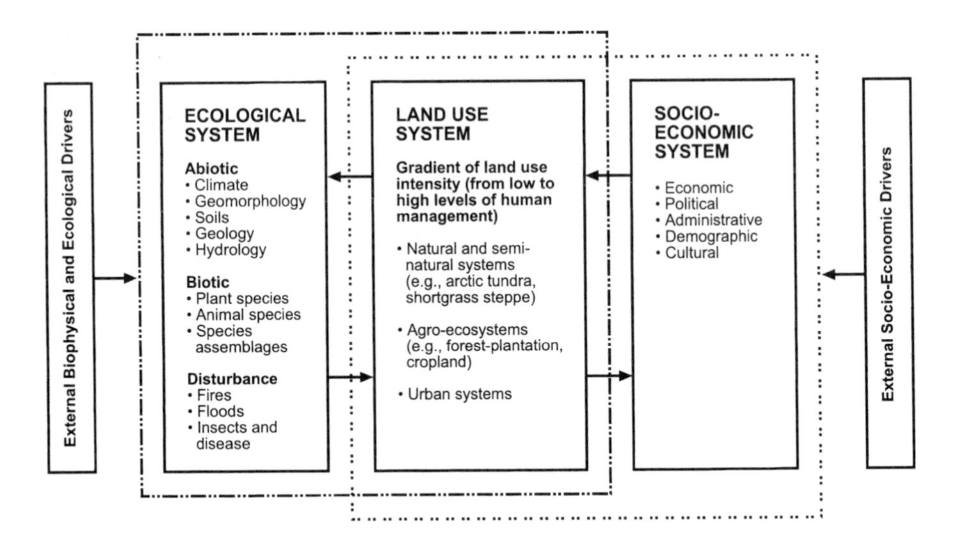

FIGURE 13.1. Simple model of relationships among regional biophysical, biological, land-use, and socio-economic systems. See text for explanation. (Adapted from Messerli and Messerli [1979] and Haber [1994].)

comparisons between levels. Therefore, during an IEA, analytical activities proceed by upward (inductive) and downward (deductive) scaling between levels of activities (Haber 1994).

Third, the spatial scales and boundaries associated with different activities of an IEA should be clearly defined (Fig. 13.3 and Section 13.2.2.4). There are six types of boundaries to consider: assessment area, characterization area, analysis area, cumulative impact area, reporting unit, and basic characterization unit. The assessment area, the area under consideration, is usually defined by socio-political concerns. It is generally smaller than the area needed for the characterization of biological and environmental ecosystem components; the characterization area should include the ecological boundaries of the ecosystem components of interest. Analyses are conducted and management recommendations directly influence the patterns and processes of interest in the analysis area. The cumulative effects of management decisions are assessed in the cumulative impact area. Reporting units are constructed to facilitate the reporting of results from different components of an assessment, e.g., terrestrial, aquatic, and socio-economic. Reporting units are important in very large assessment areas because they are the means by which the results of analyses conducted at different spatial scales are communicated to the general public, stakeholders, and decision-makers. The first five types of areas provide the basis for creating

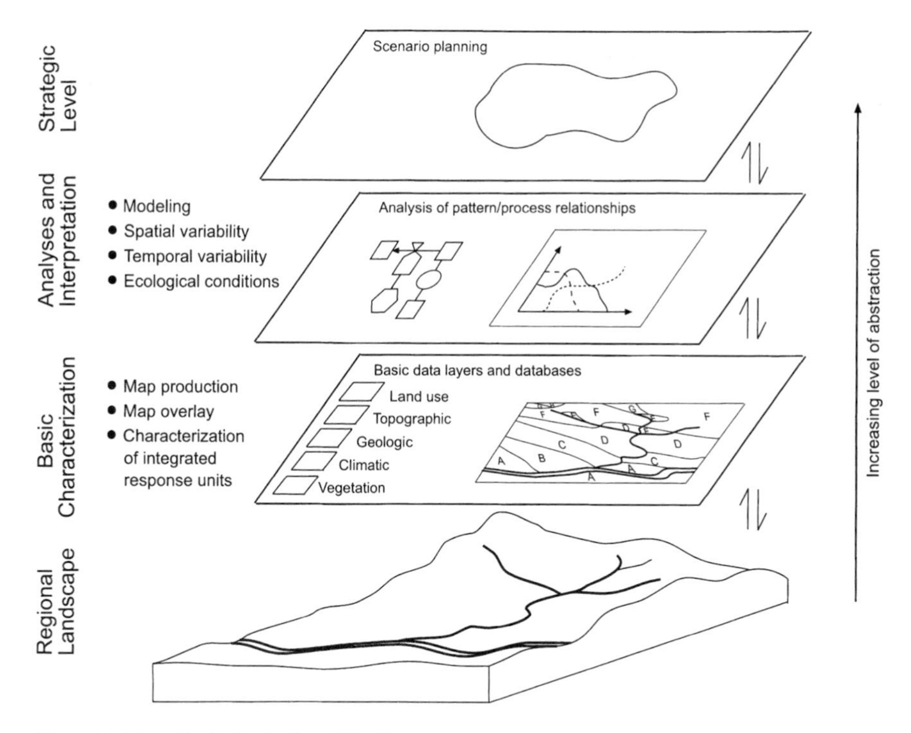

FIGURE 13.2. Scaled relationships between activities of an IEA. Examples of major tasks conducted within each level of activity indicated by bullets. Arrows indicate between-level comparisons (upward and downward scaling). See text for explanation. (Adapted from Kerner et al. [1991] and Haber [1994].)

partnerships and regional, local planning, and management frameworks (Slocombe 1993*a*, 1993*b*; Section 13.3). In a cases, some scales of conservation, political concerns, management actions, and ecological responses may be congruent and comparable, and therefore some of the types of areas may have the same boundaries. The sixth type, the basic characterization unit, defines the grain of the assessment.

13.2.2 Basic Characterization

Knowledge about the ecosystems contained in a region is needed in four areas to conduct IEAs (Bourgeron et al. 1994*a*): (1) characterization of biological component(s); (2) characterization of environmental components; (3) characterization of biological–environmental interactions; and (4) characterization of ecosystems as a whole and of ecosystem properties, including disturbances. This knowledge is acquired via the analysis and interpretation of hierarchical databases and maps that describe biophysical environments

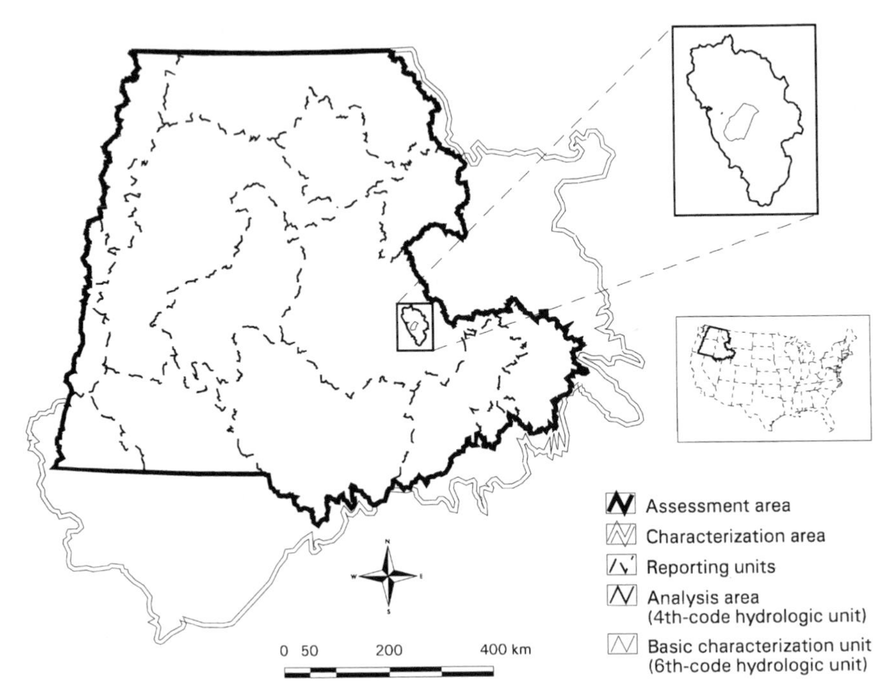

FIGURE 13.3. Map of interior Columbia River basin showing IEA boundaries: assessment area, characterization area, and reporting units, and examples of analysis area and basic characterization unit, used in Interior Columbia Basin Ecosystem Management Project.

and the current and historical status of biological and human ecosystem components (e.g., individual species, vegetation, and road density). Table 13.2 illustrates the kinds of databases, classifications, and maps needed for IEAs (Austin and Margules 1986; O'Callaghan 1996).

Complexity and scale complicate the problem of using classification systems (Austin and Margules 1986). For example, vegetation types may be distributed in a complex pattern in a landscape. At the scale of small areas (e.g., less than 100 ha), the lower levels of a vegetation classification (e.g., plant associations [*sensu* Grossman et al. 1998]) are used to represent vegetation variation. Over large areas, where broad-scale maps are needed if they are to convey useful information, more generalized vegetation classifications are necessary. Generalized vegetation classes used in mapping may have to include within-class heterogeneity to link the information gained in an assessment to finer scales, if the results of broad-scale analyses are to be used by land managers at the level of traditional planning units. Therefore, for this and other purposes required by assessment issues, classifications that are used in IEAs and land-use planning must be hierarchical.

TABLE 13.2. Examples of databases and classifications used for IEA.

Data type	Examples of variables or systems
Environmental	
Terrain	Elevation
	Slope and aspect
Rivers	River network
	River flow
	River nitrate concentration
	Catchments
Climate	Temperature
	Solar radiation
	Precipitation
	Climatic regions
Soils	Taxonomy (orders, suborders, great groups)
Disturbance	
Fire regime	Fire frequency and severity
Biological	
Floristics	Species distribution maps
	Occurrence in map grid cells or samples
Vegetation	Communities
	Alliances
	Vegetation types
Combinations of different data types	
Land classification	ECOMAP classes (Cleland et al. 1997)
Aquatic classification	Aquatic ECOMAP classes (Maxwell et al. 1995)
Land cover	Land cover characteristics (USGS 1995)
Human dimension	
Land use	Activities people undertake in a given land cover type (e.g., LULC data; USGS 1994)
Infrastructure	Roads
	Administrative boundaries

13.2.2.1 Biophysical Environments

Biophysical environments are combinations of environmental factors that drive biological responses and describe land or aquatic units that behave in a similar manner given their potential ecosystem composition, structure, and function (Zonneveld 1989; Bailey et al. 1994). The general purpose of biophysical environment maps is to stratify the environment to (1) define the potential of a site (e.g., potential vegetation, productivity, etc.), and (2) better characterize and model biological–environmental interactions and disturbances. Biophysical environments are delineated using environmental factors that do not display high temporal variability at a particular scale (e.g., regional climate, geology, landforms), and that exert primary control on the ecosystem patterns and processes of interest in an assessment. Biophysical environments have been called land types (Zonneveld 1989; Bailey et al. 1994; Cleland et al. 1997), natural landscape units (Rowe and

Sheard 1981), tesserae (Forman and Godron 1986), and landscape response units (Perez-Trejo et al. 1993). Biophysical environments provide information to address land-use guidelines 1, 3, and 5–7 (Table 13.1).

Biophysical environments have been defined at a variety of scales in nested, geographic arrangements (Bailey 1996), often organized in descending order by the driving variables that influence biological function (Perera et al. 1996). Such hierarchical frameworks are required for ecosystem management (e.g., Cleland et al. 1997; Haney and Boyce 1997). Many schemes have been proposed and widely used in conservation, forestry, agriculture, range management, landscape design, and land-use planning (e.g., Wertz and Arnold [1972], Zonneveld [1979, 1988, 1989], and Meidinger and Pojar [1991]; also see reviews in Grossman et al. [1999] and Bourgeron et al. [2001]).

However, descriptions of patterns in environmental factors or habitats is useful only if the ecological meaning of these factors is understood. The accuracy and utility of descriptions and maps of biophysical environments depend on the environmental factors selected, their relationships, the estimation procedures and mapping scales used, and the strength of the purported relationship between species and the environmental factors. Ideally, the interactions between landscape features, land use, climatic factors, and ecologically meaningful variables should be determined using a combination of geographic information systems (GIS) and simple process models (e.g., water-balance models to estimate available soil moisture [Mackey et al. 1988]), but this is seldom done. Biophysical environments have been identified as drivers of land-use structure (Verburg et al. 1997), but this relationship may change with scale (Hall et al. 1995; Verburg et al. 1997).

13.2.2.2 Current and Historical Biological and Socio-Economic Patterns

Current and historical pattern databases and maps are used to characterize basic biological, land-use, and socio-economic ecosystem components that commonly display large temporal variability at a particular scale of ecological organization. The biological component of ecosystems comprises species patterns, whether individual or species assemblages, in space and time. The precise definition of the biological component of ecosystems (e.g., which concepts and classifications to apply) depends on the objectives and the scale of the assessment. Similarly, databases and maps of land-use and socio-economic elements should be considered (Brown 2001). They include tangible aspects (i.e., those that directly influence ecosystem structure, such as transportation corridors and utilities), and intangible aspects (i.e., those that have important indirect influences, such as ownership boundaries) of human impacts. Maps of biological, land-use, and socio-economic components provide snapshots of otherwise dynamic components at the spatial and temporal scales of the assessment, and must be updated through ongoing monitoring, depending on the degree of change experienced in an assessment area. Data on current and historical biological, land-use, and socio-economic patterns provide information needed to implement guidelines 1, 3, and 5–7 (Table 13.1).

13.2.2.3 Integrated Response Units

Biophysical, biological, land-use, and socio-economic databases and maps are used to derive hierarchical integrated response units that are appropriate for the type of assessment question or issue. Some systems of integrated response units have a multipurpose intent. Integrated response unit maps are generated by overlaying basic ecosystem component pattern maps (e.g., existing vegetation maps) on similar scale biophysical map themes (e.g., geoclimatic settings, watersheds) to develop response units appropriate for the types of resource interpretations under consideration. Multiple existing pattern themes (e.g., roads and vegetation) are usually associated with a primary biophysical template to display the resource response units of concern. The units should be hierarchically arranged to be used at different scales of assessment and interpretation appropriate for multiple resources. Integrated response units can be used in addressing all guidelines (Table 13.1).

13.2.2.4 Defining Areas for IEA Activities

Characterization maps are often used to delineate the various areas associated with different IEA activities (Section 13.2.1): assessment, characterization, analysis, reporting, and basic characterization. For example, in the ICBEMP (Fig. 13.3), the assessment area was defined by an analysis of biophysical, ecological, and administrative maps in light of political directives to develop an ecosystem management strategy for forests east of the Cascade crest in the U.S. Pacific Northwest (Quigley and Arbelbide 1997*a*). Based on the objectives, a larger characterization area was delineated that encompassed the distributions of the patterns and processes of interest (Quigley and Arbelbide 1997*a*). Subbasins (4th-field hydrologic unit codes) were used as the units for many terrestrial and aquatic analyses (Jensen et al. 1997). Thirteen reporting units (Jensen et al. 1997) were defined for reporting the effects of land-management activities by landscape ecology, terrestrial, aquatic, and socio-economic groups. Subwatersheds (6th-field hydrologic unit codes [Brewer and Callahan, unpublished report]) comprised the basic characterization units for many ICBEMP studies (soil erosion, stream recovery potential, fine-scale vegetation patterns, fish distributions) (Hann et al. 1997; Jensen et al. 1997; Lee et al. 1997).

13.2.3 Analysis and Interpretation

13.2.3.1 Properties

Properties of biophysical environments and integrated response units must be defined. They may differ among assessments but should include those properties that are needed to predict the effects of global or land-use change (O'Callaghan 1996). The properties are usually derived from the multiscaled data analyses that determine biophysical environments and integrated response units (Urban et al. 1987; Bourgeron and Jensen 1994; Bourgeron et al. 1994*b*; Jensen et al. 1996). Most quantitative analyses and modeling are conducted at this stage of an assessment. Properties include hydrological, biogeochemical, ecological, and landscape characteristics, either alone or in combination. For example, biological components can be ordered along environmental gradients (e.g., Austin and Margules 1986; Faith 1992), and individual biological components can be modeled as functions of key environmental factors to determine their expected distributions within biophysical environments (e.g., Austin et al. 1990; Yee and Mitchell 1991; Lenihan 1993; Franklin 1998; Leathwick 1998). Properties that are central to IEAs include disturbances of all kinds, such as fires, regional-scale storms, and volcanic eruptions.

Natural disturbances are parts of ecosystem and landscape function. Many ecological patterns depend on disturbances for their persistence (Section 13.2.4). Therefore, natural disturbance regimes should be defined at relevant scales of occurrence (Pickett and White 1985; Pickett et al. 1989), whether the effects are regional (e.g., fires in southwestern U.S. grasslands [Bahre 1995]) or local (e.g., pocket gopher burrows, [Reichman and Smith 1985]). The definition of properties also includes spatial and temporal variability in ecosystem components (Sections 13.2.3.2 and 13.2.3.3). Modeling is routinely conducted to examine the range of biological responses expected as a result of changes in environmental attributes over space and time, and to extrapolate pattern/process-disturbance relationships from sampled to unsampled areas. Ives et al. (1998) discuss the degree of model complexity required to represent spatial patterns in biological distributions.

13.2.3.2 Spatial Variability Characterization

The spatial variability of patterns and processes of interest must be analyzed because such patterns and processes are structured by forces that have spatial components (Legendre and Legendre 1998). The spatial variability observed in many ecosystem components occurs at a variety of scales, including variability in biogeochemical cycles within and among regions (e.g., Burke et al. 1991; Fan et al. 1998), variability in the functional relationships between structure and processes within and among landscapes (Baskent 1999), and variability in the dynamic structure of single-species populations (Stenseth et al. 1999). Maps of biophysical environments and integrated response units, attributed with their ecological properties, provide the basic templates for interpreting the spatial variability of an area at multiple scales. Stratification of an area by biophysical environments or integrated response units contributes to understanding biological distributions, and to extrapolation of ecosystem pattern/process-disturbance relations from sampled to unsampled areas.

Because the strength and nature of biological–environmental interactions may change along regional environmental gradients, information should be conveyed to planners that the patterns and processes depicted on maps used for decision-making are not spatially invariant. The spatial pattern of gradual changes in landscape properties (Bourgeron et al. 1994*a*) should be incorporated into the definitions of biophysical environments and integrated response units. For example, occurrences of a regional vegetation type found at different locations along an environmental gradient would be likely to respond differently to conservation/management practices. Geographic variability should be considered as part of the land-use planning process. Characterization of spatial variability is needed to address guidelines 1–3, 5, and 8 (Table 13.1).

13.2.3.3 Temporal Variability Characterization

Biophysical environments and integrated response units provide the basic templates for interpreting the temporal variability of a landscape, i.e., its dynamics. For example, the vegetation present in a landscape changes as a result of the effects of natural disturbance regimes and resulting plant succession. Predicting the responses of individual plant species and the vegetation as a whole to disturbances is facilitated by identifying the biophysical environments or integrated response units in which they occur (e.g., Arno et al. 1986; Host and Pastor 1998). The species and type of vegetation that may occur in a biophysical environment or integrated response unit can be described, and successional pathways identified, by means of various ecological models (e.g., Huschle and Hironata 1980; Fischer and Clayton 1983; Keane et al. 1990*a*, 1990*b*; Host and Pastor 1998). For example, fire group models (Fischer and Clayton 1983; Fischer and Bradley 1987) use biophysical environments to characterize the fire regimes that drive vegetation succession in forested landscapes. The linkage of simulation modeling with biophysical environments allows testing of management practices (Pastor and Mladenoff 1992). The use of such a modeling approach has been shown to be valuable in land-use planning (e.g., Host et al. 1996; Host and Pastor 1998).

Characterization of past fluctuations in patterns and processes is also required to determine the upper and lower limits or range of variability in which an ecosystem historically operated. Knowledge of how far an ecosystem departs from its historical range of variability (HRV) at multiple scales is often useful (Morgan et al. 1994; Humphries and Bourgeron 2001). It is gained by comparison of the current state of the ecosystem with HRV (Morgan et al. 1994; Landres et al. 1999; Swetnam et al. 1999). Analyses focus on the processes and disturbance regimes operating on an ecosystem, such as the frequency and severity of specific types of disturbance (e.g., fire, floods, drought), or they focus on the states of the system resulting from the operation of processes and disturbances, such as descriptions of the size, shape, and other characteristics of patches (Swanson et al. 1994; Cissel et al. 1998; Kaufmann et al. 1998). Limitations to the power of HRV analyses are known, but careful application of the method can provide important information (Landres et al. 1999). Characterization of temporal variability is important in addressing guidelines 1–3, 5, and 8 (Table 13.1).

13.2.4 Assessment of Ecological Conditions and Persistence of Patterns and Processes

The assessment of ecological conditions and pattern and process persistence requires the interpretation and integration of ecological properties, spatial variability, and temporal variability, and should be conducted in light of the potential response expected in an area at any scale. Integrated response units provide a basic template for interpretation of change in biological patterns (e.g., existing vegetation, plant or animal abundance) and processes that commonly display changes after management or the impact of natural disturbances (e.g., fire, flood, insect outbreak). The effects of land-use practices, and natural processes and disturbances, can be assessed by contrasting the current ecological conditions of an area with conditions in other managed or unmanaged areas that occur in the same integrated response unit class. Information on ecological conditions and persistence of patterns and processes is necessary for implementing guidelines 1, 2, and 4–7 (Table 13.1).

Integrated response unit maps, coupled with models, combine information on biophysical environments, biological, land-use, and socio-economic patterns, past and present conditions, processes, and disturbances at different hierarchical scales, to provide a template for estimating spatial heterogeneity and its effects on landscape and ecosystem dynamics (Pickett and Cadenasso 1995), including the persistence of patterns at multiple scales. Within these units, the persistence of a pattern of interest depends on many factors that are spatially and temporally bounded, including landscape mosaics, ecosystem complexity, biodiversity, vegetation composition and structure, disturbance regimes, land-use patterns, road density, and fragmentation. It is at this stage of an IEA that the suitability of integrated response units is determined for specific objectives (e.g., conservation, forestry). The spatial distribution of areas targeted for specific land uses should be analyzed based on their suitability for those uses.

For example, in assessing the population viability of many species, both coarse-scale and site-level conditions are important (Glenn and Collins 1994). Across a landscape, there may be reproducing populations in high-quality habitats that are sources for populations in low-quality habitats (sinks) (Pulliam 1988). In areas with less suitable habitats, where mortality exceeds reproduction, populations are dependent on immigration from source habitats to maintain existing numbers. Integrated response units can be identified that contain strongholds, locations with strong populations of species of interest, i.e., those that are stable or increasing and have adequate population size (Lee et al. 1997), to ensure that such locations are considered for conservation. The result is the identification of a network of source and sink habitats (guideline 5). At the community level, analyses can be conducted to characterize distribution patterns and probabilities of occurrence of species assemblages (Margules and Nicholls 1987; Cox et al. 1994; Bojorquez-Tapia et al. 1995).

Analysis of ecological conditions over a range of scales may include developing habitat suitability models and determining the potential for manipulating ecosystem components (e.g., vegetation) to improve habitat (Glenn and Collins 1994). For example, population viability can be analyzed at broad landscape and local levels, and marginal habitats can be expanded, linked, or otherwise improved by using information about existing conditions and biophysical environment characteristics (Cox et al. 1994). Similar analyses can be conducted for communities.

A regional landscape analysis using long-term monitoring data was performed to assess the importance of landscape-scale factors in defining favorable wolf habitat in the northern Great Lakes region (Mladenoff et al. 1995). Mapped wolf pack territories were superimposed on maps of habitat variables such as land-cover class, land ownership, road density, and deer density. The maps were the basis for constructing logistic regression models to estimate the amount and spatial distribution of favorable wolf habitat at the regional scale. Road density and fractal dimension (land-cover patch boundary complexity) were the most important predictor variables. Models were used to create a regional map of predicted favorable wolf habitat. Wolf populations may experience source–sink dynamics as a result of the spatial distribution of habitat, with wolves moving throughout the landscape as dispersers, but successful establishment of packs occurring only in higher quality habitat areas. The spatial information generated in this kind of broad-scale analysis can be used to identify areas with high or low priority for active management and to determine the degree of connectivity among the areas to facilitate recovery of the species (guideline 5).

Restoration or maintenance of ecological systems should be conducted with knowledge of their dynamics. For example, one of the potential important contributions of IEAs to land-use planning is the characterization of large, infrequent disturbances (e.g., large severe storms, such as hurricanes, or regional fires) and their impact on ecosystems over a large area (Foster et al. 1998; Turner et al. 1998). Large, infrequent disturbances create patterns of regional spatial variability that interact with geomorphology, species dynamics, and potential global change (Turner et al. 1998). This spatial variability and its impact on patterns of recovery should be documented and incorporated into models of succession and models of responses to management in local planning units (Dale et al. 1998), e.g., integrated response units. Without an IEA, past large, infrequent disturbances and their imprint on regional landscapes (Foster et al. 1998) may be overlooked or incorrectly characterized (e.g., as stochastic processes, see discussion in Turner et al. [1998]).

The impact of management decisions or landscape conditions on natural disturbance regimes and landscape configuration can be analyzed in a hierarchical fashion (Turner et al. 1993, 1994; Milne 1994). Knowledge of ecological conditions at multiple scales is useful in devising scenarios for land-use planning that enable desired conditions to be attained and perpetuated. Conflicting resource uses (e.g., recreational activities versus no access, habitat management for rare and threatened species versus commercially important species) can be minimized by considering the effects of projects at different scales of analysis (Brenner and Jordan 1991; Milne 1994). For example, in the Boundary Waters Canoe Area Wilderness of Minnesota, forests historically consisted of stands of jack pine or aspen, which burned in frequent intense fires while still in an even-aged condition, and were replaced by the same species (Frelich and Reich 1995). Currently, in the absence of fire, small canopy openings created by windstorms and insects have led to increases in multi-aged old-growth stands consisting of a mixture of black spruce, balsam fir, paper birch, and white cedar, a successional condition rarely seen prior to the early 1900s. There is some evidence that continued lack of fire may lead to loss of jack pine from the area, severely limiting vegetation management options. In the absence of jack pine, restoration of burning would produce aspen forests, and lack of fire would lead to the old-growth mixture. In either case, a major component of the vegetation prior to fire suppression would be lost. From the standpoint of ecosystem conservation (e.g., Noss et al. 1997), allowing for conditions that did not occur prior to current land-use practices is not a good strategy in a wilderness area.

13.2.5 Strategic Scenario Planning

Scenario planning is a strategic activity and is the last step in an IEA (Fig. 13.2). It is a tool that incorporates IEA data, analyses, and products to determine the implications and tradeoffs of various possible conservation and management actions (Lessard 1995; Haynes et al. 1996). Each scenario describes a set of possible outcomes given a particular conservation or management approach. Uncertainty is explicitly included, with an emphasis on what might happen, including the potential merits and pitfalls of an approach. Usually, snapshots of probable future states are provided at designated intervals (e.g., 5, 10, 50, and 100 years into the future). Ecological modeling is required to construct scenario projections that incorporate the effects of processes influencing landscape patterns at various spatial and temporal scales. Scenario planning commonly includes displays of landscape patterns, usually as maps or graphs at different periods of time, to assist in determining the ecological consequences of specific conservation and management actions. Scenario planning can be applied at many different scales to resolve specific issues (e.g., Wear et al. 1996; Quigley et al. 1997*a*, 1997*b*, 1997*c*, 1997*d*).

Scenarios characterizing the effects of different land-use practices were implemented using a model that projects land-use changes, carbon release, and farmer turnover rates in central Rondônia, Brazil (Dale et al. 1994). Model projections were functions of initial soil and vegetation conditions, market and road infrastructure, and decision variables. Three land-use practice scenarios were simulated: (1) a scenario incorporating "typical" land-use activities employed by farmers in the area, (2) a "worst" case, in which land is cleared rapidly and not returned to fallow or used for agroforestry, and (3) a "best" case, in which innovative farming practices are employed, such as growing a diversity of perennial crops and agroforestry. An important result was the similarity between the typical scenario and the worst-case scenario, which both led to rapid deforestation and release of large amounts of carbon. The simulations suggested that the best-case scenario results in maintenance of farmers on the land, as well as retention of a mix of habitat types.

A model of land-use change was developed in the southern Appalachian highlands (part of the Southern Appalachian Man and Biosphere region) to examine issues relevant to ecosystem management of forested areas (Wear et al. 1996). The model simulated probabilities of change in forested, grassy/brushy, and unvegetated land-cover categories as functions of topographic features, distances to roads and market centers, and population density (Turner et al. 1996). The model projected future landscape development under scenarios in which land-cover dynamics in two time periods (1975–1980 and 1986–1991) were extended 100 years into the future. In addition, scenarios were developed implementing hypothetical regulatory approaches to ecosystem management, expressed as rules governing forest cutting (Wear et al. 1996). Scenario results suggested that recent (1986–1991) shifts in the dynamics of land-cover change in the southern Appalachian highlands could produce less fragmented landscape patterns and an increase in forest coverage if continued in the future. In addition, hypothetical regulatory scenario results suggested that changes in privately owned land have a much greater effect on landscape pattern in this area than changes on publicly owned land.

A promising approach to strategic scenario planning is the use of ecological/economic policy models (Carpenter et al. 1999). Carpenter et al.'s (1999) model includes a representation of decision-making to control lake pollution incorporated into three modules (ecosystem, assessment and forecast, and human behavior). The model runs as a stand-alone economic optimization or as a game in which participants in a class or workshop can initiate the year-by-year actions of various experts or stakeholders. The ecosystem representation includes a minimum number of attributes required to represent ecosystem behavior adequately, i.e., three variables functioning at three qualitatively different response rates, and also incorporates relationships that generate multiple stable states. The ecosystem module connects to an economic optimization module and to a module representing information and decision processes. An important characteristic of the model is its integration of an appropriate level of ecosystem reality with economic optimization and decision processes. The behavior of key actors (stakeholders, scientific and economic advisors, managers, and decision-makers) is represented either through model simulation, or alternatively, through explicit interventions made by an individual or group who take control of one or more of the actors' decisions. In the latter form, the model becomes a learning game in which the limits of the ecosystem and opportunities open to the key actors are discovered.

13.3 IEAs and Planning

A considerable body of work exists concerning the planning process and the role of environmental impact assessments and IEAs (Innes et al. [1994] and Ortolano [1997]; see recent reviews in Margerum [1999] and Morrison-Saunders and Bailey [1999]). The integration of IEAs within the planning process presents some specific problems regarding the size of the assessment area, multiple ownerships and jurisdictions, and multiple objectives. Assessments that cover large geographic domains can provide a common scientific foundation for coordinated planning and management among all concerned parties (Slocombe 1993*a*, 1993*b*, 1998; Margerum 1999; Haeuber 2001). Land-use planning and management within the bounds of sustainable local and regional economies require coordination among many partners, including landowners and managers (Haeuber 2001). There are substantial problems involved in conducting conservation across multiple ownerships (Landres et al. 1998), but known general solutions can be applied (Knight and Clark 1998; Landres 1998; Groom et al. 1999). One solution is to create mechanisms to facilitate broad public participation in establishing general direction (Slocombe 1998; Moore and Lee 1999; Haeuber 2001) at all stages of an IEA, especially in defining human dimensions and in scenario planning (Fig. 13.4).

A major obstacle to the integration of knowledge (Slocombe 2001) is the lack of relationship of management and planning units to ecosystems, ecosystem connections with economic and social processes, or the cultural and political identities of local people (Slocombe 1993*b*; Knight and Landres 1998). A solution is the creation of partnerships and regional and local planning and management frameworks that can be used to decide upon common strategies that will provide for ecosystem maintenance and a stable economy (e.g., Slocombe 1993*b*; Yaffee 1998) (Fig. 13.4). Other considerations include the following points (Haeuber 2001). First, legitimate representatives of the broadest currents of public perceptions and values should be

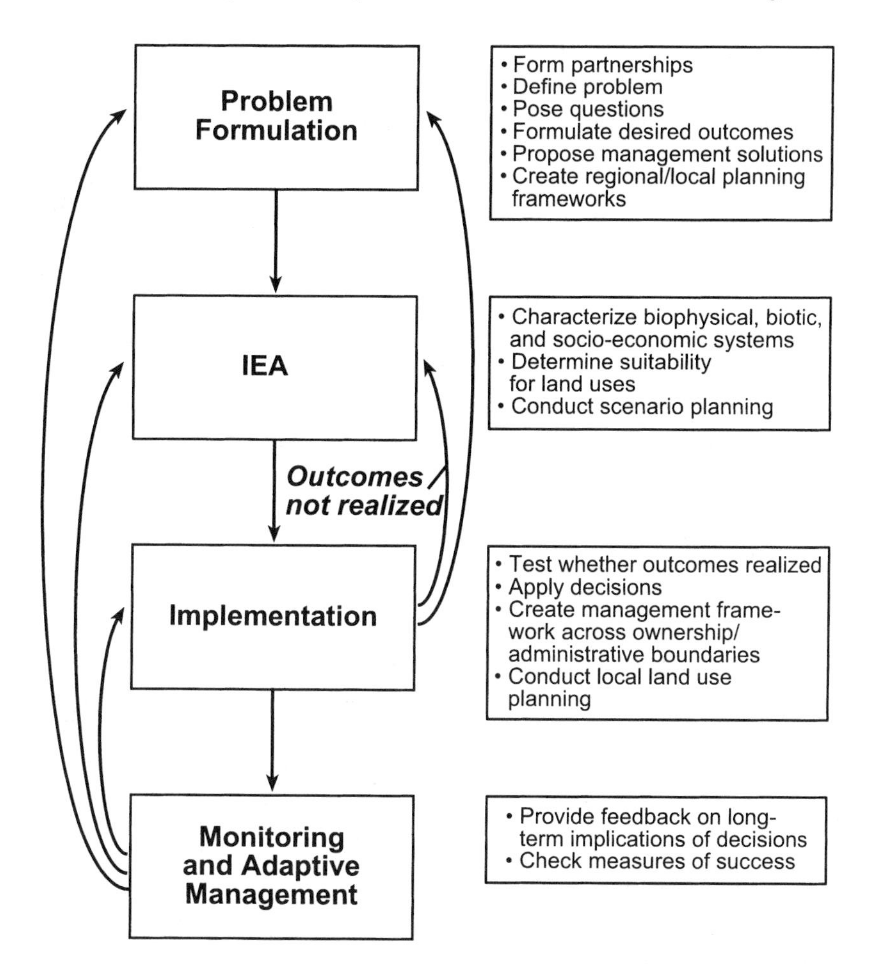

FIGURE 13.4. Simplified representation of a planning model for IEAs and ecosystem management. Model is iterative; each step is related to one or more other steps. See text for explanation. (Adapted from Lessard [1995] and Haynes et al. [1996].)

determined. Second, membership in groups is not exclusive. An individual may belong to several interest groups, which sometimes have conflicting agendas (e.g., hunting versus hiking). Third, improper conceptual models (e.g., assumption of equilibrium in large areas) may lead to socio-political struggles over ideas and perceptions (e.g., Rapport et al. 1998). Fourth, significant challenges exist in the way governmental and other institutions function, which may be impediments to IEAs. Finally, the scientific output from IEAs or any other source should be recognized as only one of many inputs into the decision-making process. In a review of six case studies, Morrison-Saunders and Bailey (1999) reported that the level of scientific rigor in assessment predictions may not result in improved environmental management. Instead, the case studies showed that early issue identification and formulation may be more important than rigorous impact predictions. Decisions are not made on sound science alone, but in relation to many other competing values. Given this fact, the resources spent on designing decision support systems should be directed at systems that are interactive, allow for real-time scenario planning during negotiations, and permit visualization of the impacts of possible land-use decisions (e.g., an interactive system for conservation in Australia [Pressey et al. 1995]; the USFS Ecosystem Management Decision Support System for ecosystem management [Reynolds et al., unpublished report]).

The choice of a flexible, adaptive-management-oriented planning process is a critical step in an IEA (Shea et al. 1998; Johnson 1999; Morrison-Saunders and Bailey 1999; Haeuber 2001). Although there are differences of opinion about the actual implementation of adaptive management and its success, the approach is generally acknowledged as important (see, in particular, a series of articles on the topic in Conservation Ecology, Volumes 1 and 2, 1999). Figure 13.4 presents a prototype of such a planning process for IEA (adapted from Lessard [1995], and Haynes et al. [1996]; see Yaffee et al. [1996] for alternative models). Regardless of the model selected, scientists should work within the planning framework to clearly formulate the problems, questions, and alternatives as hypotheses, with explicit predicted outcomes that will be tested by an IEA designed specifically to test the hypotheses (Underwood 1995). If the predicted outcomes are not realized, an evaluation should be conducted of how the process led to failure. In particular, an evaluation is required of how the outcomes were determined. The criteria for testing hypotheses should also be examined. This is an activity that is not traditionally undertaken by scientists. However, it is an area of active research and discussion (e.g., Hardin 1968; Fairfax 1978; Klapp 1992; Underwood 1995). If the predicted outcomes are realized, IEA results are implemented, and there is a shift to other scientific and planning activities. Monitoring schemes are designed at site, landscape, and regional levels to determine whether the desired outcomes are maintained over time; the results of monitoring are then incorporated into the planning process.

13.4 Examples of IEAs and Their Relationships to Land-Use Planning

Many IEAs have been completed or are currently underway. The IEA process outlined in this chapter is very flexible, allowing emphasis of some components and deletion of others. However, because each step potentially makes a significant contribution to the final assessment, the overall structure of a given IEA must be chosen carefully. Three examples of IEAs for land-use planning are presented that differ considerably in specific purposes, political context, socio-economic conditions, and spatial extent.

13.4.1 Interior Columbia Basin Ecosystem Management Project

The ICBEMP represents an example of an IEA prompted by potential natural resource conflicts and changes in institutional priorities (shift to ecosystem management by U.S. land management agencies). To date, the ICBEMP is the largest regional effort undertaken in the United States, implemented in 30.9 million hectares of U.S. Forest Service (USFS) and Bureau of Land Management (BLM) land. USFS and BLM land comprises 53% of the 58 million hectares of the interior Columbia River basin (approximately 8% of U.S. land area), which consists of mosaics of terrestrial and aquatic ecosystems, as well as a variety of land-use and ecological conditions (Quigley and Arbelbide 1997*a*, 1997*b*, 1997*c*, 1997*d*). Although it comprises only 3.6% of the U.S. economy, the area is significant politically, economically, and biologically because it contains 24% of National Forest System lands, 10% of BLM-administered land, and 29% of all wilderness in the United States. Biologically, it contains a number of significant land-cover types and habitats for many sensitive species. Despite its low human population density (approximately 20 people per square mile compared to the national average of 70 people per square mile), and despite the extent of its protected and undeveloped land, the interior Columbia River basin has experienced recent population growth and increased changes in land use (Quigley and Arbelbide 1997*a*), which have led to natural resource management conflicts (e.g., declining abundance of salmon species, controversy over old-growth forests).

The structure of the project was guided by a charter stemming from a directive from President Clinton that was issued in response to management conflicts and changes in institutional priorities. The planning process generally followed the steps shown in Figure 13.4, with some notable exceptions. The ICBEMP comprised an executive steering committee, an IEA scientific team, and two environmental impact assessment teams. Products included a scientific assessment (the IEA described in this chapter) and two environmental impact assessments. Public participation was encouraged, but because of federal regulations, only federal employees could be members of executive committees, which created problems with stakeholders. The IEA provided the information needed by the environmental impact assessment teams, including scenario planning, to address 16 major policy and process questions formulated from the charter. All IEA documents received anonymous peer review under the supervision of an independent science review board.

Results from the IEA indicate that the region contains some of the most significant land-cover types in terms of their economic and conservation contributions, yet these are among the most potentially endangered of the country (Quigley and Arbelbide 1997*a*, 1997*b*, 1997*c*, 1997*d*). Land cover in this region is susceptible to both land-use and climate-induced change (Hann et al. 1997). As a case in point, only 2% of current landscape patterns are similar to historical patterns; in particular, land-cover types have experienced recent economically and biologically significant transitions (Hann et al. 1997). For example, the interior ponderosa pine cover type, of great importance to local economies (e.g., grazing, logging), and conservation (e.g., for bird species), has shown a decline of 32% compared to its historical minimum extent (1850–1900). All IEA results, including scenario planning, were provided to the environmental impact assessment teams for planning purposes. ICBEMP results currently are being used as templates for local land-use planning. For example, individual national forests are using data and results as a context for their forest plan revisions. All data (more than 20 databases and 170 GIS data layers) and documentation have been made available to the public through the Internet (*http://www.icbemp.gov/*).

13.4.2 The United Kingdom Natural Environment Research Council/Economic and Social Research Council Land Use Programme

The Natural Environment Research Council/Economic and Social Research Council Land Use Programme (NELUP) (O'Callaghan 1996; *http://www.ncl.ac.uk/wrgi/wrsl/projects/nelup/nelup.html*) was initiated by the United Kingdom (UK) Natural Environment Research and Economic and Social Research Councils to guide national policy, and in the region, to investigate techniques for making land-use planning decisions comprising the socio-economic mechanisms of land allocation, constrained by scientific understanding of the physical and ecological environments. NELUP was designed to conduct IEAs of the catchment of the River Tyne in northeast England and the Cam basin in central England in response to a need to make decisions that conformed to new policies. Issues concerning the deintensification of agriculture and afforestation were addressed in these two regions of the UK that are mostly in private ownership. The projects were prompted by changes in agricultural and forestry policies both within the UK and in the European Union as a whole. The objectives of the assessments were to provide alternative land-use scenarios to regional and national policy makers, and their economic, biophysical, and biological implications, so that policy changes and incentives for landowners could be developed. NELUP was intended to provide a quantitative descriptions of the main impacts arising out of rural land-use change at the river basin scale. Specific program objectives were to integrate models of economics, ecology, and hydrology to describe changes in the spatial pattern of land use and the impacts of such changes; to integrate nationally available data sets within a database describing the biophysical and socio-economic conditions in the river basin; to create an interactive, user-friendly interface to the database; and to construct models to explore possible future land-use conditions.

In conducting the two assessments, existing national and regional databases (e.g., rivers, climate, land use, roads), generally maintained by governmental institutions (e.g., the Institute of Terrestrial Ecology), were available for all aspects of the project. Development of a decision support system was emphasized to provide real-time analyses that the public, policy makers, and planners could see visualized as maps. Several models installed in the decision support system were used to determine the characteristics of the river basin under a wide range of scenarios. The links between environmental–economic policy and the physical–ecological systems of the river basin were described and formalized through the integration of models, databases, and user interfaces, in a framework that allowed quantification of the interactions among water, land use, and the rural economy. The system provided visual displays of results in graphical and tabular form to illustrate the consequences of land-use change. The analyses included estimates of the potential of each site, distributions of rare species, economic effects of various types of management, and effects on water quality. An analysis was conducted of the issues associated with designating land for agricultural deintensification, and the management necessary to provide the environmental benefits required by new national and European policies. In addition, the political and social implications of making large landscape changes were analyzed. The decision support system and results of analyses were provided to regional and local planning frameworks and are currently in use for planning purposes.

13.4.3 The Malpai Borderlands Group

The Malpai Borderlands Group provides an unusual example of IEA, land-use planning, and ecosystem management in the U.S. southwest (southeastern Arizona and southwestern New Mexico). The group is a nonprofit volunteer corporation founded by ranchers in southern Arizona and New Mexico. The goal of the group is "to restore and maintain the natural processes that create and protect a healthy, unfragmented landscape to support a diverse, flourishing community of human, plant, and animal life in our Borderlands Region." The ranches are distributed in a landscape mosaic that also includes public land (USFS, BLM, and state trust land in Arizona and New Mexico). The coalition of ranchers has enlisted state and federal agencies, nongovernmental organizations (e.g., The Nature Conservancy), and universities in a partnership to assess ecological conditions and develop planning scenarios for the region. The partnership has two co-executive directors, a rancher and a Nature Conservancy staff person. The project has developed (1) a joint fire management plan for much of the 3400 km^2 in the region, (2) a "grass banking" program to help drought-stricken ranchers and protect overgrazed land, (3) programs for reseeding with native plant species and ecosystem management practices; (4) and cooperative relationships with research and management entities interested in resource management and protection. The ultimate goal of the Malpai Borderlands Group is to prevent further landscape fragmentation, and to create a region in which natural processes, including disturbances, can be used to restore and manage the productivity of natural ecological communities.

An IEA was conducted using biophysical environments, potential and existing vegetation, past and current fire regimes (used to define the planning area), and distribution of relevant plant and animal species. Current planning includes burning prescribed in the area based on historical fire regimes. As a consequence of this strategy, not all ranches are burned in a given year. To compensate for the loss of forage, cattle from ranch(es) where burning is prescribed are permitted to graze in unburned areas on other ranches. Each ranch continues to be managed according to the owner's requirements, but this is accomplished within a regional framework, based on knowledge of biophysical environments, land unit potential, ecological processes, and past and current ecological conditions. This is a unique planning case, in which private landowners were the initiators and drivers of the assessment and planning process, and governmental agencies were the stakeholders.

13.5 Conclusions

This chapter presents a framework for conducting IEAs for land-use planning. IEA is the process by which all the biophysical, biological, land-use, and socio-economic components of regional ecosystems are characterized at all spatial and ecological scales that are relevant to the objectives of an assessment. IEA is a strategic component of the planning process, and as such should be imbedded within it. Methods should be chosen that strengthen the on-the-ground implementation of land-use goals. Many of the steps described are the focus of active research and development. IEAs are being conducted and incorporated into the planning system increasingly often, as a number of national and international regulations and new management orientations trigger their implementation.

Because ecosystems do not follow political, ownership, or jurisdictional boundaries, coordination across all boundaries is necessary for effective conservation planning and management. If geopolitical and ecological boundaries do not coincide, the full range of spatial and temporal variability in the ecosystems under consideration cannot be considered. Therefore, key pattern/process–disturbance relationships may be characterized erroneously, increasing the chances of misinformed land-use decisions. Over the past several decades, ecological and environmental impact assessments have been initiated by various private, state, and federal institutions to aid in implementing a scientific approach to land-use planning. Such assessments usually have been restricted in scope and spatial extent, and therefore, their effectiveness has been limited. Furthermore, since they have differed in their specific goals, information from these projects usually has not been compatible and cannot be shared among institutions.

Data, analyses, and interpretations are needed to incorporate information about the dynamic nature of ecosystems into land-use planning at multiple scales. Human activities should be considered integral components of ecosystems, not external to them. The best applicable science and technology should be identified so that the required knowledge about ecosystems is used to make predictions about ecological conditions under different scenarios, from regional to local scales. However, the best science will not necessarily lead to the implementation of the eight land-use planning guidelines proposed by Dale et al. (Chapter 1). Final land-use decisions will not be made by scientists alone. All parties that may have an interest in these decisions (e.g., landowners, land managers, decision-makers, the general public) must be involved in all steps of the planning process. The participation of all interested parties, including scientists, is the only way to ensure that scientific input is incorporated at the appropriate stages of planning. Finally, new ways to define success should be formulated, such as the definition of testable outcomes.

The principles and proposed guidelines for land-use planning and management described by Dale et al. (Chapter 1) have implications for the types of data, the range of scales, and the kinds of analyses required for making decisions by landowners, planners, and managers. At the most basic level, characterization of ecological systems (Treweek 1999; Bourgeron et al., 2001*a*) is needed to implement these guidelines. A requirement for ecological information to identify, quantify, and evaluate the potential impact of land-use decisions on ecosystems has been recognized for some time (Everett et al. 1994; O'Callaghan 1996; Boyce and Haney 1997; Lyle 1999; Treweek 1999; Jensen et al. 2001). In response to this need for information, the process of integrated ecological assessment (IEA) has been developed to provide a comprehensive description of those ecosystem patterns, processes, and functions, including relevant socio-political factors, needed to synthesize our knowledge of ecological and human systems. IEAs are generally used to describe the biophysical, biological, and social limits of ecosystems, the interactions between the various ecosystem components, and uncertainty in ecosystem knowledge within the context of a specific set of issues (Gunderson 1998; Treweek 1999; Jensen et al. 2001).

References

Allen, T.F.H., and T.B. Starr. 1982. Hierarchy: perspectives for ecological complexity. University of Chicago Press, Chicago, Illinois, USA.

Arno, S.F., D.G. Simmermann, and R.E. Keane. 1986. Characterizing succession within a forest habitat type: an approach designed for resource managers. USDA Forest Service Research Note INT-357. Intermountain Research Station, Ogden, Utah, USA.

Austin, M.P., and C.R. Margules. 1986. Assessing representativeness. Pages 47–67 *in* M.B. Usher, editor. Wildlife conservation evaluation. Chapman & Hall, London, England.

Austin, M.P., A.O. Nicholls, and C.R. Margules. 1990. Measurement of the realized qualitative niche: environmental niches of five *Eucalyptus* species. Ecological Monographs **60**:161–177.

Bahre, C.J. 1995. Human impacts on the grasslands of southwestern Arizona. Pages 230–264 *in* M.P. McClaran and T.R. Van Devender, editors. The desert grassland. University of Arizona Press, Tucson, Arizona, USA.

Bailey, R.G. 1996. Multi-scale ecosystem analysis. Environmental Monitoring and Assessment **39**:21–24.

Bailey, R.G., M.E. Jensen, D.T. Cleland, and P.S. Bourgeron. 1994. Design and use of ecological mapping units. Pages 95–106 *in* M.E. Jensen and P.S. Bourgeron, editors. Ecosystem management, Vol. II: principles and applications. General Technical Report PNW-GTR-318. U.S. Department of Agriculture Forest Service, Pacific Northwest Research Station, Portland, Oregon, USA.

Baskent, E.Z. 1999. Controlling spatial structure of forested landscapes: a case study towards landscape management. Landscape Ecology **14**:83–97.

Beek, K.J., and J. Bannema. 1972. Land evaluation for agricultural land use planning: an ecological methodology. Department of Soil Sciences and Geology, Agricultural University, Wageningen, The Netherlands.

Belbin, L. 1993. Environmental representativeness: regional partitioning and reserve selection. Biological Conservation **66**:223–230.

Berry, M., R. Flamm, B. Hazen, and R. MacIntyre. 1996. LUCAS: a system for modeling land-use change. IEEE Computational Science Engineering **3**:24–35.

Bojorquez-Tapia, L.A., I. Azuara, E. Ezcurra, and O. Flores-Villela. 1995. Identifying conservation priorities in Mexico through geographic information systems and modeling. Ecological Applications **5**:215–231.

Bourgeron, P.S., and M.E. Jensen. 1994. An overview of ecological principles for ecosystem management. Pages 45–67 *in* M.E. Jensen and P.S. Bourgeron, editors. Ecosystem management, Vol. II: principles and applications. General Technical Report PNW-GTR-318. U.S. Department of Agriculture Forest Service, Pacific Northwest Research Station, Portland, Oregon, USA.

Bourgeron, P.S., H.C. Humphries, R.L. DeVelice, and M.E. Jensen. 1994*a*. Ecological theory in relation to landscape and ecosystem characterization. Pages 58–72 *in* M.E. Jensen and P.S. Bourgeron, editors. Ecosystem management, Vol. II: principles and applications. General Technical Report PNW-GTR-318. U.S. Department of Agriculture Forest Service, Pacific Northwest Research Station, Portland, Oregon, USA.

Bourgeron, P.S., H.C. Humphries, and M.E. Jensen. 1994*b*. Landscape characterization: a framework for ecological assessment at regional and local scales. Journal of Sustainable Forestry **2**:267–281.

Bourgeron, P.S., H.C. Humphries, and M.E. Jensen. 2001*a*. Ecosystem characterization and ecological assessments. *In* M.E. Jensen and P.S. Bourgeron, editors. A guidebook for integrated ecological assessments. Springer-Verlag, New York, New York, USA.

Bourgeron, P.S., H.C. Humphries, and M.E. Jensen. 2001*b*. Representativeness assessment. *In* M.E. Jensen and P.S. Bourgeron, editors. A guidebook for integrated ecological assessments. Springer-Verlag, New York, New York, USA.

Bourgeron, P.S., H.C. Humphries, and M.E. Jensen. 2001*c*. General data collection and sampling design considerations for integrated regional ecological assessments. *In* M.E. Jensen and P.S. Bourgeron, editors. A guidebook for integrated ecological assessments. Springer-Verlag, New York, New York, USA.

Bourgeron, P.S., H.C. Humphries, and M.E. Jensen. 2001*d*. Elements of ecological land classifications for ecological assessments. *In* M.E. Jensen and P.S. Bourgeron, editors. A guidebook for integrated ecological assessments. Springer-Verlag, New York, New York, USA.

Boyce, M.S., and A. Haney, editors. 1997. Ecosystem management: applications for sustainable forest and wildlife resources. Yale University Press, New Haven, Connecticut, USA.

Brenner, R.N., and J.K. Jordan. 1991. The role of an ecological classification system in forest plan development and implementation. Pages 70–80 *in* Proceedings of the 1991 Symposium on Systems Analysis in Forest Resources, 3–6 March 1991, Charleston, South Carolina, USA.

Brewer, C.K., and P. Callahan. Interior Columbia Basin watershed delineation guidelines. Unpublished report on file with the Interior Columbia Basin Ecosystem Management Project, Walla Walla, Washington, USA.

Brown, D. 2001. Characterizing the human imprint on landscapes for ecological assessment. *In* M.E. Jensen and P.S. Bourgeron. editors. A guidebook for integrated ecological assessments. Springer-Verlag, New York, New York, USA.

Burhenne-Guilmin, F., and L. Glowka. 1994. An introduction to the Convention on Biological Diversity. Pages 15–19 *in* A.F. Krattiger, J.A. McNeely, W.H. Lesser, K.R. Miller, Y.St. Hill, and R. Senanayake, editors. Widening perspectives on biodiversity. IUCN/IAE, Gland, Switzerland.

Burke, I.C., G.F. Kittel, W.K. Lauenroth, P. Snook, C.M. Yonker, and W.J. Parton. 1991. Regional analysis of the central Great Plains: sensitivity to climate variability. BioScience **41**:685–692.

Carpenter, S., W. Brock, and P. Hanson. 1999. Ecological and social dynamics in simple models of ecosystem management. Conservation Ecology **3**(2):4 [online at *http://www.consecol.org/vol3/iss2/art4*].

Christensen, N.L., J.K. Agee, P.F. Brussard, J. Hughes, D.H. Knight, G.W. Minshall, J.M. Peek, S.J. Pyne, F.J. Swanson, J.W. Thomas, S. Wells, S.E. Williams, H.A. Wright. 1989. Interpreting the Yellowstone fires of 1988. BioScience **39**:678–685.

Christensen, N.L., A.M. Bartuska, J.H. Brown, S.R. Carpenter, C. D'Antonio, R. Francis, J.F. Franklin, J.A. MacMahon, R.F. Noss, D.J. Parsons, C.H. Peterson, M.G. Turner, and R.G. Woodmansee. 1996. The report to the

Ecological Society of America committee on the scientific basis for ecosystem management. Ecological Applications **6**:665–691.

Cissel, J.H., F.J. Swanson, G.E. Grant, D.H. Olson, S.V. Gregory, S.L. Garman, L.R. Ashkenas, M.G. Hunter, J.A. Kertis, J.H. Mayo, M.D. McSwain, S.G. Swetland, K.A. Swindle, and D.O. Wallin. 1998. A landscape plan based on historical fire regimes for a managed forest ecosystem: the Augusta Creek study. General Technical Report PNW-GTR-422. U.S. Department of Agriculture Forest Service, Pacific Northwest Research Station, Portland, Oregon, USA.

Cleland, D.T., P.E. Avers, W.H. McNab, M.E. Jensen, R.G. Bailey, T. King, and W.E. Russell. 1997. National hierarchical framework of ecological units. Pages 181–200 *in* M.S. Boyce and A. Haney, editors. Ecosystem management: applications for sustainable forest and wildlife resources. Yale University Press, New Haven, Connecticut, USA.

Cooney, R.T. 1995. SEA94: sound ecosystem assessment (SEA)—an integrated science plan for the restoration of injured species in Prince William Sound, Exxon Valdez Oil Spill Restoration Project Final Report. Restoration Project 94320. Alaska Department of Fish and Game, Habitat and Restoration Section, Anchorage, Alaska, USA.

Costanza, R., L. Wainger, C. Folke, and K. Maler. 1993. Modeling complex ecological economic systems. BioScience **43**:545–555.

Cox, J., R. Kautz, M. MacLaughlin, and T. Gilbert. 1994. Closing the gaps in Florida's wildlife habitat conservation system. Office of Environmental Services, Florida Game and Fresh Water Fish Commission, Tallahassee, Florida, USA.

Dale, V.H., and R.V. O'Neill. 1999. Tools to characterize the environmental setting. Pages 62–93 *in* V.H. Dale and M.R. English, editors. Tools to aid environmental decision making. Springer-Verlag, New York, New York, USA.

Dale, V.H., R.V. O'Neill, F. Southworth, and M. Pedlowski. 1994. Modeling effects of land management in the Brazilian Amazonian settlement of Rondonia. Conservation Biology **8**:196–206.

Dale, V.H., A.E. Lugo, J.A. MacMahon, and S.T.A. Pickett. 1998. Ecosystem management in the context of large, infrequent disturbances. Ecosystems **1**:546–557.

Dale, V.H., S. Brown, R.A. Haeuber, N.T. Hobbs, N. Huntly, R.J. Naiman, W.E. Riebsame, M.G. Turner, and T.J. Valone. 2000. Ecological principles and guidelines for managing the use of land. Ecological Applications **10**:639–670.

English, M.R., V.H. Dale, C. Van Riper-Geibig, and W. Hudson Ramsey. 1999. Overview. Pages 1–31 *in* V.H. Dale and M.R. English, editors. Tools to aid environmental decision making. Springer-Verlag, New York, New York, USA.

Everett, R.L., P.F. Hessburg, M.E. Jensen, and B.T. Bormann. 1994. Vol. I: Executive summary. General Technical Report PNW-GTR-317. U.S. Department of Agriculture Forest Service, Pacific Northwest Research Station, Portland, Oregon, USA.

Fairfax, S.K. 1978. A disaster in the environmental movement. Science **199**:743–748.

Faith, D.P. 1992. Conservation evaluation and phylogenetic diversity. Biological Conservation **61**:1–10.

Faith, D.P., and P.A. Walker. 1996. Environmental diversity: on the best-possible use of surrogate data for assessing the relative biodiversity of sets of areas. Biodiversity and Conservation **5**:399–415.

Fan, W., J.C. Randolph, and J.L. Ehman. 1998. Regional estimation of nitrogen mineralization in forest ecosystems using geographic information systems. Ecological Applications **8**:734–747.

FAO. 1976. A framework for land evaluation, 32. United Nations Food and Agriculture Organization, Rome, Italy.

FAO. 1993. Guidelines for land use planning. United Nations Food and Agriculture Organization, Rome, Italy.

Fischer, W.C., and A.F. Bradley. 1987. Fire ecology of western Montana forest habitat types. General Technical Report INT-223. U.S. Department of Agriculture Forest Service, Intermountain Research Station, Ogden, Utah, USA.

Fischer, W.C., and B.D. Clayton. 1983. Fire ecology of Montana forest habitat types east of the Continental Divide. General Technical Report INT-141. U.S. Department of Agriculture Forest Service, Intermountain Forest and Range Experiment Station, Ogden, Utah, USA.

Forman, R.T.T., and M. Godron. 1986. Landscape ecology. John Wiley and Sons, New York, New York, USA.

Foster, D.R., D.H. Knight, and J.F. Franklin. 1998. Landscape patterns and legacies resulting from large, infrequent forest disturbances. Ecosystems **1**:497–510.

Franklin, J. 1998. Predicting the distribution of shrub species in southern California from climate and terrain-derived variables. Journal of Vegetation Science **9**:733–748.

Frelich, L.E., and P.B. Reich. 1995. Spatial patterns and succession in a Minnesota southern-boreal forest. Ecological Monographs **65**:325–346.

Glenn, S.M., and S.L. Collins. 1994. Species richness and scale: the application of distribution and abundance models. Pages 135–142 *in* M.E. Jensen and P.S. Bourgeron, editors. Ecosystem management, Vol. II: principles and applications. General Technical Report PNW-GTR-318. U.S. Department of Agriculture Forest Service, Pacific Northwest Research Station, Portland, Oregon, USA.

Graham, R.L., C.T. Hunsaker, R.V. O'Neill, and B.L. Jackson. 1991. Ecological risk assessment at the regional scale. Ecological Applications **1**:196–206.

Groom, M., D.B. Jensen, R.L. Knight, S. Gatewood, L. Mills, D. Boyd-Heger, L.S. Mills, and M.E. Soulé. 1999. Buffer zones: benefits and dangers of compatible stewardship. *In* M.E. Soulé and J. Terborgh, editors. Continental conservation: scientific foundations of regional reserve networks. Island Press, Washington, D.C., USA.

Grossman, D.H., D. Faber-Langendoen, A.S. Weakley, M. Anderson, P. Bourgeron, R. Crawford, K. Gooding, S. Landaal, K. Metzler, K.D. Patterson, M. Pyne, M. Reid, and L. Sneddon. 1998. International classification of ecological communities: terrestrial vegetation of the United States, Vol. I: The national vegetation classification standard. The Nature Conservancy, Arlington, Virginia, USA.

Grossman, D.H., P.S. Bourgeron, W.-D.N. Busch, D. Cleland, W. Platts, G.C. Ray, C.R. Robins, and G. Roloff. 1999. Principles for ecological classification. Pages 353–393 *in* N.C. Johnson, A.J. Malk, W.T. Sexton, and R.C. Szaro, editors.

Ecological stewardship, Vol. II: a common reference for ecosystem management. Elsevier Science, Amsterdam, The Netherlands.

Grumbine, R.E. 1993. What is ecosystem management? Conservation Biology **8**:27–38.

Gunderson, L.H. 1998. Stepping back: assessing for understanding in complex regional systems. Pages 27–42 *in* K.N. Johnson, F. Swanson, M. Herring, and S. Greene, editors. Bioregional assessments. Island Press, Washington, D.C., USA.

Haber, W. 1994. System ecological concepts for environmental planning. Pages 49–67 *in* F. Klijn, editor. Ecosystem classification for environmental management. Dordrecht, The Netherlands: Kluwer Academic Publishers, Dordrecht, The Netherlands.

Haeuber, R. 2001. Ecological assessments and implementing ecosystem management: challenges, opportunities, and the road ahead. *In* M.E. Jensen and P.S. Bourgeron, editors. A guidebook for integrated ecological assessments. Springer-Verlag, New York, New York, USA.

Hall, C.A.S., H. Tian, Y. Qi, G. Pontius, and J. Cornell. 1995. Modelling spatial and temporal patterns of tropical land use change. Journal of Biogeography **22**:753–757.

Haney, A., and M.S. Boyce, 1997. Introduction. Pages 1–17 *in* M.S. Boyce and A. Haney, editors. Ecosystem management: applications for sustainable forest and wildlife resources. Yale University Press, New Haven, Connecticut, USA.

Hann, W.J., J.L. Jones, M.G. Karl, P.F. Hessburg, R.E. Keane, D.G. Long, J.P. Menakis, C.H. McNicol, S.G. Leonard, R.A. Gravenmier, and B.G. Smith. 1997. Landscape dynamics of the basin. Pages 337–1055 *in* T.M. Quigley and S.J. Arbelbide, editors. An assessment of ecosystem components in the Interior Columbia Basin and portions of the Klamath and Great Basins, Vol. II. General Technical Report PNW-GTR-405. U.S. Department of Agriculture Forest Service, Pacific Northwest Research Station, Portland, Oregon, USA.

Hardin, G. 1968. The tragedy of the commons. Science **62**:1243–1248.

Haynes, R.W., R.T., Graham, and T.M. Quigley. 1996. A framework for ecosystem management in the Interior Columbia Basin. General Technical Report PNW-GTR-374. U.S. Department of Agriculture Forest Service, Pacific Northwest Research Station, Portland, Oregon, USA.

Herring, M. 1998. Introduction. Pages 1–8 *in* K.N. Johnson, F. Swanson, M. Herring, and S. Greene, editors. Bioregional assessments. Island Press, Washington, D.C., USA.

Holling, C.S., editor. 1978. Adaptive environmental assessment and management. John Wiley and Sons, New York, New York, USA.

Holling, C.S. 1986. The resilience of terrestrial ecosystems: local surprise and global change. Pages 292–320 *in* W.M. Clark and R.E. Munn, editors. Sustainable development in the biosphere. Oxford University Press, Oxford, England.

Host, G., and J. Pastor. 1998. Modeling forest succession among ecological land units in northern Minnesota. Conservation Ecology **2**(2):15 [online at *http://www.consecol.org/vol2/iss2/art15*].

Host, G.E., P.L. Polzer, M.A. Mladenoff, M.A. White, and S.J. Crow. 1996. A quantitative approach to developing regional ecosystem classifications. Ecological Applications **6**:608–618.

Humphries, H.C., and P.S. Bourgeron. In Press. Methods for determining historical range of variability. *In* M.E. Jensen and P.S. Bourgeron, editors. A guidebook for integrated ecological assessments. Springer-Verlag, New York, New York, USA.

Hunsaker, C.T., R.T. Graham, G.W. Suter II, R.V. O'Neill, L.W. Barnthouse, and R.H. Gardner. 1990. Assessing ecological risk on a regional scale. Environmental Management **14**:325–332.

Huschle, G., and M. Hironata. 1980. Classification and ordination of seral plant communities. Journal of Range Management **33**:179–182.

Innes, J., J. Gruber, M. Neuman, and R. Thompson. 1994. Coordinating growth and environmental management through consensus building. Proceedings of the California Policy Seminar, Berkeley, California, USA.

Ives, A., M.G. Turner, and S.M. Pearson. 1998. Local explanations of landscape patterns: can analytical approaches approximate simulation models of spatial processes? Ecosystems **1**:35–51.

Jensen, M.E., and P.S. Bourgeron, editors. 2001. A guidebook for integrated ecological assessments. Springer-Verlag, New York, New York, USA.

Jensen, M.E., P.S. Bourgeron, R. Everett, and I. Goodman. 1996. Ecosystem management: a landscape ecology perspective. Journal of American Water Resources Association **32**:208–216.

Jensen, M., I. Goodman, K. Brewer, T. Frost, G. Ford, and J. Nesser. 1997. Biophysical environments of the basin. Pages 99–314 *in* T.M. Quigley and S.J. Arbelbide, editors. An assessment of ecosystem components in the Interior Columbia Basin and portions of the Klamath and Great Basins, Vol. II. General Technical Report PNW-GTR-405. U.S. Department of Agriculture Forest Service, Pacific Northwest Research Station, Portland, Oregon, USA.

Jensen, M.E., N.L. Christensen, Jr., and P.S. Bourgeron. 2001. An overview of ecological assessment principles and applications. *In* M.E. Jensen and P.S. Bourgeron, editors. A Guidebook for integrated ecological assessments. Springer-Verlag, New York, New York, USA.

Johnson, B.L. 1999. The role of adaptive management as an operational approach for resource management agencies. Conservation Ecology **3**(2):8 [online at *http://www.consecol.org/vol3/iss2/art8*].

Johnson, K.N., F. Swanson, M. Herring, and S. Greene, editors. 1998. Bioregional assessments. Island Press, Washington, D.C., USA.

Kaufmann, M.R., L.S. Huckaby, C.S. Regan, and J. Popp. 1998. Forest reference conditions for ecosystem management in the Sacramento Mountains, New Mexico. General Technical Report RMRS-GTR-19. U.S. Department of Agriculture Forest Service, Rocky Mountain Research Station, Fort Collins, Colorado, USA.

Kay, J.J. 1991. A nonequilibrium thermodynamic framework for discussing ecosystem integrity. Environmental Management **15**:483–495.

Keane, R.E., S.F. Arno, and J.K. Brown. 1990*a*. Simulating cumulative fire effects in ponderosa pine/Douglas-fir forests. Ecology **71**:189.

Keane, R.E., S.F. Arno, J.K. Brown, and D.F. Tomback. 1990*b*. Modelling stand dynamics in whitebark pine *Pinus albicaulis* forests. Ecological Modelling **51**:73–95.

Kerner, H.F., L. Spandau, J.G. Koppel, and T. Wachter. 1991. Methoden zur angewandten kosystemforschung, entwickelt im MAB-Projekt 6 kosystemforschung Berchtesgaden' (Werkstattbericht). *In* Deutsches nationalkomitee

'Der Mensch und die Biosphere' (MAB). MAB-Mitteilungen 35, hrsg.v. Bundesministerium fur Umwelt, Naturschutz und Reaktorsicherheit, Bonn, Germany.

Kessler, J.J., editor. 1997. Strategic Environmental Analysis, AIDEnvironment. The Hague, Amsterdam, The Netherlands.

Kessler, J.J., and M. Van Dorp. 1998. Structural adjustment and the environment: the need for an analytical methodology. Ecological Economics **27**:267–281.

Klapp, M.G. 1992. Bargaining with uncertainty: decision-making in public health, technological safety, and environmental quality. Auburn House, New York, New York, USA.

Klingebiel, A.A., and D.H. Montgomery. 1961. Land capability classification. Agricultural Handbook 210. Soil Conservation Service, US GPO, Washington, D.C., USA.

Knight, D.H., and L.L. Wallace. 1989. The Yellowstone fires: issues in landscape ecology. BioScience **39**:700–706.

Knight, R.L., and T.W. Clark. 1998. Boundaries between public and private lands: defining obstacles, finding solutions. *In* R.L. Knight and P.B. Landres, editors. Stewardship across boundaries. Island Press, Washington, D.C., USA.

Knight, R.L., and P.B. Landres, editors. 1998. Stewardship across boundaries. Island Press, Washington, D.C., USA.

Landres, P.B. 1998. Integration: a beginning for landscape-scale stewardship. *In* R.L. Knight and P.B. Landres, editors. Stewardship across boundaries. Island Press, Washington, D.C., USA.

Landres, P.B., R.L. Knight, S.T.A. Pickett, and M.L. Cadenasso. 1998. Ecological effects of administrative boundaries. Pages 39–64 *in* R.L. Knight and P.B. Landres, editors. Stewardship across boundaries. Island Press, Washington, D.C., USA.

Landres, P.B., P. Morgan, and F.J. Swanson. 1999. Evaluating the utility of natural variability concepts in managing ecological systems. Ecological Applications **9**:1179–1188.

Leathwick, J.R. 1998. Are New Zealand's *Nothofagus* species in equilibrium with their environment? Journal of Vegetation Science **9**:719–732.

Lee, D.C., J.R. Sedell, B.E. Rieman, R.F. Thurow, J.E. Williams, D. Burns, J. Clayton, L. Decker, R. Gresswell, R. House, P. Howell, K.M. Lee, K. MacDonald, J. McIntyre, S. McKinney, T. Noel, J.E. O'Connor, C.K. Overton, D. Perkinson, K. Tu, and P. Van Eimeren. 1997. Broadscale assessment of aquatic species and habitats. Pages 1057–1496 *in* T.M. Quigley and S.J. Arbelbide, editors. An assessment of ecosystem components in the Interior Columbia Basin and portions of the Klamath and Great Basins, Volume III. General Technical Report PNW-GTR-405. U.S. Department of Agriculture Forest Service, Pacific Northwest Research Station, Portland, Oregon, USA.

Legendre, P., and L. Legendre. 1998. Numerical ecology. Elsevier Science B.V., Amsterdam, The Netherlands.

Lenihan, J.M. 1993. Ecological response surfaces for North American boreal tree species and their use in forest classification. Journal of Vegetation Science **4**:667–680.

Lessard, G. 1995. A national framework for integrated ecological assessments. Federal InterAgency Working Group, Washington, D.C., USA.

Lessard, G., M.E. Jensen, M. Crespi, and P.S. Bourgeron. 1999. A national framework for integrated ecological assessments. *In* H.K. Cordel and J.C. Bergstrom, editors. Integrating social sciences with ecosystem management: human dimensions in assessment policy and management. Sagamore Press, Champaign-Urbana, Illinois, USA.

Lyle, J.T. 1999. Design for human ecosystems: landscape, land use, and natural resources. Island Press, Washington, D.C., USA.

Lyndon, M.L. 1999. Characterizing the regulatory and judicial setting. Pages 130–156 *in* V.H. Dale, and M.R. English, editors. Tools to aid environmental decision making. Springer-Verlag, New York, New York, USA.

Mackey, B.G., H.A. Nix, M.F. Hutchinson, J.P. MacMahon, and P.M. Fleming. 1988. Assessing representativeness of places for conservation reservation and heritage listing. Environmental Management **12**:501–514.

Mackey, B.G., H.A. Nix, J.A. Stein, S.E. Cork, and F.T. Bullen. 1989. Assessing the representativeness of the wet tropics of Queensland World Heritage Property. Biological Conservation **50**:279–303.

Margerum, R.D. 1999. Integrated environmental management: the foundations for successful practice. Environmental Management **24**:151–166.

Margules, C.R., and A.O. Nicholls. 1987. Assessing the conservation value of remnant habitat 'islands': mallee patches on the western Eyre Peninsula, South Australia. Pages 89–102 *in* D.A. Saunders, G.W. Arnold, A.A. Burbidge, and A.J.M. Hopkins, editors. Nature conservation: the role of remnants of native vegetation. Commonwealth Scientific and Industrial Research Organization Division of Water and Land Resources, Canberra, Australia.

Maxwell, J.R., C.J. Edwards, M.E. Jensen, S.J. Paustian, H. Parrott, and D.M. Hill. 1995. A hierarchical framework of aquatic ecological units in North America. General Technical Report NC-176. U.S. Department of Agriculture Forest Service, North Central Forest Experiment Station, St. Paul, Minnesota, USA.

May, R. 1976. Simple mathematical models with very complicated dynamics. Nature **261**:459–467.

Meidinger, D., and J. Pojar. 1991. Ecosystems of British Columbia. Special Report Series 6. Research Branch, Ministry of Forests, Victoria, British Columbia, Canada.

Messerli, B., and P. Messerli. 1979. Wirtschaftliche Entwicklung und ökologische belastbarkeit im berggebiet. Fachbeiträzur Schweizerischen MAB-Information (Bern) Nr.1 [also *in* Geographica Helvetica **33**:203–210, 1978].

Milne, B.T. 1994. Pattern analysis for landscape evaluation and characterization. Pages 121–134 *in* M.E. Jensen and P.S. Bourgeron, editors. Ecosystem management, Vol. II: principles and applications. General Technical Report PNW-GTR-318. U.S. Department of Agriculture Forest Service, Pacific Northwest Research Station, Portland, Oregon, USA.

Mladenoff, D.J., T.A. Sickley, R.G. Haight, and A.P. Wydeven. 1995. A regional landscape analysis and prediction of favorable gray wolf habitat in the northern Great Lakes region. Conservation Biology **9**:279–294.

Moore, S.A., and R.G. Lee. 1999. Understanding dispute resolution processes for American and Australian public wildlands: towards a conceptual framework for managers. Environmental Management **23**:453–465.

Morgan, P., G.H. Aplet, J.B. Haufler, H.C. Humphries, M.M. Moore, and W.D. Wilson. 1994. Historical range of variability: a useful tool for evaluating ecosystem change. Journal of Sustainable Forestry **2**(3–4):57–72.

Morrison-Saunders, A., and J. Bailey. 1999. Exploring the EIA/environmental management relationship. Environmental Management **24**:281–295.

National Research Council (NRC). 1994. Science and judgment in risk assessment. National Academy Press, Washington, D.C., USA.

Noss, R.F., M.A. O'Connell, and D.D. Murphy. 1997. The science of conservation planning: habitat conservation under the Endangered Species Act. Island Press, Washington, D.C., USA.

O'Callaghan, J.R. 1996. Land use: the interaction of economics, ecology and hydrology. Chapman & Hall, London, England.

Olson, G.W. 1974. Land classifications. Cornell University, Ithaca, New York, USA.

O'Neill, R.V., and A.W. King. 1998. Homage to St. Michael; or, why are there so many books on scale? *In* D.L. Peterson and V.T. Parker, editors. Ecological scale: theory and applications. Columbia University Press, New York, New York, USA.

Ortolano, L. 1997. Environmental regulation and impact assessment. John Wiley and Sons, New York, New York, USA.

Ortolano, L., and A. Shepherd. 1995. Environmental impact assessment. Pages 3–31 *in* F. Vanclay and D.A. Bronstein, editors. Environmental and social impact assessment. John Wiley and Sons, Chichester, England.

Pastor, J., and D.J. Mladenoff. 1992. Modeling the effects of timber management on population dynamics, diversity, and ecosystem processes. Pages 16–29 *in* D.C. LeMaster and R.A. Sedjo, editors. Modeling sustainable forest ecosystems. Forest Policy Center, Washington, D.C., USA.

Perera, A.H., J.A. Baker, L.E. Band, and D.J.B. Baldwin. 1996. A strategic framework to eco-regionalize Ontario. Pages 85–96 *in* R.A. Sims, I.G.W. Corns and K. Klinka, editors. Global to local ecological land classification. Kluwer Academic Publishers, Dordrecht, The Netherlands.

Perez-Trejo, F., N. Clark, and P. Allen. 1993. An exploration of dynamical systems modelling as a decision tool for environmental policy. Journal of Environmental Management **39**:305–319.

Peterson, G., C.R. Allen, and C.S. Holling. 1998. Ecological resilience, biodiversity, and scale. Ecosystems **1**:6–18.

Pickett, S.T.A., and M.L. Cadenasso. 1995. Landscape ecology: spatial heterogeneity in ecological systems. Science **269**(5222):331–334.

Pickett, S.T.A., and P.S. White. 1985. The ecology of natural disturbance and patch dynamics. Academic Press, London, England.

Pickett, S.T.A., J. Kolasa, J.J. Armesto, and S.L. Collins. 1989. The ecological concept of disturbance and its expression at various hierarchical levels. Oikos **54**:129–136.

Poiani, K.A., J.V. Baumgartner, S.C. Buttrick, S.L. Green, E. Hopkins, G.D. Ivey, K.P. Seaton, and R.D. Sutter. 1998. A scale-independent, site conservation planning framework in The Nature Conservancy. Landscape and Urban Planning **43**:143–156.

Pressey, R.L., S. Ferrier, C.D. Hutchinson, D.P. Sivertsen, and G. Manion. 1995. Planning for negotiation: using an interactive geographic information system to explore alternative protected area networks. Pages 23–33 *in* D.A. Saunders, J.L. Craig and E.M. Mattiske, editors. Nature conservation, 4: the role of networks. Surrey Beatty and Sons, Sydney, Australia.

Pulliam, H.R. 1988. Sources, sinks, and population regulation. American Naturalist **132**:652–661.
Quigley, T.M., and S.J. Arbelbide, editors. 1997*a*. An assessment of ecosystem components in the interior Columbia Basin and portions of the Klamath and Great Basins, Vol. I. General Technical Report PNW-GTR-405. U.S. Department of Agriculture Forest Service, Pacific Northwest Research Station, Portland, Oregon, USA.
Quigley, T.M., and S.J. Arbelbide, editors. 1997*b*. An assessment of ecosystem components in the interior Columbia Basin and portions of the Klamath and Great Basins, Vol. II. General Technical Report PNW-GTR-405. U.S. Department of Agriculture Forest Service, Pacific Northwest Research Station, Portland, Oregon, USA.
Quigley, T.M., and S.J. Arbelbide, editors. 1997*c*. An assessment of ecosystem components in the interior Columbia Basin and portions of the Klamath and Great Basins, Vol. III. General Technical Report PNW-GTR-405. U.S. Department of Agriculture Forest Service, Pacific Northwest Research Station, Portland, Oregon, USA.
Quigley, T.M., and S.J. Arbelbide, editors. 1997*d*. An assessment of ecosystem components in the interior Columbia Basin and portions of the Klamath and Great Basins, Vol. IV. General Technical Report PNW-GTR-405. U.S. Department of Agriculture Forest Service, Pacific Northwest Research Station, Portland, Oregon, USA.
Rapport, D.J., C. Gaudet, J.R. Karr, J.S. Baron, C. Bohlen, W. Jackson, B. Jones, R.J. Naiman, B. Norton, and M.M. Pollock. 1998. Evaluating landscape health: integrating societal goals and biophysical process. Journal of Environmental Management **53**:1–15.
Reichman, O.J., and S.C. Smith. 1985. Impact of pocket gopher burrows on overlying vegetation. Journal of Mammalogy **66**:720–725.
Reynolds, K., J. Slade, M. Saunders, and B. Miller. NetWeaver for EMDS version 1.0 user guide: a knowledge base development system. Unpublished report. U.S. Department of Agriculture, Forest Service, Pacific Northwest Research Station, Portland, Oregon, USA.
Robinson, N. 1992. International trends in environmental impact assessment. Boston College Environmental Affairs Law Review **19**:591–621.
Romme, W.H., and D.G. Despain. 1989. Historical perspective on the Yellowstone fires of 1988. BioScience **39**:695–699.
Rowe, J.S., and J.W. Sheard. 1981. Ecological land classification: a survey approach. Environmental Management **5**:451–464.
Scott, J.M., and M.D. Jennings. 1998. Large-area mapping of biodiversity. Annals of the Missouri Botanical Garden **85**:34–47.
Shea, K., and the NCEAS Working Group on Population Management. 1998. Management of populations in conservation, harvesting, and control. Trends in Ecology and Evolution **13**:371–375.
Sierra Nevada Ecosystem Project (SNEP). 1996. Sierra Nevada Ecosystem Project: final report to Congress, assessments and scientific basis for management options. Center for Water and Wildlands Resources, University of California, Davis, California, USA.
Slocombe, D.S. 1993*a*. Implementing ecosystem-based management. BioScience **43**:612–622.

Slocombe, D.S. 1993*b*. Environmental planning, ecosystem science, and ecosystem approaches for integrating environment and development. Environmental Management **17**:289–303.

Slocombe, D.S. 1998. Defining goals and criteria for ecosystem-based management. Environmental Management **22**:483–493.

Slocombe, D.S. 2001. Integration of physical, biological and socio-economic information. *In* M.E. Jensen and P.S. Bourgeron, editors. A guidebook for integrated ecological assessments. Springer-Verlag, New York, New York, USA.

Southern Appalachian Man and the Biosphere (SAMAB). 1996. The Southern Appalachian assessment summary report. U.S. Department of Agriculture, Forest Service, Southern Region, Atlanta, Georgia, USA.

Stenseth, N.C., K.-S. Chan, H. Tong, R. Boonstra, S. Boutin, C.J. Krebs, E. Post, M. O'Donoghue, N.G. Yoccoz, M.C. Forchhammer, and J.W. Hurrell. 1999. Common dynamic structure of Canada lynx populations within three climatic regions. Science **285**:1071–1073.

Swanson, F.J., J.A. Jones, D.O. Wallin, and J.H. Cissel. 1994. Natural variability: implications for ecosystem management. Pages 80–94 *in* M.E. Jensen and P.S. Bourgeron, editors. Ecosystem management, Vol. II: principles and applications. General Technical Report PNW-GTR-318. U.S. Department of Agriculture Forest Service, Pacific Northwest Research Station, Portland, Oregon, USA.

Swetnam, T.W., C.D. Allen, and J.L. Betancourt. 1999. Applied historical ecology: using the past to manage the future. Ecological Applications **9**:1189–1206.

Thérivel, R., and M. Rosário Partidário. 1996. The practice of strategic environmental assessment. Earthscan Publications, London, England.

Thérivel, R., and S. Thompson. 1996. Strategic environmental assessment and nature conservation. English Nature, Peterborough, England.

Thérivel, R., E. Wilson, S. Thompson, D. Heaney, and D. Pritchard. 1992. Strategic environmental assessment. Earthscan Publications, London, England.

Treweek, J. 1999. Ecological impact assessment. Blackwell Science, Oxford, England.

Turner, M.G., and R.H. Gardner. 1991. Quantitative methods in landscape ecology: an introduction. Pages 3–14 *in* M.G. Turner and R.H. Gardner, editors. Quantitative methods in landscape ecology: the analysis and interpretation of environmental heterogeneity. Springer-Verlag, New York, New York, USA.

Turner, M.G., W.H. Romme, R.H. Gardner, R.V. O'Neill, and T.K. Kratz. 1993. A revised concept of landscape equilibrium: disturbance and stability on scaled landscapes. Landscape Ecology **8**:213–227.

Turner, M.G., R.H. Gardner, R.V. O'Neill, and S.M. Pearson. 1994. Multiscale organization of landscape heterogeneity. Pages 73–79 *in* M.E. Jensen, and P.S. Bourgeron, editors. Ecosystem management, Vol. II: principles and applications. General Technical Report PNW-GTR-318. U.S. Department of Agriculture Forest Service, Pacific Northwest Research Station, Portland, Oregon, USA.

Turner, M.G., D.N. Wear, and R.O. Flamm. 1996. Land ownership and land-cover change in the southern Appalachian highlands and the Olympic peninsula. Ecological Applications **6**:1150–1172.

Turner, M.G., W.L. Baker, C.J. Peterson, and R.K. Peet. 1998. Factors influencing succession: lessons from large, infrequent natural disturbances. Ecosystems **1**:511–523.

Underwood, A.J. 1995. Ecological research and (and research into) environmental management. Ecological Applications **5**:232–247.

Urban, D.L., R.V. O'Neill, and H.H. Shugart, Jr. 1987. Landscape ecology: a hierarchical perspective can help scientists understand spatial patterns. BioScience **37**:119–127.

U.S. Environmental Protection Agency (US EPA). 1992. Framework for ecological risk assessment. EPA/630/R-92/001. Risk Assessment Forum, U.S. Environmental Protection Agency, Washington, D.C., USA.

U.S. Environmental Protection Agency (US EPA). 1998. Guidelines for ecological risk assessment. EPA/630/R-95/002F. Risk Assessment Forum, U.S. Environmental Protection Agency, Washington, D.C., USA.

U.S. Geological Service (USGS). 1994. USGS land use and land cover (LULC) data. Map. U.S. Geological Survey, Sioux Falls, South Dakota, USA.

U.S. Geological Service (USGS). 1995. U.S. land cover characteristics data set 1990 prototype. Map. U.S. Geological Survey, Sioux Falls, South Dakota, USA.

Vanclay, F., and D.A. Bronstein, editors. 1995. Environmental and social impact assessment. John Wiley and Sons, Chichester, England.

Verburg, P.H., G.H.J. de Koning, K. Kok, A. Veldkamp, L.O. Fresco, and J. Bouma, 1997. Quantifying the spatial structure of land use change: an integrated approach. Proceedings of the Conference on Geo-information for Sustainable Land Management, ITC-journal, special edition.

Wear, D.N., M.G. Turner, and R.O. Flamm. 1996. Ecosystem management with multiple owners: landscape dynamics in a southern Appalachian watershed. Ecological Applications **6**:1173–1188.

Wertz, W.A., and J.A. Arnold. 1972. Land systems inventory. U.S. Department of Agriculture Forest Service, Intermountain Region, Ogden, Utah, USA.

Wiens, J.A. 1989. Spatial scaling in ecology. Functional Ecology **3**:385–397.

Yaffee, S.L. 1998. Cooperation: a strategy for achieving. Stewardship across boundaries. Pages 299–324 *in* R.L. Knight and P.B. Landres, editors. Stewardship across boundaries. Island Press, Washington, D.C., USA.

Yaffee, S.L., A. Phillips, I. Frentz, P. Hardy, S. Maleki, and B. Thorpe. 1996. Ecosystem management in the United States: an assessment of current experience. Island Press, Washington, D.C., USA.

Yee, T.W., and N.D. Mitchell. 1991. Generalised additive models in plant ecology. Journal of Vegetation Science **2**:587–602.

Zonneveld, I.S. 1979. Landscape science and landscape evaluation. Enschede, The Netherlands.

Zonneveld, I.S. 1988. Basic principles of land evaluation using vegetation and other land attributes. Pages 499–517 *in* A.W. Kuchler and I.S. Zonneveld, editors. Vegetation mapping. Kluwer Academic Publishers, Boston, Massachusetts, USA.

Zonneveld, I.S. 1989. The land unit-a fundamental concept in landscape ecology, and its applications. Landscape Ecology **3**:67–89.

14
New Directions in Land Management: Incorporation of Ecological Principles

Richard A. Haeuber and Virginia H. Dale

Several themes resonate in the chapters of this volume. There is a clear call for integrating perspectives, disciplines, and approaches to land use and management. Longer temporal and broader spatial scales, as well as human dimensions, must be considered in this integration. To manage and use land in an ecologically sound manner, land-use planning efforts should recognize the significance of natural processes, the dynamic nature of ecological systems, the uncertainty and inherent variability of ecological systems, and the importance of cumulative effects. Advances in land-management tools and technology should be balanced by an understanding of ecological processes and sources of change. The new directions in land management and use discussed in this chapter explicitly incorporate ecological principles and guidelines.

> A land ethic changes the role of *Homo sapiens* from conqueror of the land-community to plain member and citizen of it. It implies respect for his fellow-members, and also respect for the community as such.
>
> Aldo Leopold

Aldo Leopold sought to develop and adopt a land ethic that would foster an appreciation of the land and its values and become a fundamental part of our relation to nature. The environmental movement that began in the 1960s has taken up some of these themes. Even though our children learn much more about ecological relationships than we did, most of them believe that they are doing their part to protect the earth by recycling. The education of a whole generation of recyclers has tremendous benefits, yet there is still much more to accomplish. The roots of human impacts on the earth run deep and require broader recognition and realistic approaches to contain them. This volume aims to further that recognition and promote development of such approaches.

The initial chapter in this volume described ecological principles and guidelines for land use and management (Dale et al., Chapter 1). Subsequent chapters explored the diverse possibilities for applying and further developing these principles and guidelines toward the goal of implementing land-use and land-management practices that are ecologically sound, economically productive, and socially and politically viable. The purpose of this chapter is twofold. First, we briefly review perspectives presented in earlier chapters, discussing common themes and contrasting viewpoints presented by various authors. Given the many perspectives and approaches found throughout this volume, exploring contrasts and syntheses is important for advancing the development and implementation of ecological approaches to land-use policy and management.

Second, we suggest new directions in land management that combine the principles and guidelines presented in Chapter 1 with the approaches and examples found throughout other chapters of this volume. The future for incorporation of ecological ideas in land-use decisions is bright. The necessary tools and technology for implementing these concepts are advancing. Moreover, knowledge of ecological systems is now at a stage where ecological concepts and ideas can be incorporated into the decision-making process. The big challenge, however, lies in taking concepts developed by the research community and making them accessible and useful to land managers.

14.1 Contrasts and Common Ground

14.1.1 Integrating Perspectives, Disciplines, and Approaches

The chapters in this volume come from a wide variety of perspectives and describe various approaches to making land-use decisions. The authors differ according to discipline, substantive concern, methodology, and occupation. The decision to include such a variety of perspectives was conscious and, we feel, represents one of the strengths of this volume. The reality of land use and management itself is reflected in this intellectual, substantive, and methodological breadth. Landowners, land managers, and land-use policymakers are themselves a varied lot, as many of the chapters in this volume demonstrate. If heterogeneity is a defining characteristic of land use, management, and policy, then it is essential that a volume devoted to exploring the subject be heterogeneous as well.

However, it also important to acknowledge that heterogeneity and diversity of perspectives, or rather the *integration* of diverse perspectives and approaches, is one of the most significant obstacles to effective land management and policy. This integration is necessary to understand land-use issues and to develop new approaches for addressing them. The difficulty of integration is apparent whenever attempts are made to understand the interaction between ecological and human dimensions of systems at any scale, from local-level urban systems (Pickett et al. 1997) to subcontinental regional systems (Slocombe 1993; Blood 1994). The integration task is one of the most difficult challenges confronting the social, natural, and physical sciences and is at the forefront of the scientific endeavor.

The challenge, of course, is not new. Over the last few decades, some attempts have been made to bridge the gap between the "two cultures," as (Snow 1993) referred to the social and natural sciences. Such efforts have met with varying levels of success, though research that truly integrates the perspectives of diverse disciplines is rare (Miller 1994). In the last decade, however, integration efforts have become more intense as the scientific community realized the necessity of building bridges between disciplines to answer challenging intellectual questions and address pressing societal problems, such as land-use change (Lubchenco et al. 1991; CIESIN 1992; Stern et al. 1992; Dale et al., 2000).

The chapters in this volume demonstrate that efforts to integrate natural, physical, and social sciences in developing a comprehensive understanding of how systems function confront difficulties that are epistemological, technical, substantive, and process-oriented in nature. Perhaps the most basic obstacle is that individuals from different disciplines often possess very different cognitive "maps" or "schema." As one scholar of cognitive theory observes, "Every individual acquires during the course of his development a set of beliefs and personal constructs about the physical and social environment. These beliefs provide a relatively coherent way of organizing and making sense of what would otherwise be a confusing and overwhelming array of signals and cues picked up from the environment from his senses" (George 1980). Cognitive maps provide mental constructs that assist individuals in absorbing, processing, and interpreting new information and making sense of the world around them (Steinbruner 1974; Fiske and Taylor 1991; Lau and Sears 1986; Rosati, 2000).

To the extent that disciplinary training constitutes an important influence on intellectual development, it provides a paradigmatic and perceptual superstructure through which researchers view, understand, and interpret the world. Disciplinary superstructures consist of basic concepts, modes of inquiry, observational categories, representation techniques, standards of proof, and types of explanation (Petrie 1976; Vedeld 1994). As a result, individual members of different disciplines often observe, understand, and communicate about phenomena in ways that are alien to one another, creating distinct and difficult barriers to interaction.

A careful reading of the chapters in this volume uncovers many obvious examples of the clash of perceptual superstructures. Take the concept of "community," for instance. Haufler and Kernohan (Chapter 4) use the term community to describe an association of species living in a particular geographic area that is relevant for natural resource management purposes. For Botteron (Chapter 7) or Weiss (Chapter 5), on the other hand, community refers to collections of humans, and their social, economic, political, and cultural patterns of interaction. The same word is used with two very different meanings, both of which are essential for understanding land-use policy and management issues and moving toward an ecological approach to land use.

A single word—community—elegantly captures the complexity of the integration challenge. Language is only one example of conflict in perceptions. Indeed, such tension between competing perspectives pervades much of the discussion between disciplines. The effort to formulate principles and guidelines for land management provides one means to address the need for better communication between disciplines. Our intent is that an ecological perspective can be combined with other views about the land, be they economic, social, political, religious, or aesthetic in emphasis. As the chapters in this volume demonstrate, principles and guidelines derived from an explicitly ecological perspective provide a useful starting point—but only a starting point—for developing a broad, interdisciplinary approach to land management.

14.1.2 Broadening the Temporal and Spatial Scale of Conservation

Several authors in this volume suggest that successful future conservation efforts will require approaches that address land-management issues at longer temporal and broader spatial scales (e.g., see Hulse and Gregory, Chapter 9; Haeuber and Hobbs, Chapter 12). Frequently referred to as "ecosystem management," attempts to address scale issues in the context of conservation initiatives have consumed scientists and managers alike over the past two decades. The importance and challenge of undertaking conservation at broader spatial and longer temporal scales is a central message of contemporary conservation science (e.g., see Noss 1983; Grumbine 1994, 1997; Christensen et al. 1996; Committee of Scientists 1999).

Because ecological systems are dynamic in space and time, issues of scale pose a significant land-use and land-management challenge (Haufler and Kernohan, Chapter 4; Bourgeron et al., Chapter 13). Patterns and processes are manifest at many spatial scales, from a stand of trees to a watershed to a continent, and temporal scales may range from a single event to centuries or eons. It is increasingly apparent that understanding and interpreting ecosystem structure and process depends on the scale at which measurements are made. Indeed, the issue of scale in understanding ecological pattern and process has been referred to as the "fundamental conceptual problem in ecology" (Levin 1992). Thus, in addressing questions of land use and management, the appropriate scale largely depends on the management or policy issue in question. Scaling from the leaf to the ecosystem to the landscape to the region, and understanding the transfer of information, matter, and energy between these scales, not only is a central question for ecology, but constitutes a fundamental management challenge as well (Haufler and Kernohan, Chapter 4; Bourgeron et al., Chapter 13).

The issue of scale highlights one of the central challenges explored throughout this volume—the general mismatch between ecosystems and institutions that commonly creates difficulties for conservation and land management. Fundamental differences exist between the spatial and temporal scales of resource- and land-management issues and the spatial and temporal context in which institutions designed to address these issues must operate (Meidinger 1997; Folke et al. 1998; Hulse and Gregory, Chapter 9). For example, regional-scale ecological systems are broader than established administrative and jurisdictional boundaries, but must be managed in the context of a governmental system that seldom empowers regional-scale institutions (Wuichet 1995).

As several chapters in this volume demonstrate, temporal scale mismatches between ecological systems and resource management institutions frequently exist as a function of the short-term time horizons of planners and politicians (e.g., Steinitz and McDowell, Chapter 8; Hulse and Gregory, Chapter 9; Santelmann et al., Chapter 11). However, a more basic temporal disjunct involves the underlying foundation of traditional resource management philosophies, which aim to reduce the variability in natural systems over short time periods. Recent experiences with both fire and flooding demonstrate that reduction of variability of endogenous disturbances leads to intense long-term disruptions of natural and social systems (Holling and Meffe 1996; Haeuber and Michener 1998). Not only do the ecological principles and guidelines discussed in Dale et al. (Chapter 1) address these issues, but several of the chapters in this volume describe the impacts of such mismatches on wildlife (Hobbs and Theobald, Chapter 2), how existing tools can be used to address scale issues (Haeuber and Hobbs, Chapter 12), and the ways in which planning efforts can be guided in directions more consonant with ecologically relevant spatial and temporal scales (Russell, Chapter 6; Hulse and Gregory, Chapter 9; Santelmann et al., Chapter 11).

14.1.3 Incorporating the Human Dimension in Conservation

It may make ecological sense to broaden the spatial context of land-use and land-management decisions. However, there are few areas—some would argue none—that are "natural" enough to undertake conservation efforts at landscape to regional scales without somehow incorporating humans and human systems as an integral element of conservation (Vitousek et al. 1997). Under these circumstances, the interface between human and natural systems becomes an area of significant debate and uncertainty in terms of intellectual inquiry and practical application. In a landscape to regional-scale conservation context, in fact, the potential for conflict between the needs of human systems and ecological systems has become a central defining issue of contemporary conservation. The importance of this issue is particularly evident in the context of conservation efforts undertaken in heavily populated, less-developed regions of the world (e.g., see Huntley et al. 1991).

Although not presented as starkly as it might be, the tension between "biocentric" and "anthropocentric" conservation approaches is an area of significant contrast found in the chapters of this volume. Perhaps the clearest instance of a biocentric conservation approach found here is the discussion of nature reserve design offered by Hansen and Rotella (Chapter 3). Their discussion reflects broad-scale biocentric approaches, such as efforts to set aside large areas of wilderness in the northern Rocky Mountains, which have been a mainstay of conservation dialogues throughout the 1980s and 1990s (Noss 1992). On the other hand, some authors in this volume place humans and human systems at the center of conservation, maintaining that approaches that do not acknowledge this reality are doomed to failure (e.g., Weiss, Chapter 5; Botteron, Chapter 7). In a sense, they advocate approaches emphasizing "conservation of social and cultural interactions" as an essential prerequisite of successful conservation and land management.

In general, however, the authors in this volume take a decidedly practical approach to conservation, exploring the middle ground between biocentric and anthropocentric perspectives. Several chapters in this book make it clear that ecological principles must be integrated into a broader set of land-use and land-management ideas that cut across several perspectives. For example, Santelmann et al. (Chapter 11) suggest new principles dealing with human values and cultural practices, as well as the importance of communicating uncertainty to decision makers. The case study by Dale (Chapter 10) illustrates the difficulties inherent in applying ecological principles alone to understand and evaluate land-use practices of small farmers in Rondônia, Brazil, since social, economic, and political conditions inevitably influence land-use decisions and their impacts. Weiss (Chapter 5) takes the analysis of Santelman et al. (Chapter 11) a step further, illustrating how ecologically sound land use and management in tribal communities in southern India actually contributes to community building and empowerment. Other authors, such as Haufler and Kernohan (Chapter 4), explore how humans and their socio-political, economic, and cultural systems constitute distinct concrete constraints on the types of conservation efforts that are viable. Also recognizing this constraint, Haeuber and Hobbs (Chapter 12) explore how existing land-use tools and approaches can be used creatively to find middle ground between the needs of human systems and ecological systems. Finally, Hulse and Gregory (Chapter 9), Santelmann et al. (Chapter 11), Steinitz and McDowell (Chapter 8), and Bourgeron et al. (Chapter 13) all explore the decision-making methods and tools through which this common ground can be defined.

14.2 Key Concepts

An important lesson demonstrated by the chapters in this volume is that certain key concepts and ideas must be understood in order to manage and use land in an ecologically sound manner. It is important that land-use planning recognize the dynamic nature of ecological systems, the significance of natural processes, the uncertainty and inherent variability of ecological systems, and the importance of cumulative effects (Dale et al. 1999). At the same time, as Santelmann et al. (Chapter 11) observe, acknowledgment of errors and uncertainties in decisions must be factored into all land-use and land-management activities.

14.2.1 Ecological Systems Are Dynamic

A central, dominant reality of ecological systems is that they are not in equilibrium; instead, ecological systems are inherently dynamic and constantly changing over some range of temporal and spatial scales. For example, ecological systems are regularly subjected to episodic, natural disturbances that shape their states and dynamics. As described throughout this volume, it is essential to recognize that ecological systems are hierarchical structures, best evaluated at a variety of spatial scales (Haufler and Kernohan, Chapter 4; Bourgeron et al., Chapter 13). Unique processes, features, and dynamics become apparent when viewed from different scales. The only way to ensure ecological diversity and productivity for future generations is to sustain ecological processes within the expected bounds of variation at each scale. However, this poses a significant management dilemma, revolving around the potentially conflicting needs and decision timeframes of human and natural systems described above. As Botteron (Chapter 7) notes, for example, human activities can alter land-cover patterns and ecological processes, jeopardizing the species that depend on existing landscapes. And, as Haeuber and Hobbs (Chapter 12) point out, traditional conservation approaches frequently involve attempts to freeze ecological systems in time, including the types of anthropogenic alterations to ecological patterns and processes that Russell (Chapter 6) describes in her case studies.

14.2.2 Natural Processes Are Significant

The fact that ecological systems change over both time and space is at the heart of the dynamism described above. Such changes include succession, disturbance, changes in climate, loss of site productivity related to land-use activities, establishment and spread of nonnative species, and the loss of native species diversity. Anthropogenic disturbances to ecological processes must be considered against the background of natural dynamics. Thus, as

Russell (Chapter 6) demonstrates, a simple return to more natural conditions after human-induced landscape change is often difficult or, in fact, may be impossible. As the Hulse and Gregory (Chapter 9) case study illustrates, natural processes must be considered in defining desired future conditions, as well as in developing land-management strategies. The observed range of environmental variation in natural processes must be compared to what would have been expected in the absence of recent human changes to the landscape. If the degree of variation exceeds expectations, then it is likely that recent human activity is changing the frequency or magnitude of disturbance processes.

14.2.3 Ecological Systems Are Inherently Uncertain

Uncertainty arises from an incomplete understanding of how ecological systems work and from insufficient information. However, even if these sources of uncertainty could be removed through more research and better theory, the fact remains that ecological systems are inherently variable. Thus, variability must be factored into expressions of desired future conditions as well as into expectations related to management actions and strategies. In fact, Santelmann et al. (Chapter 11) call for a principle that deals explicitly with including uncertainty in communications with decision-makers. Uncertainty and variability are present in nearly all stewardship actions and are best acknowledged and incorporated into stewardship through monitoring, adaptive management, and communication.

14.2.4 Cumulative Effects Are Important

Cumulative effects involve the impact on the environment resulting from the incremental impact of an action when added to other past, present, and reasonably foreseeable future actions. As Haeuber and Hobbs (Chapter 12) point out, however, many land-use impacts are cumulative, making it difficult to define a single causal act. In addition, because of the wide variation in site-specific practices and local environmental conditions, impacts of past and present management practices may not always be well understood or predicted (e.g., see Russell Chapter 6). Cumulative effects generally reach beyond specific parcel boundaries, and thus there is a need to coordinate with landowners in adjacent parcels and throughout the region when undertaking cumulative-effects analysis and monitoring ongoing changes.

There are few analytical methods available that effectively address cumulative impacts. However, techniques discussed in this volume, such as integrated regional ecological assessment (Bourgeron et al., Chapter 13) and alternative futures analysis (e.g, Steinitz and McDowell, Chapter 8; Hulse and Gregory, Chapter 9; Santelmann et al., Chapter 11) allow proposed actions to be considered in terms of their cumulative effects. This type of "real-time" cumulative-effects analysis significantly enhances our ability to address the future consequences of specific decisions. Nevertheless, only active and ongoing monitoring can detect unanticipated changes and the introduction of new elements to the system. Therefore, monitoring and adaptive management must remain a part of land management and policy.

14.3 Tools and Technology for Land Management

The success of future efforts to apply ecological principles to land-use decisions partially depends on the creative use of existing tools and the development of new tools, as many of the chapters in this volume demonstrate. Tools for applying ecological approaches to land-use and land-management decisions include spatially-explicit models, geographic information systems, and visualization tools. With recent advances in computers, networks, electronic transfer, data management, data collection, and measurement technology, these tools are becoming increasingly sophisticated. For example, technological advances are apparent in terms of speed, user friendliness, accessibility, and the connectivity of models and information.

Increasing data availability is clearly a key development. For example, modern dairy farmers regularly receive detailed data and information on individual cows based on submitted samples of blood and milk. These data provide the basis for decisions about feeding, heath care, and the expected productivity of each animal. In a similar manner, data about the land can be used to inform decision-makers. For example, satellites collect reflectance information that can be interpreted to infer land-cover types and even vegetation characteristics. Moreover, one can now download information from Internet sites that is pertinent to a particular land use and place (e.g., see Hobbs and Theobald, Chapter 2). As we move toward the increasingly integrated and electronically connected community of the future, however, the vastly increased amount of available information ensures that a critical challenge will be learning how to manage and understand the flow of information.

We believe that a key advance in managing and interpreting information involves models that allow landowners to visualize the implications of specific land actions on their own property (as described by Hulse and Gregory, Chapter 9; Haeuber and Hobbs, Chapter 12). A challenge to the use of visualization tools is that projections become so real that they cannot be distinguished from photographs, and thus landowners may believe that model simulations are predictions of actual future conditions. Therefore, as several chapters in this volume describe, it is useful to project a set of future scenarios so that decision-makers can compare alternative futures (Steinitz and McDowell, Chapter 8; Hulse and Gregory, Chapter 9; Santelmann et al., Chapter 11).

The future use of models for land management depends on their usability, the development of software architecture that allows ready access to these models, real-time geographic information systems data, spatial analysis that relates to a particular landowner's property, advanced three-dimensional models and visualization, image processing, cartography, and Internet deployment of the models. Such advances must occur with relatively few advances in the computer literacy of landowners; instead, the tools must become easier to use.

We can expect the savvy landowner to own a computer, have access to the Internet, and be able to follow easy directions. However, we should never expect landowners or land managers to create models or to manipulate complex technological systems. What we do hope is that there may be a time in the relatively near future when a landowner can readily access this information. To illustrate this possibility, we can imagine such a land manager in the future.

14.3.1 An Example of Future Land-Management Decision Making

Frank Forest owns 20 acres in western Kentucky that are largely forested. He wants to maintain the trees on part of the land near his house, for he enjoys observing wildlife and wants his home to be in a forest setting. He also wants to pass this land onto his children, which lends a temporal component to his management practices. In addition, however, he needs to obtain income from his land. His goals for the land thus have aesthetic, ecological, and economic aspects. His decisions about the use of the land lead to a set of land-use and land-management questions. Where should he situate the house? In which forested areas should he harvest trees, and what timber-management techniques and technologies should be employed?

To make these decisions, Frank Forest goes to his computer and downloads vegetation, soils, and geology maps for his land and surrounding parcels from the local community server. He then downloads a computer model that allows him to examine the impacts of particular land-use decisions. He checks the boxes indicating the model output that he would like to see. He selects aerial visualization of the land, bird diversity, and economic value (current and projected by decade). He also could have explored myriad other variables, such as contribution to carbon sequestration, seed production output of the land, expected growth of the trees, or animals and plants that may occur on his land. Based on the selected decision parameters, the model provides several management options for harvesting his land. To select the sites for harvesting, however, the model first directs him to the soil analysis that suggests sites where growth of commercial tree species would be most appropriate and makes some suggestions about harvesting principles.

At this stage, the model provides visualizations that explore ecological concerns, such as the importance of the temporal setting, the presence of individual species and networks of interacting species, the uniqueness of each site or region, natural disturbances, and landscape relationships. Based on these considerations, the model suggests particular management strategies. Frank is offered numerous options to explore, enabling him to examine visualizations of what the site may look like under future conditions. He selects his preferred option and then is led by the program through a procedure to examine how this option fits into the ecological guidelines for land use. For example, he is encouraged to examine how his local decisions fit into the regional context. The model allows him to examine his land in conjunction with his neighbors and, therefore, consider critical habitat and rare species supported by his land in relation to the larger region or connected areas. A pop-up screen provides him with the e-mail addresses of his neighbors, allowing him to describe and discuss various land-management options to see how they fit with the land-management schemes of adjacent areas.

The program already has demonstrated potential trajectories for long-term change and future events, but it also encourages him to consider unexpected impacts, such as global climate change, or disturbances, such as fire or flooding. The program also alerts him to the presence of rare landscape features and provides alternative options for managing those features. It reminds him of deed restrictions, zoning conditions, and relevant federal and state regulations. It suggests alternatives to avoid land uses that might deplete natural resources on the land, whether they be natural features or the presence of unique species. The program also suggests ways to minimize the introduction or spread of nonnative invasive species, providing a list of native species that are appropriate for the site conditions if the landowner chooses to perform some planting. When selected land-management practices call for active development, the program suggests ways to avoid or compensate for the effects of development on the ecological processes at the site. Finally, the program explores land-use and land-management practices that are compatible with the natural potential of the area.

In summary, the land manager of the future will be able to readily use a variety of tools to make decisions about how to manage and use his land. However, he must have some basic knowledge to make these decisions in an ecologically informed manner.

14.4 Building Ecological Knowledge

The example explored above sounds like a lot for a computer model to accomplish, and it is. Clearly, such a model can only produce recommendations that are encoded within its memory. Ultimately, it is the individual landowner's responsibility to make actual decisions about what is in his best interest, as well as considering the best interests of the land and the region. To arrive at the best-informed decisions, however, landowners must know what questions to ask and how to interpret the answers to their questions. Therefore, even though the tools and technology for land management may become more sophisticated and useful, advances in the knowledge of individual landowners are necessary as well. Equipped with information about the implications of land-use choices, the landowner also must possess ecological knowledge and understanding in order to interpret this information and make decisions that are appropriate for the ecological conditions of the land and the socioeconomic conditions of the community and region.

If such models are to become practical and useful tools for ordinary landowners, a central challenge involves taking concepts from the research community and making them available and accessible to land managers. The chapters by Stenitz and McDowell (Chapter 8), Hulse and Gregory (Chapter 9), and Santelmann et al. (Chapter 11) provide examples of the transfer of such information and tools. Part of the challenge in applying ecological principles to decisions about land use involves the diversity of landowners and their varying levels of ecological knowledge and understanding. New tools and technology should be available to owners of a few hectares of land as well as those individuals, corporations, or government agencies that manage large areas. A key need is to make the tools accessible to the numerous landowners who own relatively small areas. Individual landowners frequently would like to manage their land in an ecologically appropriate way, but lack the information or tools to do so. In the state of Tennessee, for example, there are over 300,000 parcels of forestland larger than 15 acres and only 33 state foresters to assist with the management of these lands.

One way to meet this need for knowledge is to distribute information electronically. For example, the South Carolina Forestry Commission has produced a CD-ROM on "Forest Stewardship" that interactively explores management options targeted toward aesthetics, timber, wildlife, recreation, and soil and water integrity. The information is developed for forestland owners and natural resource professionals. Using the interactive CD-ROM provides a way to bring people to the same level of understanding about the definition, components, and options for sustainable forest management.

Therefore, we believe that the future is bright for efforts to incorporate ecological understanding in land-use decisions, but success depends on individual knowledge, as well as the availability of tools and technology. Scientists and educators confront significant educational challenges if they are to help ensure that future land management considers the broad, ecological impacts of changes resulting from land-use decisions. Schools already are relatively successful in imparting this knowledge. Third and fourth graders understand what ecosystems are and discuss issues such as interactions among ecological communities, but it is not clear that they have a practical understanding of key ecological concepts and their implications. Land-management agencies at both state and federal levels now provide information in pamphlets, books, and CD-ROMS regarding land-management alternatives. Yet, landowners do not always know that this information is available and, too frequently, the information is not presented in ways that landowners can understand and use. In moving information and concepts from the research community to the land-user community, it is essential that everyday language be employed and that applications are relevant to landowners' needs. Challenges remain in terms of the substance and presentation of ecological knowledge. However, we can be certain of one thing—The Internet will be critical to the future of land use and management through data distribution, public notification, and awareness raising as we strive to consider the broader impacts of our decisions both in space and time on ecological and human systems.

14.5 Conclusion

If ecology is to make a difference in future land-use and land-management decisions, decision-making processes must be changed to include an ecological perspective. This change in viewpoint can be facilitated by advances in visualization tools; the further development and application of ecological principles; organizational efforts to increase knowledge and technology; attention to data sets, schematics, and standards; and leadership among educators and the research community.

By themselves, however, technological advances and increased availability of information will not be enough to stimulate the necessary evolution in land-use planning and decision-making. The need to focus on how humans affect the land led Aldo Leopold to call for a land ethic more than a half-century ago (Leopold 1949). Leopold was not asking for the government to intervene and establish protected areas. Rather, he challenged each of us to examine our relationship with the land. He sought the means to develop and embrace an ethic that recognized the land and its inherent value as an integral component of our relationship with nature. We hope that the development and application of ecological principles and guidelines for action explored in this volume is another small step toward realizing Leopold's dream.

Acknowledgments. Some of the ideas presented in this paper evolved while V. Dale served on the Committee of Scientists for the Secretary of Agriculture. Thus, she acknowledges thought-provoking discussions with other members of that Committee, particularly Robert Bescha and Barry

Noon. Comments on the paper by Lisa Olsen and Tom Ashwood are appreciated. While one of the co-authors, Richard Haeuber, is an employee of the U.S. EPA, this research was completed on his own time. It was conducted independent of EPA employment and has not been subjected to the Agency's peer and administrative review. Therefore, the conclusions and opinions drawn are solely those of the authors and should not be construed to reflect the views of the EPA.

References

Blood, E. 1994. Prospects for the development of integrated regional models. *In* P.M. Groffman and G.E. Likens, editors. Integrated regional models: interactions between humans and their environment. Chapman & Hall, New York, New York, USA.

Christensen, N.L., A.M. Bartuska, J.H. Brown, S. Carpenter, C. D'Antonio, R. Francis, J.F. Franklin, J.A. MacMahon, R.F. Noss, D.J. Parsons, C.H. Peterson, M.G. Turner, and R.G. Woodmansee. 1996. The report of the Ecological Society of America Committee on the Scientific Basis for Ecosystem Management. Ecological Applications **6**:665–691.

CIESIN. 1992. Pathways of understanding: the interactions of humanity and global environmental change. CIESIN, University Center, Michigan, USA.

Committee of Scientists. 1999. Sustaining the people's lands: recommendations for stewardship of the national forests and grasslands into the next century. USDA, Washington, D.C., USA.

Dale, V.H., J. Agee, J. Long, and B. Noon. 1999. Ecological sustainability is fundamental to managing the national forests and grasslands. Bulletin of the Ecological Society of America. **80**:207–209.

Dale, V.H., S. Brown, R.A. Haeuber, N.T. Hobbs, N. Huntly, R.J. Naiman, W.E. Riebsame, M.G. Turner, and T.J. Valone. 2000. Ecological principles and guidelines for managing the use of land. Ecological Applications **10**:639–670.

Fiske, S.T., and S.E. Taylor. 1991. Social cognition. McGraw-Hill, New York, New York, USA.

Folke, C., L. Pritchard, Jr., F. Berkes, J. Calding, and U. Svedin. 1998. The problem of fit between ecosystems and institutions. IHDP Working Paper No. 2. IHDP, Bonn, Germany.

George, A.L. 1980. Presidential decisionmaking in foreign policy: the effective use of information and advice. Westview Press, Boulder, Colorado, USA.

Grumbine, E.R. 1994. What is ecosystem management? Conservation Biology **4**:27–38.

Grumbine, E.R. 1997. Reflections on "what is ecosystem management?" Conservation Biology **11**:41–47.

Haeuber, R.A., and W.K. Michener. 1998. Policy implications of recent natural and managed floods. Bioscience **48**:765–772.

Holling, C.S., and G.K. Meffe. 1996. Command and control and the pathology of natural resource management. Conservation Biology **10**:328–337.

Huntley, B.J., E. Ezcurra, E.R. Fuentes, K. Fujii, P.J. Grubb, W. Haber, J.R.E. Harger, M.M. Holland, S.A. Levin, J. Lubchenco, H.A. Mooney, V. Neronov, I. Noble, H.R. Pulliam, P.S. Ramakrishnan, P.G. Risser, O. Sala, J. Sarukhan, and W.G. Sombroek. 1991. A sustainable biosphere: the global imperative: the International Sustainable Biosphere Initiative. Ecology International **20**:1–14.

Lau, R.R., and D.O. Sears, editors. 1986. Political cognition. Lawrence Erlbaum Associates, Hillsdale, New Jersey, USA.

Leopold, A. 1968. A Sand County almanac and sketches here and there. Oxford University Press, New York, New York.

Levin, S.A. 1992. The problem of pattern and scale in ecology. Ecology **73**:1943–1967.

Lubchenco, J., A.M. Olson, L.B. Brubaker, S.R. Carpenter, M.M. Holland, S.P. Hubbell, S.A. Levin, J.A. MacMahon, P.A. Matson, J.M. Melillo, H.A. Mooney, C.H. Peterson, H.R. Pulliam, L.A. Real, P.J. Regal, and P.G. Risser. 1991. The sustainable biosphere initiative: an ecological research agenda. Ecology **72**:371–412.

Meidinger, E.E. 1997. Organizational and Legal Challenges for Ecosystem Management. *In* K.A. Kohm and J.F. Franklin, editors. Creating a forestry for the 21st century: the science of ecosystem management. Island Press, Washington, D.C., USA.

Miller, R.B. 1994. Interactions and collaboration in global change across the social and natural sciences. Ambio **23**:19–42.

Noss, R.F. 1983. A regional approach to maintain diversity. Bioscience **33**:700–706.

Noss, R.F. 1992. The Wildlands Project: land conservation strategy. Wild Earth (Special Issue):10–25.

Petrie, H.G. 1976. Do you see what I see? The epistemology of interdisciplinary inquiry. The Journal of Aesthetic Education **10**:29–43.

Pickett, S.T. A., W.R. Burch, Jr., S.E. Dalton, T.W. Foresman, J.M. Grove, and R. Rowntree. 1997. A conceptual framework for the study of human ecosystems in urban areas. Urban Ecology **1**:185–199.

Rosati, J. 2000. Bounded rationality, cognitive actions, and the study of world politics: the power of human cognition and policymaker belief. International Studies Review **2**(3).

Slocombe, D.S. 1993. Implementing ecosystem-based management. Bioscience **43**:612–622.

Snow, C.P. 1993. The two cultures. (Canto edition) Cambridge University Press, Cambridge, England.

Steinbruner, J.D. 1974. The cybernetic theory of decision. Princeton University Press, Princeton, New Jersey, USA.

Stern, P.C., O.R. Young, and D. Druckman. 1992. Global environmental change: understanding the human dimensions. National Academy Press, Washington, D.C., USA.

Vedeld, P.O. 1994. The environment and interdisciplinarity: ecological and neo-classical economical approaches to the use of natural resources. Ecological Economics **10**:1–13.

Vitousek, P.M., H.A. Mooney, J. Lubchenco, and J.M. Melillo. 1997. Human domination of earth's ecosystems. Science **277**:494–499.

Wuichet, J.W. 1995. Toward an ecosystem management policy grounded in hierarchy theory. Ecosystem Health **1**:161–169.

Index